Sasquatch
in
British Columbia

Sasquatch in British Columbia

A Chronology of Incidents
&
Important Events

Christopher L. Murphy

in association with

Thomas Steenburg

hancock

house

ISBN 978-0-88839-721-8
Copyright © 2012 Christopher L. Murphy

Library and Archives Canada Cataloguing in Publication

Murphy, Christopher L. (Christopher Leo), 1941–
    Sasquatch in British Columbia : a chronology of incidents &
    important events / Christopher L. Murphy in association with
    Thomas Steenburg.

Includes bibliographical references and index.
ISBN 978-0-88839-721-8

    1. Sasquatch—British Columbia. 2. Sasquatch—British
Columbia —Anecdotes. I. Steenburg, Thomas N. (Thomas Nelson)
II. Title.

QL89.2.S2M8779 2012   001.944   C2012-901055-3

Printed in South Korea — PACOM

*We acknowledge the financial support of the Government of Canada through
the Canada Book Fund for our publishing activities.*

Published simultaneously in Canada and the United States by

HANCOCK HOUSE PUBLISHERS LTD.
19313 Zero Avenue, Surrey, B.C. Canada V3S 9R9
(604) 538-1114 Fax (604) 538-2262

HANCOCK HOUSE PUBLISHERS
1431 Harrison Avenue, Blaine, WA U.S.A. 98230-5005
(604) 538-1114 Fax (604) 538-2262

Website: www.hancockhouse.com
Email: sales@hancockhouse.com

CONTENTS

Acknowledgements

John Green: Most of the material I present in this work was originally found and researched by John Green. His dedication and perseverance in the field of sasquatch studies has no equal. I met John in 1994 and he has provided me with counsel, assistance, advice, photographs, and artifacts for all of the projects I have undertaken. John remains the most knowledgeable person in the field, and I express my sincere gratitude to him for all he has done for me.

René Dahinden: A day does not go by without René entering my thoughts. I met him in 1993 and we became very close friends. I spent more time talking with René about sasquatch than I have with any other person on a specific subject. He was my mentor and confidant, and although we parted company a few years before he died, I am very grateful to him for the knowledge he imparted to me.

John Fuhrmann: I never personally met John. He left this world before I became involved in the sasquatch/bigfoot issue. He was a highly meticulous researcher who collected and neatly filed all the information he could find on sasquatch/bigfoot, as well as that of possible related creatures. His massive collection found its way to me by way of Dr. Grover Krantz and John Green. It was John Fuhrmann's collection that provided me with the inspiration to compile this work…many thanks, John.

(Photos: Top and Center, C. Murphy; Lower, J. Green.)

Introduction

When the 49th parallel was used to essentially cut North America in two, forming Canada and the United States (US), the latter certainly got the "more comfortable" half. From this parallel north it gets wetter and colder, and to the south dryer and warmer. However, if one is looking for the sasquatch, not comfort, then the top half in the West is significantly better.

Based on sighting statistics and current populations in British Columbia (BC), Washington, Oregon and California, one's chance of finding a sasquatch in BC is five times greater than in these US states, although access in BC to remote areas is more difficult.

The math is based on the fact that with just 4.2 million people, (2006) BC has about 31% of the reported sasquatch incidents in the four regions cited. A total population of about 46 million people in the US regions makes up the other 69%. Of course, what I am implying here is that the more people there are in a region, then the more likelihood of a sasquatch sighting. Naturally, without people there cannot be a sasquatch sighting.

There is also another factor that comes into the equation. Most of BC's population hugs the 49th parallel. Indeed, the number of people who get into remote BC

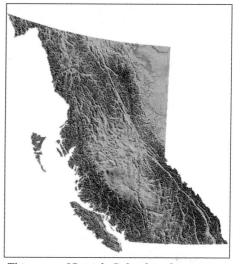

This map of British Columbia shows its vast forests and mountainous regions. Most of the province is highly inaccessible.

The province was given the name "British Columbia" in 1858 by Queen Victoria. The "Columbia" part refers to what was known as the "Columbia District"— territory drained by the Columbia River (origins and upper reaches are in Southwestern BC.). The Queen chose "British" to distinguish the region from that of the US ("American Columbia" or "Southern Columbia"). The name, however, had been in use previously by the Hudson's Bay Company, but it was definitely the Queen's call.

BC's approximate population is shown here by year.

Year	Population
1850	55,000
1900	179,000
1950	1,200,000
2000	3,900,000

regions is actually fewer than it was 100 years ago—the First Nations population in such regions has decreased, and trappers no longer work the forests as they did in the early years.

Most of BC is highly inaccessible, except by plane, and even then it can be "tough going." One has to simply check Google Earth™ to see just how rugged this region is. Indeed, when one "zeros in" to, say, a remote lake

Somewhere in Northern British Columbia from about 20 miles in the air.

on a satellite image, it is likely few (if any) people have been (physically or otherwise) to that spot. Generally most people don't wander more than about 500 feet from a forest service road.

Many sightings in BC occur on islands, and here it needs to be mentioned that the province has some 40,000 islands. Few have official names or human habitation. For certain there are more islands in BC than there are sasquatch.

BC's coast, seen here only partially, has about 40,000 islands. Few are inhabited by humans.

That BC has so many sasquatch-related incidents with so few people appears to indicate that there are considerably more sasquatch in this province than anywhere else in North America.

However, in one respect it is surprising that BC has any sasquatch-related incidents at all. By far, most incidents certainly take place within

50 miles of some sort of human habitation. With the vast expanse of the province (365,900 square miles), it is hardly necessary that the creatures need to get anywhere near a human (or a road for that matter). Generally speaking, to say that humans are invading the creature's domain is incorrect. It is the other way around. Thomas Steenburg states that from his experience, "Sightings mostly occur when the creatures wander close to people, not when people wander deep into their wilderness home."

Be that as it may, the fact is that hundreds of people in BC have reported seeing the creature, and/or finding evidence of its passage, for at least 150 years (not counting legend and speculation based on prehistoric artifacts).

From what I have been able to determine, for every sasquatch-related incident that is reported and made public (media or internet), eight (8) are not reported. In other words, most people who see a sasquatch, or find associated evidence, don't make their experience known beyond a few relatives and friends. Based on that theory, this work should be many times larger with regard to the number of incidents provided.

With regard to the *verification* of material I present in this book, all I can say is that I have provided everything as accurately as I can in accordance with the sources shown. Absolutely no additional credibility can be assigned because the incidents appear in this work. Also, I make no claim that I have included all sasquatch-related incidents that were reported in BC. Some were so brief that they did not warrant inclusion (incidental reports), or were sadly lacking as to the time frame and reasonable location aspects. Furthermore, although I believe I have "tapped" the most common and available resources for sighting information, I am sure there are many reports I failed to find. If the reader has, or knows of, a report that should be included, then I ask that he or she contact Thomas Steenburg >sasquatch@telus.net< and we will look at including it in an update of this work.

Concerning report credibility, I will mention that I have great faith in the investigative skills of researchers such as John Green, René Dahinden, Bob Titmus, Thomas Steenburg, and Bill Miller. They were personally involved in many of the incidents cited, and have done extensive field work. I worked with Dahinden (d. 2001) for some 6 years, and have worked fairly closely with Green and

Steenburg for at least 10 years. As to material that did not involve these researchers, I have reasoned that if the information has been used in various media, then regardless of its level of truth, (notwithstanding obvious fabrications) it has still become a part of sasquatch history. In all cases, I cannot prove that the story was wholly or partially fictitious. I don't wish to act as judge and jury in such cases. I have pointed out what I think, or have been led to believe by others. The rest is up to the reader.

The scope of the entries in this work is not limited to just sasquatch sightings and directly associated incidents. I have included major events, milestones, historical references, and human interest stories. In some cases, the event did not occur in BC, however, it was deemed appropriate in the course of events, or involved one or more major BC researchers. Also, some incidents involve ordinary "wild men" (or men gone wild) who were definitely not sasquatch. I believe that at a distance such individuals could be mistaken for sasquatch, and have included them for that reason.

The order of the entries is chronological. However, as I may be missing parts of dates, or only have approximate time frames, I have used what I consider to be a logical sequence as follows using a fully numerical date.

The date entries for accounts prior to 1800 simply show an approximated year reference which is self explanatory.

The date entries for accounts that have occurred since 1800 are arranged as follows:

1. Year, month, day, known (e.g., 1985/01/01)
2. Year and month known (e.g., 1985/01/00 – Entry placed at the last day of the month.
3. Year and season known:
   1985/SP/00 (Spring) – Entry placed at April 30
   1985/SU/00 (Summer) – Entry placed at July 31
   1985/AU/00 (Autumn) – Entry placed at October 31
   1985/WI/00 (Winter) – Entry placed at January 31
4. Year only known (e.g., 1985/YR/00) – Entry placed at December 31 ("YR" means "year")
6. Decade only known (e.g., 1989/DE/00) – Entry placed at December 31 of the last year in the decade. In this case the decade was the 1980s ("DE" means "decade")

5. Date is approximate (e.g., 1985/CI/00) – Entry placed at December 31 ("CI" means "circa"). As "CI" could be the following year, it is shown last in all cases.

It should be noted that, generally speaking, the actual seasons of the year in the Northern Hemisphere are as follows:

> Spring starts on March 21
> Summer starts on June 21
> Autumn starts on September 23
> Winter starts on December 21

The "entry dates" I have chosen for the seasons are arbitrary. They are simply representative dates that I feel best define the season.

All entries start with the date in numerical format as shown. *This coding is for referencing only.* In all cases, the date information is provided in normal date format in the first part of the entry. The date is followed by the geographical *area* of the incident, generally the closest and largest town or "registered" community (distances will vary considerably). However, in some cases the specific location is not known or is so remote that showing the closest town is not logical. Here there is no choice but to show what is given as the location (lake, inlet, canal and so forth). More exacting location information, when available, is provided in the first part of the entry. Unfortunately, Canada does not have the "county system" as used in the US, so providing location information is cumbersome.

To distinguish incidents on Vancouver Island and the Queen Charlotte Islands (as opposed to the mainland or a small island) I have shown (VI) or (QCI) after the location name (town or other indicator).

The geographical area is follow by a short title that serves to identify the entry. Example:

1985/01/01 – Terrace. Picking Salmonberries: Two men reported that they saw a sasquatch near Terrace along Highway 22 on January 1, 1985. *(Full details of the incident are then provided. In this case the creature was seen picking salmonberries.)*

It is seen here that the first sentence provides: 1) the witness or witnesses; 2) the location, which is expanded as necessary; 3) the date of the incident (not the date of the report). I have purposely repeated the information in the title (date and location) for clarity and to provide a "summary" statement.

The reader needs to take note that even when exact dates are referenced, there can be discrepancies of a few days. Newspaper articles sometime reference "two weeks ago" or other similar statements that have been worked out to provide a proper date.

Lists are provided in the Appendix showing all entries in date order, geographical area order (alphabetical), and title order (alphabetical). Nevertheless, if one is looking to see if an incident that occurred at a location within a geographical area (small creek, valley, local name for a section of land, company property, and so forth), then it is best to check the General Index.

Notes on the Entries

Witness Names: When actual witness names are known and the person involved has not asked that his or her name be deleted, then names have been shown. However, this does not imply that names are known and have been withheld in all cases.

Witness Descriptions: Unless a witness specifically says that he or she thinks the creature observed was a sasquatch, we really have no choice but to call what was seen as an *unusual creature, a strange creature*, an *oddity* or another non-specific name. In this way, the creature is identified as not being a common animal that would be immediately recognized (e.g., a bear). Whether or not the description given for the creature matches what is believed to be a sasquatch is left up to the reader.

Terms and Word Usage: The term "sasquatch" has been used in this work except where the term "bigfoot" is in quoted or reference material. Also, I consider the word "sasquatch" to be both singular and plural (e.g., one sasquatch, two sasquatch, and so forth). Furthermore, I have elected not to to show the word with a capital letter.

With regard to the term "First Nations," it refers to native or aboriginal people. Only in cases where another term for these people is in quoted material is that other term used.

Victoria Newspapers: References to Victoria newspapers shown in books or other media can be confusing. The following is from Wikipedia:

> The *Times Colonist* is an English-language daily newspaper in Victoria, British Columbia, Canada. It was formed by the merger, in 1980, of the *Victoria Daily Times,* established in 1884, and the *British Colonist* (later the *Daily Colonist),* established in 1858. The *British Colonist,* was also called the *Daily British Colonist,* the *Daily Colonist,* and other variants from its first issue in 1858 to June of 1910.

In many cases, the *Colonist* is referenced in works I have used for material presented. From the foregoing, this would be either the *British Colonist* or the *Daily Colonist* up to 1979, and the *Times Colonist* from 1980 onward. I have used this reasoning for providing source entries in this book.

Entry "inferences": Some of the entries in this work imply that the sasquatch may do things, have characteristics, or have abilities many researchers do not agree with. It must not be construed that inclusion of this material signifies concurrence by either my associate Thomas Steenburg or myself. It is impossible to prove that such claims or occurrences did not take place, and equally impossible to prove that they did. Their inclusion is strictly a matter of "record," not a matter of "fact."

Missing Evidence: Throughout this work some entries indicate the photographs of footprints and/or plaster casts were taken of such prints. Also, that even movie footage or photographs were taken of the creature sighted. The obvious question is, of course, why have I not shown images of (or from) this evidence.

The first part of the answer is that I have not seen it myself and *to my knowledge,* it did not end up with any of the major researchers in BC active around 1993. Trying to trace this material down at this late date is very difficult, if not impossible. For certain, researchers at the time of the incident would have tried to obtain it if they thought it was important. That these researchers may have simply obtained and "filed away" photographs or other evidence is possible, and in the case of René Dahinden, highly likely. The bottom line is that I do not have any of this material at this time.

The second part of the answer, which concerns strictly photographs, is that although I might have the images (or have access to them), I do not have permission to provide them in a book. I will mention here that there are other types of photographs I would like to include, but can't do so for the same reason. Using a photograph on the internet is totally different from using it in a book (different rules apply).

In cases where I do not show that photographs were taken or plaster casts made, it is *probable* that such were not taken or made. However, I don't know the answer.

John Green's Data Base: Many reports provided in this work are from John Green's posted data base www.sasquatchdatabase.com/. Please access this data base using the Incident Identification Number shown, date, or location for additional information on the incident (location statistics or witness personal information). Also, access the data base to see incidental reports not provided in this work. Please note that not all of John Green's material is in his published data base. Some material is only in his books and his private Sasquatch-Incidents File (which includes everything he has). In cases where I show the source as "Sasquatch-Incidents File," this is material that has not been published either in Green's books or in his data base. Please note that the data base has been given priority when minor details differ.

Note on the Term "Stride": This term as used in this work *generally* means a "pace," or the distance from the heel of a footprint to the heel of the next footprint. I realize the actual definition is from the heel of say the left foot to the heel of the next left foot (i.e., two paces). However, most people use the term "stride" to mean a "pace,"and consequently used "stride" in their reports. Unfortunately, the matter gets more complex because it is not known exactly how the measurement was made. Some people might measure just the space between prints, rather than including the space occupied

14

by the first print. The adjacent chart illustrates the official (proper) definition of "stride" and "pace." However, please note that a "pace" can also be called a "step."

In looking at the measurement provided by witnesses or researchers, it is perhaps prudent to keep in mind that the creature in the Patterson/Gimlin film had a 14.5-inch foot with a STRIDE of 81.5 inches or 6.8 feet. The PACE was therefore 40.75 inches or 3.4 feet (Krantz, *Bigfoot Sasquatch Evidence,* p. 93). For comparison purposes, a man about 6 feet tall with 11.25-inch long feet would have a normal walking pace of about 22 inches. His stride would therefore be about 44 inches.

In this work, I have shown "[pace]" after the word "stride" in cases where I believe "pace" is correct.

Note on Images: I have provided as many images as possible in this work to support the information in the entries and give the reader something to reflect upon in evaluating what has been stated. Many images are from Google Earth, whose kind permission to use their material is greatly appreciated. The Google Earth program is an astounding research facility for all Earth sciences studies. I urge sasquatch researchers to use it to better understand circumstances with regard to reported sasquatch-related incidents. I recommend that you use

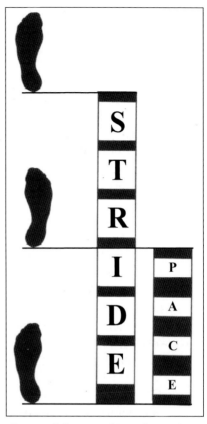

Proper definition of "stride" and "pace." The only reason I can see for a "stride" measurement is to take into account the length of the other foot. In some cases, a human can have one foot slightly larger than the other (up to ¼ inch). Sasquatch footprint casts also sometimes show a discrepancy. The "stride" is therefore a more accurate measurement of the distance covered than simply doubling the pace.

the program to "visit," as it were, the location of sightings or other incidents presented in this book. You will be able to see geographical aspects at different angles in great detail—far beyond what can be provided in print.

Where available, I have provided actual photographs of sasquatch sighting locations. Certainly, photographs of roads, hillsides, sections of forest, and so forth are not overly exciting unless you are highly involved in sasquatch studies. The objective is to simply provide the reader as much information as possible on the incident. Such photographs facilitate an actual "mental picture" of what happened, rather than depending upon a totally imagined picture.

Furthermore, I have gone to great lengths to provide images of the people who reported sasquatch sightings or were highly involved in the sasquatch issue. Photographs that appeared in newspaper articles or other media were obtained, and those taken by Thomas Steenburg, John Green, other researchers, and myself were used. I believe one can gain valuable insights from "people photos." If anything, one will appreciate that the people seen here were all ordinary people like you and me. There is no reason, in my opinion, to doubt their honesty as to the information they have provided.

Note on Measurements: This book uses the Imperial/USA measurement units. The metric conversions are as follows:

$$
\begin{array}{rcl}
1 \text{ Inch} & = & 2.54 \text{ centimeters} \\
1 \text{ Foot} & = & 30.48 \text{ centimeters} \\
1 \text{ Yard} & = & 0.91 \text{ meters} \\
1 \text{ Mile} & = & 1.61 \text{ kilometers} \\
1 \text{ Pound} & = & 0.45 \text{ kilograms}
\end{array}
$$

Source: General Knowledge and Wikipedia. (Photos: Map of BC, Wikipedia Commons, refer to the Wikipedia website for details; Section of BC, Image from Google Earth, © 2010 TeleMetrics; ©, 2010 Google; BC's islands, Image from Google Earth, Image © 2011 Province of British Columbia; © 2011 Cnes/Spot Image; Image © 2011 Terra Metric, Data Living Oceans Society; Stride/Pace illustration, C. Murphy.

Chronological Overview

Beginning – 1799: Strange footprints locked in time, and First Nations art, provide possible testimony for the existence of sasquatch, or sasquatch-like creatures, going back long before recorded history. When Europeans first came to North America, they wrote of strange occurrences involving a, "beast of unknown species." (Photo: W.G. Burroughs.)

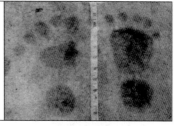

1800 – 1849: With increased settlement along the East Coast of North America, reports of a strange, man-like creature were detailed in early pioneer newspapers. The only explanation settlers could provide was that the creature was a man who had gone wild and had grown a complete covering of hair. (Photo/Artwork: David T. King.)

1850 – 1899: With cities, towns and small settlements now in place across the continent, "wild man" sightings and other related incidents increased. Movement west had indeed started to penetrate into the main domain of the creature. Explorers, prospectors, soldiers, hunters, farmers, and even "townsfolk" reported encounters, and credible captures were reported. (Photo/Artwork: RobRoy Menzies.)

1900 – 1949: With continued sightings and other incidents, the notion that the creature might indeed be real— and not myth or hallucination—started to take hold. With greatly improved communications, newspaper reports are now current and first-hand. Moreover, with the automobile generally available, sightings start to occur to people when they are driving. (Photo/Artwork: Gary Krejci.)

1950 – DATE: The sasquatch moves from myth to the fringes of science. Greater and faster encroachment into wilderness areas reveals numerous footprints which are photographed, cast in plaster, and provided to scientists. The first motion picture of a sasquatch was taken in California in 1967. With the advent of the internet, the true extent of the phenomenon is finally revealed. (Photo: R. Patterson, Public Domain.)

Distribution of Sasquatch-related Incidents
in British Columbia

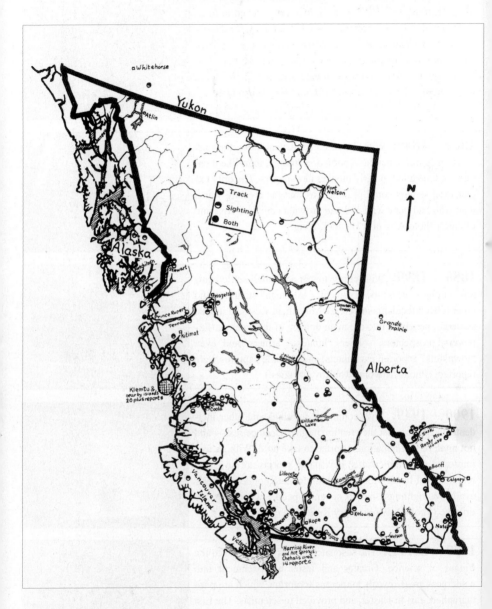

This map prepared by John Green shows the distribution of sasquatch-related incidents up to about the year 1980. For the 110-year period ending December 31, 2009 there were 379 incidents reported in British Columbia. It is felt that in reality there are about 8 times this number as most incidents are not reported.

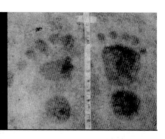

The sasquatch in British Columbia really begins with the sasquatch in North America, long before political boundaries carved up the continent.

Whatever the creature is, or might be, it probably came from Eurasia by crossing a land bridge (now the Bering Strait) that linked North America to what is now Russia. This "bridge" was usable for at least 20,000 years before the sea reclaimed the land about 10,000 years ago. The "transmigration routes" are shown in the following illustration.

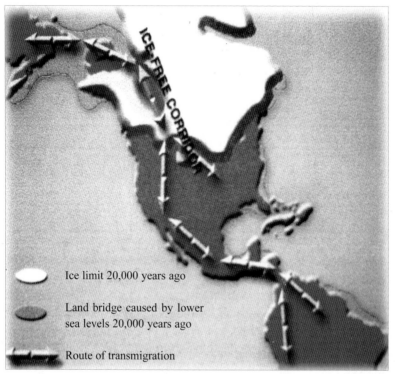

Ice limit 20,000 years ago

Land bridge caused by lower sea levels 20,000 years ago

Route of transmigration

Early transmigration route of people and animals in the Americas.

Early people and everything that walked or crawled on land trickled over to what is now Alaska, and slowly migrated down the entire length of North and Central America into South America.

It appears most of the "migrants" stayed west, and this might account for the predominant number of sasquatch incidents in BC and the Western US.

Source: General knowledge. (Photos: Stone feet, W.G. Burroughs; Map, Author's collection and Y. Leclerc.)

0000 – North America. Footprints In Stone: The earliest evidence of what could be sasquatch are bipedal footprint impressions in stone. The prints were originally made in a soft substrate which over thousands of years turned to stone retaining the impressions (fossilized footprints). Although, to my knowledge, none of these prints have been found in BC, there have been several findings in the US, particularly near Glen Rose, Texas.

Some fossilized prints show a very distinct toe configuration which effectively matches that seen in recently found, probable sasquatch footprints.

That something bipedal and with human-like feet (although generally larger) lived in North America in prehistoric times appears evident. As footprints similar to this "something"

Fossilized footprint found in Glen Rose, Texas.

are found today in ordinary soft soil or substrate, then some speculation is justified that the prints are made by the same type of creature.

The obvious question in all of this is, of course, "Where are the bones of the footprint maker?" Although there have been claims of unusual bones being found and associated with possible sasquatch bones, none to my knowledge have "survived" for proper scientific analysis. That they could be sitting in dusty museum basements with incorrect labels is a possibility. To quote Steenburg on this point: "It wouldn't surprise me if the day happens and the physical remains of a sasquatch are laid out on a table for all to see; they won't be found by a hunter or researcher who came upon them in the bush, but by some student who found them in some long forgotten museum box or drawer somewhere."

Source: General knowledge: (Photos: One stone print: J. Green; Two stone prints: W.G. Burroughs.)

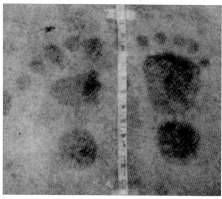

Stone prints found in Kentucky.

300,000 BC – Asia. *Gigantopithecus blacki:* The only bones we have of what might be a candidate for the fossilized footprints are those of a prehistoric Asian ape called *Gigantopithecus blacki*. The bones (all jaw bones) were found in China and India. The creature is believed to have become extinct about 300,000 years ago. It can be speculated that this creature crossed into North America over the Bering Strait land bridge. We might also speculate that for some reason, in North America it was able to outlive its predecessors in Asia. In other words, "avoid" extinction about 300,000 years ago. Could it have become, or evolved into, the creature we now call the sasquatch?

The early "hard evidence," as it were, that may be connected to the sasquatch ends here. Nevertheless, evidence in the form of First Nations artwork picks up the thread.

Bill Munns with a model he created of a Gigantopithecus blacki *using scientific information.*

Source: Dr. Grover Krantz, 1999. *Bigfoot/Sasquatch Evidence*. Hancock House Publishers, Surrey, BC, Canada, pp. 188-193. (Photo: W. Munns.)

AD 500 or earlier – North America. Petroglyphs & Pictographs: Petroglyphs (etchings in stone) and pictographs (paintings on stone) created by First Nations people in North America depict what they generally call, or describe as, the "hairy man" (many different names are used). Just when they were created is uncertain, however, definitely a long time ago (at least 500 years).

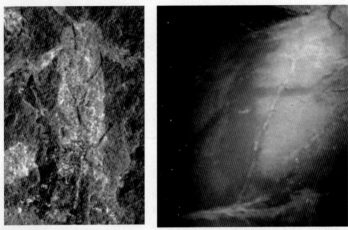

A petroglyph (left) from California, and a pictograph (right), from the Harrison, BC area. Both depict the "hairy man," or what we believe to be the sasquatch.

Petroglyphs in Bella Coola, BC. The enlarged head could be sasquatch-related. The images were not created by the Bella Coola First Nations people. We do not know exactly who made them.

Source: General knowledge. (Photos: Petroglyph from California, K.Strain; Pictograph from Harrison, author's file; Petroglyph from Bella Coola, C. Murphy.)

AD **500 or earlier – Columbia River Valley, USA. Stone Heads:** Along with pictographs and petroglyphs, very early First Nations people created stone heads that, according to anthropologists, appear to depict ape-like creatures. Several of these heads were found in the Columbia River Valley (between Oregon and Washington in the late 1800s). They are between 1,500 and 3,000 years old. As the heads definitely do not depict humans, and a carving of a known animal head was also found, we are left to wonder what (other than a sasquatch) the ape-like heads were intended to represent.

Stone head.

Source: Address by Q.C. Marsh, 1877, American Association for the Advancement of Science, Nashville, Tennessee, USA; Halpin.Ames, 1980. *Manlike Monsters on Trial*, University of BC Press, p. 229. (Photo: C. Murphy.)

AD **500 or earlier – Lillooet. Stone Foot:** Perhaps the most intriguing ancient First Nations artifact is what is commonly called the "stone foot." It was probably created about the same time-frame as the stone heads. It was found in Lillooet, BC, in or about 1947 and donated to the Vancouver (BC) Museum that year. The foot does not appear to represent a human foot, but it is "human-like."

The size of the foot as seen (next page) is 8.81 inches long and 7 inches wide. The big toe is missing, and the foot itself is broken off, indicating that the original foot was considerably longer. probably by 3 or 4 inches. It does not appear to represent the foot of a bear, as there are no claws.

As the top of the foot (right photo) is hollowed-out, museum officials classified it as "Ceremonial bowl: Medicine man's ceremonial stone." I have reasoned that this might provide a bit of a clue as to a "sasquatch" connection. In some First Nations cultures, the sasquatch was considered sacred. If the foot was patterned after a sasquatch footprint, then the resulting "bowl" might be said to have spiritual or "magical" properties. Something mixed in the bowl

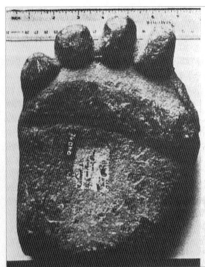

The "stone foot," a First Nations carving found in Lillooet, BC. The first image shown is from underneath, the second from above.

would therefore carry this quality.

Sources: John Green, 1973 *The Sasquatch File*. Cheam Publishing, Agassiz, BC, pp. 66–67. (Photos: J. Green)

AD 1700 and earlier – Pacific Coast. Totem Poles & Masks: Cer-

tainly, during this early time First Nations people depicted what are believed to be sasquatch-like creatures (utilizing several names) on their totem poles, wooden masks and other works. We know this because current totem poles and masks show such creatures. Naturally, stories handed down from generation to generation would result in such expressions. However, because such material is generally highly ornate and often very colorful, it gives a greater sense of mythology than reality. That it was based on an actual "sighting" of some sort in the distant past might be true, but there is no way to confirm this.

A recent Kwakiutl (BC) heraldic pole showing D'sonoqua, the "canni-bal woman" and her offspring.

Although these Kwakiutl "buck'was" (wild man of the woods) masks are recent, for certain the basic designs go back many hundreds of years. I am told that the mask on the right is a direct copy of a very old original.

Having said that, it is perhaps important to note that the sasquatch is, to my knowledge, the only prehistoric First Nations "mythological" creature that was subsequently seen by non-First Nations people.

Source: General Knowledge. (Photos: Heraldic pole and mask on the left, J. Green; Mask on the right, C. Murphy.)

1792/YR/00 – Nootka Sound (VI). First Nations People Fear Unusual Creatures: Other than what has been expressed in the First Nations art mentioned, we have no further possible sasquatch-related information up until 1792. In that year Jose Mariano Mocino, a naturalist on a Spanish voyage of exploration, heard of an unusual creature in BC. Mocino, reported in his book, *Noticias de Nutka: An Account of Nootka Sound in 1792,* that First Nations people on the Pacific Coast of North America (Nootka Sound, BC area) have a strong belief in terrifying creatures called "matlox" that are said to inhabit the vast forests. Mocino stated:

> I do not know what to say about Matlox, inhabitant of the mountainous district, of who all have an unbelievable terror. They [First Nations people] imagine his body as very monstrous, all covered with stiff black bristles; a head similar to a human one, but with much greater, sharper, and stronger fangs than those of the bear; extremely long arms; and toes and fingers armed with long curved claws. His shouts alone

25

(they say) force those who hear them to the ground, and any unfortunate body he slaps is broken into a thousand pieces.

Source: John Green, 1981. *Sasquatch, The Apes Among Us.* Hancock House Publishers, Surrey, BC, p. 25, from Jose Mariano Mocino, 1792. *Noticias de Nutka: An Account of Nootka Sound in 1792,* McClelland and Steward Ltd, Toronto/Montreal, pp. 27, 28.

General Information and Some Speculation: There have been claims that other early (prior to 1800) explorers of BC's coast heard of possible sasquatch-related experiences, however, I have not been able to substantiate such.

Nevertheless, some years ago I noticed that a depiction of Jacques Cartier (d. 1557) on a Canadian postage stamp (issued in 1984) showed him with a clay smoking pipe that has what appears to be an image of D'sonoqua, the wild woman of the woods. I thought that perhaps the stamp designer had found in a museum a pipe of Cartier's time and used in for his artwork. If so, this would at least provide an early and reasonably firm possible reference date for non-First Nations recognition of D'sonoqua. However, clay pipes were not officially invented (by Europeans) until the latter part of the 1500s, so I doubt that Cartier had one. Be that

Top: A Canadian postage stamp depicting Jacques Cartier. Lower left, a detail of the smoking pipe Cartier is holding. Lower right, a First Nations carving of D'sonoqua.

what it may, if the pipe dated from even the late 1500s or 1600s it would still say something (i.e., some late 1500s non-First Nations pipe maker was aware of D'sonoqua).

This, of course, is highly speculative. The only thing I do know is that stamp designers are usually very exacting and would indeed search out actual artifacts for their art work.

Source: General Knowledge. (Photos: Stamp, © Canada Postal Corp., 1984; D'sonoqua, Author's collection.)

The Period from 1800 to 1849

During this period in what we now know as British Columbia, the fur trade and establishment of fur trading posts took place. The posts became the major cities in the province. Although there were numerous sasquatch-related incidents in other parts of Canada and the United States, the record for the BC region is silent.

One has to take into consideration that First Nations people would have kept sasquatch-related incidents to themselves, and the few non-First Nations people in the region would not have considered a sasquatch sighting "big news." Indeed, the best one could hope for at this time was a diary or journal entry.

1811/01/07 – Jasper, Alberta. Unusual Footprints: Very close to BC, in what is now Jasper, Alberta, the geographer and explorer, David Thompson (1770–1857), reported in his journal the finding of unusual large footprints in 1811 (journal entry for January 7, 1811). The prints indicated just four toes which had claws. He did not indicate whether the prints were made by a bipedal or four-legged creature; however, the First Nations people in his party would not accept that the prints were made by a bear. As sasquatch usually have five toes and no claws, the indication is that whatever made the prints was not a sasquatch. Nevertheless, it is possible that the little toes did not register enough in the prints to see them, and we can suppose that some sasquatch have long toenails.

David Thompson: There is no known drawing or painting that shows Thomson's actual likeness. This painting was created by the author from a possible likeness based on descriptions of Thompson.

Source: John Green, 1981. *Sasquatch: The Apes Among Us.* Hancock House Publishers, Surrey, BC, pp 35-37, from *David Thompson's Narrative*: Entry for January 7, 1811. Unpublished. (Photo/Artwork, C. Murphy.)

1847/03/26 – Mt. St. Helens, Washington, USA. Race of a Different Species: Again close to BC, we have the unusual incident that occurred at Mount St. Helens, Washington, on March 26, 1847, as related in a book by the noted artist and explorer Paul Kane. The book, *Wanderings of an Artist Among the Indians of North America,* provides the following entry for the date March 26, 1847:

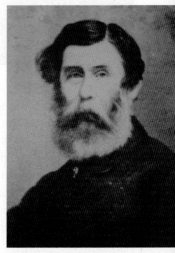

Paul Kane

> This mountain has never been visited by either Whites or Indians; the latter assert that it is inhabited by a race of beings of a different species, who are cannibals, and whom they hold in great dread; they also say that there is a lake at its base with a very extraordinary kind of fish in it, with a head more resembling that of a bear than any other animal. These superstitions are taken from a statement of a man who, they say, went to the mountain with another, and escaped the fate of his companion, who was eaten by the "Skoocooms," or evil genii. I offered a considerable bribe to any Indian who would accompany me in its exploration, but could not find one hardy enough to venture.

Source: Paul Kane. *Wanderings of an Artist Among the Indians of North America.* The Radisson Society of Canada Ltd., Toronto, Ontario. 1925. (Photo/Artwork: Public Domain.)

General Information:
We can, of course, presume that First Nations people in the Pacific Northwest continued to create sasquatch-related masks and other carvings during this period. The display shown on the right, next page, provided some years ago at the British Columbia Museum of Anthropology contains some very old "sasquatch" masks, although I can't confirm which, if any, were created in the 1800 to 1849 period.

Although First Nations people have predominantly used wood for artistic expression, there are 2 regular drawings (shown below) that possibly depict sasquatch. They were created in the 1820s by David Cusick, a Tuscarora First Nations artist in North Carolina. This is certainly a long way from BC, I realize, however, the nature of the drawings is very interesting and might provide some insights.

In the first Cusick drawing, the creature is depicted along the lines of most conventional sasquatch descriptions—gorilla-like, no neck, very long arms. In the second drawing, we see very human-like, hair-covered giants. I find this interesting because some 180 years later we still have sighting descriptions that fall in, and between, these 2 boundaries. Essentially, there

Some First Nations masks depicting D'sonoqua at the British Columbia Museum of Anthropology.

are those who believe the sasquatch is a North American ape, and those who believe it is a human of some sort.

Drawing of a "cannibal monster" by David Cusick (1820). The ominous creature is seen watching a First Nations woman parching acorns. It was frightened away because it thought the woman was eating red hot coals.

Drawing of "Stonish Giants" by David Cusick (1828). Based on legend, in AD 242 the creatures were said to be starting to overrun the country, and so ravenous that they had devoured the people in almost every settlement. With the help of the "Holder of the Heavens," the giants were eventually defeated and forced to seek asylum in the regions of the north.

Source: David Cusick, 1828. *Sketches of Ancient History of the Six Nations*, Tuscarora Village, Lewiston, Niagara, County, New York, USA. (Photos: First Nations masks, C. Murphy; Artwork, Both images by David Cusick, Public Domain.)

The Period from 1850 to 1899

During this time, the Canadian Pacific Railway pushed into British Columbia. News of gold in the province brought about a "gold rush," swelling the populations of many small towns.

1850/YR/00 – Nass River Area. The Monkey Mask: At some time in about 1850, a Niska (Tsimshian) First Nations artist living in the Nass River area of northern BC created a ceremonial mask that appears to depict an ape or monkey of some sort. As there are no wild creatures of this nature in North America, we are left to wonder as to the source of the imagery. A possible explanation might lie in sightings or stories of the giant, hairy, ape-like creature that is said to inhabit the forests—the sasquatch.

The mask was collected by Lieutenant George Thornton Emmons in about 1914, and is said to date back to the mid 1800s. In doc-

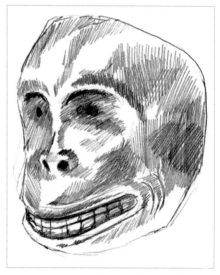

What is commonly called the "monkey mask." Drawing by Peter Travers.

umentation accompanying the mask, Emmons wrote that it represented, "a mythical being found in the woods and called today a monkey." It has been determined (2004) that the Niska (or Nishga) First Nations people hold a strong belief in "mountain monkeys." This is odd because BC's climate is hardly suitable for any known monkey. Certainly few monkeys, if any, had been brought into the area back in the mid-1800s as pets, and thereby served as an inspiration for the artwork.

Marjorie Halpin, assistant professor in the anthropology department at the University of British Columbia, prepared a paper on the mask that was published in 1980. She is quoted as follows:

George Thornton Emmons

> A good way to begin is by attempting a zoological classification of the creature represented in the Peabody mask [reference here is to the museum that has the artifact]. According to physical anthropologist R.D.E. Mashpee, it does not resemble any known species of primate closely enough to be identified as a zoologically verifiable animal. It does have primate attributes, four of which are especially noteworthy: the brow ridges, the shape of the nose, the distance between the base of the nose and the upper lip, and the prognathic, chinless lower face. The brow ridges appear from the photographs to be bow-shaped and sharply delineated from the forehead, a form replicated in many monkey species and quite unlike those of man, fossil or recent. The dished-out nose with its rounded, somewhat laterally directed nares is striking and is perhaps the best single monkey-like feature of the mask. While the large mobile upper lip is prominent in apes, it is also characteristic of numerous monkeys. The prognathism of the lower face and its lack of a chin also strongly evoke the non-human primates. Such identification is, however, counteracted by the lack of large canines, which are characteristic of every other primate except man. Comparing these and other attributes with those of both Old and New World primates, McPhee concludes that, except for the absence of prominent canines, if the mask represents a monkey, the most likely model would be found among the short-faced monkeys of the Old World.

Source: Marjorie Halpin and Michael M Ames, editors, 1980. *Manlike Monsters on Trial; Early Records and Modern Evidence.* University of British Columbia Press, Vancouver, BC, pp. 211, 212. Information on "belief in mountain monkeys" was communicated by Niska elders in 2004 when the mask was requested to be displayed in a sasquatch exhibit at the Vancouver Museum, BC. The request was denied because of the religious significance given the artifact by the Niska people. (Photos/Artwork, Sasquatch in water, RobRoy Menzies; "Monkey mask," Peter Travers; George Thornton Emmons, Public Domain.)

"Monkey Mask" Significance and Other Similar Masks: Because the "monkey mask" is not "mythological" in nature (i.e., lifelike, not painted in various colors), it can be said to provide a good argument for the existence of sasquatch. Remarkably, we have yet other examples of unpainted First Nations masks with ape-like characteristics from the same region, and probably the same time-frame.

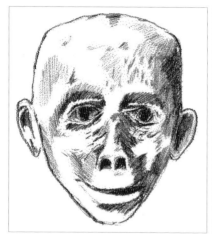 

Nishga masks showing ape-like characteristics. Drawing by Peter Travers.

Source: General Knowledge. (Photos/Artwork: Drawing (left) Peter Travers; Mask (right), C. Murphy.)

1864/YR/00 – Fraser River Canyon. The Caulfield Incident: For the year 1864 there is an unsubstantiated report that Hudson Bay Company inspector, Alexander Caulfield Anderson and his party of fur traders, were attacked with a bombardment of rocks hurled by "wild giants of the mountains." The incident is said to have occurred along the Fraser River canyon. The source is stated as the "Journal of Alexander Caulfield Anderson." However, there is no reference to the incident in Anderson's journal. Evidently, either the source stated is incorrect or the incident is a fabrication.

33

Source: Account of the attack: John W. Burns, 1954. "My Search for B.C.'s Giants." *Liberty* magazine, Canadian edition, December 1954; shown as from the journal of Alexander Caulfield Anderson, 1864. Other information: John Green, communications.

1871/CI/00 – Harrison Mills (Chehalis Reservation). Abduction of Serephine Long:

The earliest alleged abduction by a sasquatch in BC happened on the Chehalis First Nations reservation in about 1871. The story was not published until 1954, and as it is said to have come from Serephine Long, the person who had been abducted, it might have some credibility.

Whatever the case, in 1872 Serephine Long, a 17-year-old Chehalis girl who disappeared nearly one year

Serephine Long in about 1941 (age about 87).

previous, arrived back at her reservation. She was in a state of utter exhaustion and too weak to talk. She was put to bed, and that same night gave birth; but the infant died within a few hours.

At the time of her disappearance, she was about to marry a young brave named Qualac, and was last seen going to gather cedar roots. Serephine stated that she had been abducted by one of the hairy giants said to inhabit the region. The creature smeared tree gum over her eyelids so that she could not see, hoisted her on his shoulder, and raced off with the struggling woman to a cave on Mount Morris. Here she was kept a prisoner with her abductor and his elderly parents. "They fed me well," she said. After almost one year, she grew sick and pleaded with her captor, "I wish to see my own people before I die." He reluctantly gave in and after again putting tree gum on her eyelids, carried her back to her reservation. When interviewed in 1925 she said she was glad her baby died and hoped that she would never again see a hairy giant.

The glaring "oddity" in the account is the reference to "Mount Morris." This mountain is really not much more than a hill and is located right next to the main (populated) area of the Chehalis reser-

I believe what was known as "Mount Morris" at the time of Serephine Long's abduction was the large hill (or mountain) seen in the background of the photo (left). At this time, one of the hills seen in the foreground apparently has that name. I have not been able to find any official documentation in this regard. Nevertheless, that Serephine was taken to one of the smaller hills does not make sense. However, even the large hill or "mountain" I believe is Mount Morris does not make a lot of sense either, but I suppose she could have been taken to its other side. The artwork on the right shows Serephine being carried away by the sasquatch. It was featured in the published article on the incident.

vation. To believe the story, we would have to concede that either Serephine gave the wrong name for the mountain she was taken to, or the person (John W. Burns) who documented the story provided the wrong name. I have one account that states that Serephine's last name was Leon, not Long so there may have been other errors.

Source: John W. Burns, 1954. "My Search for B.C.'s Giant Indians." *Liberty* magazine, December, 1954. (Photos: Serephine Long, Public Domain; Mountains, C. Murphy; Artwork–Abduction drawing, author's collection.)

1875/12/00 – Stump Lake. Martin the Wild Man: A wild man was reported sighted at Stump Lake in December 1875. British Columbia, along with the rest of North America also has stories of "wild men," or what appear to be "men gone wild." It is likely some of these individuals have been mistaken for sasquatch.

The following is the newspaper account (1875) of the Stump Lake wild man.

A wild man in British Columbia

A wild man identified as Martin, who at one time had a store in Victoria, was sighted here [Stump Lake] a few days ago. He asked a shepherd for something to eat, and the latter gave him the remnants of his dinner. Martin was described as having a strip of what appeared to be a piece of old trousers around his neck, and not another rag or hat. His hair is gray and matted, and hangs down to his shoulders, and his body has become thickly covered with hair like an animal of the gorilla species. He carried a rifle, which was very rusty. He did not answer the many questions asked by the shepherd, but did remark after finishing his food, "That is the first bread I have eaten in five years." The shepherd offered him clothes, but he seemed unconscious of anything and simply walked away.

Obviously it appears Martin was simply a man who had "gone back to nature," however, how he managed to grow thick hair all over his body is a mystery. At that time, people believed that such did happen if a man "went wild," but we now know this is not true.

Source: Robert Colombo, 2004. *Canadian Monsters.* George A Vanderburgh, Publisher, Shelburne, Ontario, pp. 31, 32, from "A Wild Man in British Columbia," The *Times,* Ottawa, Ontario, December 4, 1875; reprinted from The *Daily Colonist,* Victoria, BC, date not known.

1884/06/30 – Yale. Jacko the Ape-boy: A boy resembling an ape was said to have been captured near Yale on June 30, 1884. The story of the boy, nicknamed Jacko, is one of the classic tales in BC's sasquatch lore. The story was made public in a *Daily Colonist,* Victoria, BC, newspaper article on July 4, 1884 under the heading: "What is it? A Strange Creature Captured Above Yale."

The article relates that workers on a train headed for Yale, BC saw what appeared to be a young boy laying unconscious near the railroad tracks, close to Tunnel No. 4, on June 30, 1884. They blew the train whistle and stopped the train. In this process, the boy woke up and scurried up a cliff side. The train men pursued him and captured him. They now saw that he was an ape-boy of some sort, totally covered in black hair.

The men tied him up, placed him in the train's baggage car, and then proceeded toward Yale. They dropped the boy off at the railway machine shops, about three-quarters of a mile north of Yale proper (actual town).

During this time, he apparently got his nicknamed. He was put in the care of a man by the name of George Tilbury, and later probably seen by a medical doctor, Dr. Hanington, from Yale.

Jacko bound and in the baggage car. Drawing by Duncan Hopkins.

The following information from the article gives a full description of the boy.

"Jacko," as the creature has been called by his captors, is something of the gorilla type standing about four feet seven inches in height and weighing 127 pounds. He has long, black, strong hair and resembles a human being with one exception, his entire body, excepting his hands, (or paws) and feet are covered with glossy hair about one inch long. His forearm is much longer than a man's forearm, and he possesses extraordinary strength, as he will take hold of a stick and break it by wrenching or twisting it, which no man living could break in the same way.

We are told that people in Yale were notified of the capture by telephone and apparently waited to see Jacko, not knowing that he had been taken off the train at the machine shops.

A newspaper report a few days later stated that Jacko had been placed in the Yale jail. Many people went to see him, but he was not there and had never been there. Another paper implied that the entire *Daily Colonist* article was a hoax.

Nevertheless, subsequent information revealed that Jacko had been quietly taken to the East or England to be exhibited in a sideshow. As far as we know, he did not arrive at either destination.

When sasquatch researchers learned of the incident in the 1950s, they did extensive research. They found that all of the people mentioned in the *Daily Colonist* article were real people. In other words, the names were not fabricated. They found people in Yale who remembered the incident, or whose relatives remember it, but no one had actually seen Jacko. Naturally, if they had not gone up to the machine shops, they could not have seen him. However, even if someone had gone up to the shops, we have to wonder if Tilbury would have allowed anyone, other than Dr. Hanington, to see the creature.

Further research revealed that an ape-boy of some sort had been exhibited in Vancouver in 1884, but this could not be substantiated (see next entry). It was also learned that a creature of the same nature had been taken up Burrard Inlet where its hair was cut off to see what was underneath, and it thereupon died. There is marginal evidence that suggests these were both different creatures, and nothing really suggests that either were Jacko.

The Number 4 Tunnel (background) and the cliffs up which Jacko was probably pursed by the train workers.

The actual railway machine shops where Jacko was said to have been dropped off. The shops were three-quarters of a mile north-east of Yale proper. Dr. Hanington was apparently the only person from Yale who went up to the shops to see the creature.

I believe that the most intriguing aspect of this incident is the detail and "quality," as it were, of the *Daily Colonist* newspaper article. Although there are a few anomalies, they can be rationalized. If the story were a hoax, as is generally believed, the person who wrote the article certainly did his or her homework. Also, given it was a hoax and the *Daily Colonist* knew this, it is odd that the full story was picked up and provided in 2 other newspapers (another BC paper and one in Manitoba). I have been given to be-

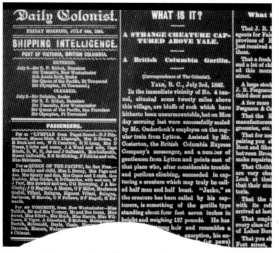

Part of the actual Daily Colonist *newspaper with the Jacko article shown in the center. The article date (1882) is probably a typesetting error (should be 1884), and the "twenty miles" distance from Yale for the No. 4 tunnel is also wrong—it is only about 2 miles away.*

lieve that newspapers at the time prevented their hoaxes from being carried by other papers—sort of professional etiquette.

Also odd is the fact that the Yale Historic Site museum has absolutely nothing on the incident in their public archives. A search by several researchers and myself came up empty. All we found was a document that referenced Dr. Hanington, who the *Daily Colonist* implies, saw Jacko. As Hanington was a "man of science," as it were, one would think he would have recorded something for posterity if he actually saw such a creature.

Source (Main Information): "What is it? A Strange Creature Captured Above Yale; A British Columbia Gorilla." Correspondence of the *Daily Colonist*, Victoria, BC, July 3, 1884. Also featured in *The Columbian*, New Westminster, BC, July 5, 1884 and *The Free Press*, Winnipeg, Manitoba, July 15, 1884. Also, John Green, 1981. *Sasquatch, The Apes Among Us.* Hancock House Publishers, Surrey, BC, p. 86, from the *Daily Colonist* (previously mentioned) and *The Columbian*, New Westminster BC, July 12, 1884 and the *Mainland Guardian*, New Westminster, BC, July 9, 1884. Also John Green, Sasquatch Incidents Data Base, Incident Number 1000128. (Photos/Artwork, Drawing of Jacko in boxcar, Duncan Hopkins; Tunnel No. 4, T. Steenburg; Machine shops, Public Domain; Newspaper, Public Domain.)

1884/00/00 – Vancouver. Ape-boy Displayed?: In 1884, it is said that an ape-boy was displayed in what is now Vancouver. Ellen Neal, a noted First Nations artist (carver) said she was told by Chief August Jack Khahtsalano (1867–1967) that a creature of this nature did reach Burrard Inlet in Vancouver and was exhibited there. We are even told that Chief Khahtsalano actually saw the creature.

There is a natural tendency to tie in this information with the story of Jacko (previous entry). However, there are discrepancies, so we cannot say that Jacko was involved.

There is also the lack of a newspaper story or other verifying documentation on the "exhibit" information provided. All we have is word-of-mouth.

Source: John Green, 1981.
Sasquatch, The Apes Among Us.
Hancock House Publishers, Surrey, BC, p. 368. (Photo: C. Murphy.)

A painting of Chief Khahtsalano by Charles H. Scott, 1943. It hangs in Kitsilano (English version of name) High School, Vancouver. Remarkably, below it hangs a D'sonoqua or "Wild Woman of the Woods" mask. It was presented to the school in 1993.

1884/CI/00 – Spuzzum. Sasquatch Killed & Buried: In about 1884, probably earlier, it is said that a sasquatch was accidentally killed "above" Spuzzum. The following remarkable account was written by Annie York (1904–1991), a Spuzzum First Nations historian. It is provided in a book, published in 1998, written by Annie York and Andrea Laforet (see Source). The story was provided to Annie by either her father or grandfather.

Chief Petek was the chief as far as Five Mile Creek, and he ruled the band. When the CPR was laying its track through this area, there was a construction camp up at the long tunnel above Spuzzum. The contractors who lived in the camp used to miss a lot of their stuff from the outside meathouse. Two cowboys had an idea about what to do. They took some long rope, the two of them, and they stayed up all nigh to watch this meathouse.

Along came a sasquatch to the meathouse to take the food away. "So there," the cowboys thought to themselves, "that's the chance for us to catch that monster that's been taking our food away." So they lassoed him, and of course when they lassoed him they had the string around his neck. Then he jumped, and he snapped his neck and died.

One of the Indians above the tunnel there, from the reservation, came along and saw these whites looking at this monster lying on his back. The man came all the way down from there to Spuzzum to the chief and told the chief what had happened to the monster, which the Indians call "sasquatch." So the chief called together his retainers, his warriors. He put his robe on—his robe is made of weasel and his banners were made of buckskin with beautiful pictures on them. He took these things and went with this warriors to the construction camp. When he got there his interpreter asked what they had done with the sasquatch. "Oh well," said one of the men, "we'll do something about it. We'll bury it."

The chief insisted he would claim the body because the Indians have always reverenced these sasquatches. The Indians claim the sasquatch is a human being, and they always claim the body, and they bury it or put it on a scaffold, if they have that kind of system. So finally these men gave up, and they gave him the body. He took the body all the way from the tunnel right down to Spuzzum. He gave it his blessing and buried it as a human being.

The Indians claim the sasquatches are human beings because they are the people who practiced to be medicine men when they were young. When boys or girls are young and want to be medicine men, their father or grandfather takes them up to the mountains and leaves them with very little to eat. They had to sleep and pray and stay alone, and some of them never returned. They got wild in the woods and never came home again. The Indians claim that that is where the sasquatch come from.

It needs to be mentioned that the term "sasquatch" was not created until about 1926, so First Nations people in the 1880s would not have called the creature by this name. It appears Annie York wrote the account after the word was in common use and simply used it for convenience. However, I still find this a little odd.

The fact this this sasquatch was captured nearby a railway tunnel might lead one to believe that it could be somehow tied in with the story of Jacko. As the tunnel is said to be above Spuzzum, not Yale (Spuzzum is above Yale), this rules out any connection. The mention of a tunnel might appear coincidental, but such is not the case. Construction camps were likely specifically for the construction of railway tunnels, so it is inevitable that a tunnel would be mentioned. What I find a little bothering is that nowhere in the book is mention made of the Jacko incident. If Annie's father or grandfather provided this story, then one would think the Jacko story, given it were true, would also have been provided.

As to the "nature" of sasquatch, we can't draw anything from the statement, "The Indians claim the sasquatch is a human being," because Annie later provides an explanation for this—they are humans who turned into sasquatch, which we know is not physically possible.

Chief Mel Bobb of the Spuzzum First Nations people, right, with author.
Chief Bobb mentioned that his people talk of the sasquatch.

Source: Andrea Laforet and Annie York, 1998. *Spuzzum: Fraser Canyon Histories, 1808–1939.* UBC Press, Vancouver, BC, pp. 87, 88. (Photo: C. Murphy.)

1886/YR/00 – St. Alice Springs (now Harrison Hot Springs. Do Hot Springs Attract Sasquatch?: In 1886, Joseph Charles Armstrong opened the St. Alice Hotel in what is now Harrison Hot Springs. The hotel drew a lot of people to the area to bathe in the natural hot springs and enjoy the magnificent scenery. At the time, it was thought that to even breath the vapors from the hot pools could cure rheumatic aliments. There is no doubt that hot springs are therapeutic. Remarkably, in Japan, macaque monkeys (or snow monkeys) love to bathe in hot springs. I have pondered if the hot springs might have something to do with the number of sasquatch sightings in the Harrison region.

The St. Alice Hotel was destroyed by fire in 1920 and was replaced by the Harrison Hotel.

The St. Alice Hotel is seen in the background. The hot springs are in the foreground. Apparently, at the time this photograph was taken, people walked up to the hot springs to bathe. The water is now piped to the hotel where it fills a large indoor and outdoor pool.

The buildings in the foreground have been replaced by a simple well into which one can peer and see the hot water bubbling to the surface.

Source: Harrison Hotel website and general knowledge. (Photo: Public Domain.)

1895/YR/00 – Metchosin (VI). The Metchosin Monster—An Expensive Hoax: A giant of some sort was said to have been unearthed near Metchosin in 1895. Called the "Metchosin Monster," it was an armless and legless giant displayed in a coffin-like box in Victoria that year. The exhibitor, Mr. Dubois, said that the creature had been dug up by a Mr. Gilbert on his farm at Happy Valley. For an entrance fee of 25 cents, one could view the oddity.

Captain Alistair McDougall, an amateur anthropologist, was so taken by the exhibit that he purchased it for $1,500, intending to donate it to a British museum. When this information was made public, a customs officer recalled a crate with a vague humanoid figure coming through customs and allowed to enter for the nominal fee of $1. Now knowing the value, he went after McDougall for a duty of 30%, or $450.

The issue went to court and a lady who lived next to Mr. Gilbert in Happy Valley came forward and said that months before she had seen people at night burying something on his farm. It was thereupon discovered that the "relic" was a fake (apparently made in the USA). Remarkably, the judge ruled that the customs office had acted on "reasonable supposition," so McDougall was liable for the duty.

Source: *Times Colonist,* Victoria, BC, (after 1980; no details).

1898/CI/00 – Yale. A Strange Cave: A cave that appeared to be inhabited by something unusual was discovered by four men near Yale in about 1898. This interesting discovery came from Charlie Victor, one of the men who lived on a reserve near Chilliwack. I don't have an exact time reference. I do know that in a photograph of Victor (seen here) taken in about 1928 he appears about 60 years old, and that he had the experience when he was a "young man." If we say that was when he was about 30, then the event took place in about 1898.

Charlie Victor

Victor and three other men were berry picking on a mountain side some five or six miles from the town of Yale. They stumbled upon a large cave in the side of the mountain. A big bolder was located to the side at the opening of the cave. The men made torches of pitchwood and explored the cave and came across a crude "sort of stone house or enclosure." As their torches kept going out, they did not explore any further and left the area, intending to come back again later.

They discussed the find with "Old Indians" who warned them not to go back as the cave was probably that of a "hairy mountain man," but they did so anyway. When they arrived at the cave they found that the large bolder had been rolled into the mouth of the cave "so nicely that you might suppose it had been made for that purpose." As a result, they were unable to go back into the cave.

Source: John Green, 1973. *The Sasquatch File*. Cheam Publishing, Agassiz, BC, Canada, p. 10, 11, from J.W. Burns, 1929. "Introducing B.C.'s Hairy Giants." *MacLean's* magazine, April 1, 1929. The story provided is one of several in Burns' article and has been adapted for this work. (Photo: Public Domain.)

1899/DE/00 – Vancouver. Loggers & Sasquatch: By the turn of the 19th century, logging operations in the Vancouver area and elsewhere were well underway. Massive forests of fir trees had no "enemies" except forest fires and man. First Nations people naturally took trees, however, such was nothing compared to the scourge of the newcomers. We can only wonder about the number of sasquatch sightings by loggers as they leveled extensive areas. However, even today there are relatively few sightings reported by loggers, and there has been speculation that now such are suppressed to avoid public reaction to possible sasquatch habitat destruction.

Logging in 1894. A steam engine tractor made the job much easier and increased the "take."

Source: General Knowledge. (Photo: Public Domain.)

The Period From 1900 to 1949

With the dawn of the 20th century, there were many more non-First Nations people in British Columbia. Whereas Natives were inclined to accept the sasquatch as simply a part of the natural or spiritual world, the newcomers were not of that persuasion. If they saw something unusual, they were curious and many provided their stories to the media which was now fully established.

1900/CI/00 – Harrison Mills (Chehalis Reservation). First Report Known of Large Footprints in BC: In about 1900, the Olds brothers found unusually large, barefoot, human-like footprints near Harrison Mills on the Chehalis Reservation. The prints were on a creek sand bar behind their home in Morris Valley.

This information came to light about 1960 in a letter to John Green received from C.H. Olds, one of the brothers, who now lived in Prince George, BC. The 2 brothers were boys at the time the prints were found.

This appears to be the first report of large footprints found in BC.

This whole area would be considered Morris Valley. Where I took the photograph, there is a house essentially to the left, and likely a few others further in. Not much has changed in the last 110 years.

Source: John Green, 1973 *The Sasquatch File.* Cheam Publishing, Agassiz, BC, from a letter by C.H. Olds, Sr. (Photos: Sasquatch, G. Krejci; Morris Valley, C. Murphy.)

1900/CI/00 – Lillooet. The Fossilized Man Hoax: Around the turn of the century (1900) what was said to be a "fossilized man" was displayed in Lillooet. The oddity had his arms folded neatly across his chest and was housed in a coffin-like wooden crate.

The great find, however, was short-lived as it was soon discovered that the "fossilized man" was actually made of Portland cement by someone who lived in the Fairview district of Vancouver, BC.

We can reason that hoaxes of this nature, or any nature, became "attractive" because of media coverage. In other words, for a hoax to be "fun" or profitable, it must have publicity, and newspapers served the purpose. The more people one can "access" and fool, then the more successful the hoax. Many, if not most people, have a "flair" for the unusual, especially if it involves something that is human-related (giants, human fossils, hominoids, etc.) so these subjects were, and continue to be, lucrative for fabrications.

The "fossilized man" of Lillooet, another hoax.

Source: Probably the *Times Colonist,* Victoria, BC, (after 1980). (Photo: Public Domain.)

1901/CI/00 – Campbell River (VI). The Monkey Man: What might be considered the first "truly credible" reported sasquatch sighting in BC occurred near the town of Campbell River in 1901.

Mike King, a timber cruiser, reported that he had an encounter with what he referred to as a "monkey man" in a forest area known as Forbidden Plateau. King had gone into the area alone, as his First Nations packers refused to accompany him for fear of such creatures. It was late afternoon when he saw a "man beast" bending over a water hole washing some edible grass roots, which it placed in 2 neat piles—uncleaned and cleaned. When it became aware of King's

Mike King is the center person in this period photograph. The identity of the other men is not known.

presence, it uttered a very human cry of terror and started running up a hillside. It stopped at some distance and looked back at King, who kept it covered with his rifle. King described the creature as covered with reddish-brown hair, and with peculiarly long arms. When he examined the ground where it had trod, he found human-like footprints of phenomenal length, with spreading toes.

What I have provided here is the generally accepted version of the story. Another version differs with regard to the creature's cry, hair color, and footprints. Also, more creature details are provided. We are told that the cry was a half-human sort of grunt. The hair was black, long and coarse. On some portions of its body the hair was 12

Forbidden Plateau is seen here. It is a small, hilly plateau in the east of the Vancouver Island mountain ranges.

inches long, and on its hands the hair hung below the finger ends. In addition, the forehead was low and retreating, and the eyes small. Footprints were very short and broad, the heel came almost to a point. In this version, we are told that King was about 25 feet from the oddity, and that he estimated it was over 6 feet tall. We also learn that the roots the creature was washing were a kind of wild onion.

There is no way to rationalize these differences at this time. However, it needs to be mentioned that the first version was written when King was out of the country. The second, after he returned and is said to be quoting him.

Source: John Green, 1973. *The Sasquatch File.* Cheam Publishing, Agassiz, BC, p. 9, from information provided by historian Bruce McKelvie. Also John Green, Sasquatch Incidents Data Base, Incident Number 1000130. (Photo: Mike King and other loggers: Image No. 1245 courtesy of the museum at Campbell River, BC; Forbidden Plateau, Wikipedia Commons, refer to the Wikipedia website for details.)

1904/AU/00 – Port Alberni (VI). Some Sort of Wild Man—First Multiple Witness Sighting: The first reported sighting in BC by multiple witnesses occurred near Port Alberni in the fall of 1904.

Residents A.R. Crump, Jim. Kinkaid, T. Hutchins and W. Buss claim they sighted some sort of wild man while hunting near Horne Lake, which lies midway between Great Central Lake and Comox Lake. The area is an uninhabited and little a explored section of the interior of Vancouver Island.

The men described the oddity as a living, breathing and intensely interesting modern "Mowgli." He was apparently young, with long matted hair, a beard, and a profusion of hair all over his body. When he saw the hunters, he ran like a deer through the seemingly impenetrable tangle of undergrowth. Pursuit was utterly impossible.

The images of Kinkade and Buss seen on the right were taken from group photographs taken in 1896 provided by the Qualicum Beach Museum & Archives. The actual photographs, shown on the next page, provide some interesting insights as to life in the forests of BC during the late 1800s. I have mused as to what other eyes might have been on the groups as they stared stern-faced at the photographer.

Jim Kinkaid in 1896.

W. Buss in 1896.

These photographs were taken on Vancouver Island in 1896. Jim Kinkaid is shown first on the left in the top photo. W. Buss is seen eighth from the left in the lower photo (he is holding an ax). Their sighting did not take place until 1904, eight years later.

Source: John Green, 1973. *The Sasquatch File.* Cheam Publishing, Agassiz, BC, p. 9, from the *Daily Colonist,* Victoria, BC, December 4, 1904. Also, John Green, Sasquatch Incidents Data Base, Incident Number 1000129. (Photos: Courtesy of the Qualicum Beach Museum & Archives.)

1905/04/00 – Comox (VI). First Known Sasquatch Shooting:
Several men reported that they saw and shot at a sasquatch in the Comox area at Union Bay in April 1905. The men were in a canoe and saw the creature on the shore. They thought it was a bear. One of them fired at it with a shotgun whereupon it straightened up, screamed, and ran upright into the woods. The men said that it appeared like a naked human covered in hair, and that it had been digging clams when it was first noticed.

A view of Union Bay as it presently appears. There are about 1,200 people now living in the area. Union Bay was first established as "Union Wharf" in 1887.

Source: John Green, Sasquatch Incidents Data Base, Incident Number 1000131, from the *Daily Colonist,* Victoria, BC, May 2, 1905. The information had been provided by a Captain Owens, a steamboat pilot, who had apparently heard of the incident from the men who saw the creature. Also Wikipedia. (Photo: Image from Google Earth; © 2011 IMTCAN, Data Living Oceans Society; © 2011 Cnes.Spot Image; Image © 2011 Province of British Columbia.)

1905/06/18 – Little Qualicum (VI). First Urban Area Sighting:
James Kinkaid, who along with three other men reported a "wild man" sighting near Horne Lake in 1904, had another sighting near Little Qualicum on June 18, 1905.

Kinkaid reported that he came quite close to a "wild man" near the Little Qualicum schoolhouse. Kinkaid was on a bicycle and thought the oddity was a man. When Kinkaid was about 10 yards away, the creature turned its head and looked at Kinkaid. It then jumped into the bush and ran away. Kinkaid said the creature was about 6 feet tall and very stoutly built, pointing out that it looked much the same as the oddity he and three others saw near Comox in the fall of 1904 (see 1904/AU/00 – Port Alberni (VI). Some Sort of Wild Man—First Multiple Witness Sighting – creature described as having long matted hair, a beard, and a profusion of hair all over its body).

Up to this time in BC, to our knowledge, the creature had stayed clear of highly populated or "urban" areas. This incident was the first reported in the province whereby it invaded the white man's domain.

Source: John Green, Sasquatch Incidents Data Base, Incident Number 1000133, from the *Daily Colonist*, Victoria, BC, June 21, 1905.

1905/07/29 – Cowichan (VI). Prospector Sees Wild Man: A prospector reported that he saw a "wild man" near Cowichan at Cowichan Lake on July 29, 1905. The prospector thought he was seeing a bear and approached the animal with this rifle raised. All of a sudden, the creature straightened up and he could then see that it was a wild man. He shouted to him, but he sprang into the bushes and disappeared. The prospector said that the undergrowth was too dense to try and follow the wild man.

Source: John Green, Sasquatch Incidents Data Base, Incident Number 495, from the *Daily Colonist,* Victoria, BC, July 20, 1905, originally from a report in the *Leader,* Cowichan, BC.

1905/10/00 – Comox (VI). Creatures Doing Sort of Sun Dance: James Johnson reported that he saw several "wild men" near Comox in October 1905. The following is a reprint of an October 1905 newspaper article on the incident. It is very interesting from the standpoint that the creatures were possibly seen doing a dance. Also, we see that there was an inquiry to the government as to the legality of killing a sasquatch.

British Columbia Mowglis
Tribe of Wild Men Roaming Woods and Frightening People

October 28, 1905

James Johnson, a rancher living near Comox, seven miles from Cumberland, B.C., reports several Mowglis, or wild men, who have been seen in that neighborhood by ranchers, says a Nanaimo (B.C.) correspondent of the *San Francisco Call*. Johnson asserts that they were performing what seemed to be a sort of "sun dance" on the sand. One of them caught a glimpse of Johnson who was viewing the proceedings from behind a big log. The Mowgli disappeared as if by magic into a big cave.

Thomas Kinkaid [believe this should be Jim Kinkaid], a rancher living near French creek, while bicycling from Cumberland, also reports seeing a Mowgli, who he describes as powerfully built man, more than six feet in height and covered with long black hair. The wild man upon seeing Kinkaid uttered a shriek and disappeared into the woods. Upon arriving home Kinkaid wrote Government Agent Bray of Nanaimo, inquiring if it would be lawful to shoot the Mowgli, as he was terrorizing that vicinity.

The government agent replied that there was no law permitting such an act. It is reported that on a recent hunting expedition up the Qualicum River an Indian saw a Mowgli and, mistaking him for a bear, shot at and wounded him. During the past month no less than eleven persons coming to Nanaimo from Cumberland have seen the wild men. Parties have been organized, and every effort is being made to capture the Mowgli.

The fact that Kinkaid took the trouble to check on the legalities of shooting the creature implies to me that he was not sure if it was simply another "animal." Back in 1905, I don't believe there were any restrictions on shooting animals in BC, such as those in place at this time. Anyway, as far as I can see, this was the first time it was reported in a newspaper that someone had inquired on the matter.

Source: Scott McClean Collection. "British Columbia Mowglis: Tribe of Wild Men Roaming Woods and Frightening People." The *Daily Bulletin,* Van Wert, Ohio, USA, October 26, 1905.

1905/12/00 – Comox (VI). Carried a Lantern: In December 1905 residents on Valdes Island reported seeing a "hairy giant." The following newspaper article published in December 1905 follows up on the article in the previous entry. Oddly, it is stated that the creature "carried a lantern."

Hairy Giant Roams Woods
Island of Valdes Sound terrorized by a "Mowgli" in Real Life

December 31, 1905

The residents of Valdes Island, near Comox, B.C., have been terrorized of late by visits from Vancouver Island's now famous Mowgli.

A rancher from Valdes Island reported that the wild man of Vancouver Island is at present on Valdes, and by the description of the creature which he solemnly asserts he saw, including the corroboration of half a dozen others, the "Mowgli" is undoubtedly the same as seen near Horne Lake two months ago.

A family named Pitcock, residing on Valdez, last week was frightened one night about 10 o'clock while sitting at supper by seeing a ghastly face looking through the window into the room.

The face was covered with long black coarse hair, only a small portion of the skin being visible. Directly the creature saw he was noticed he uttered a diabolical scream, which was heard by ranchers nearly a mile away, and vanished.

A number of young farmers with rifles searched the vicinity next day without success. Large prints of feet upon the soft earth of a flower bed under the window was the only sign of the monster, who according to those who have seen him, is nearly seven feet tall.

Twice during the next night the family heard a scream similar to that uttered by the wild man. All that day, a careful watch was kept, but there were no signs of the "Mowgli," although there were a number of footprints corresponding with those seen before, which were found along the creek that passed the Pitcock household.

Valdes Island, as seen here, is about 1 mile wide and 10 miles long. There is evidence of human habitation from at least 5,000 years ago. There is a large cave in the middle of the island which, according to legend, runs under the sea and emerges on Thetis Island, about 2.4 miles to the south west. It appears the distance for the cave "passage" would be about 5 miles. Recent explorers have found that rockfalls in the passage have made it too narrow for humans to go through.

The next night a dozen well-armed men concealed themselves near a stack of hay where the footprints of the "Mowgli" were traceable. About 10:30 p.m., horrible yells were heard in the direction of the creek and after waiting for an hour some of the posse determined to go in that direction.

They had gone up the stream about half a mile when the light of a lantern carried by the "Mowgli" came into view. Hiding behind a tree the posse had a splendid chance to view this denizen of the woods.

They describe him as very tall and powerfully built and with the exception of a few rags hanging from a belt at his waist he was entirely naked. His body was covered with long black hair and the face was identified as the same man as that seen at the Pitcock home the evening previous.

The men, after recovering from their fright, broke from

cover and one raised his rifle, but before he could find range the "Mowgli," who had by this time discovered the farmers, threw himself into the icy waters and swam for the other side of the creek.

Nothing more has been seen of him around the Pitcock residence, although two half-breeds of Alert Bay claim to have seen the same man a few days after.—Seattle Times.

The same article with a different heading and introduction was featured the next day in a different paper. The heading was "Hairy Giant in Woods: He is a "Mowgli" in Real Life and Terrorizes an Island." This article introduces the subject with:

According to arrivals from Comox today by the steamer *City of Nanaimo,* the residents of Valdez Island, near Comox, B.C., have been terrorized of late by visits from Vancouver Island's famous "Mowgli."

Comox is about 67 air miles from Valdes Island, so it is hardly "near" the island. It appears the reporter wanted to imply a strong connection between the 2 locations. Much closer towns such as Nanaimo (13 miles) or Ladysmith (8 miles) should have been referenced.

Source: Scott McClean collection. "Hairy Giant Roams Woods." The *Sunday Globe,* Boston, Massachusetts, USA, December 31, 1905, from *The Times, Seattle, USA,* (no details). Also, "Hairy Giant in Woods: He is a "Mowgli" in Real Life and Terrorizes an Island." The *Evening News,* Lincoln, Nebraska, USA, January 1, 1906. Also Wikipedia. (Photo: Image from Google Earth; Image © 2011 IMTCAN © 2011 CNES/Spot Image; Image © 2011 Province of British Columbia; Data Living Oceans Society.)

1905/YR/00 – Nahanni Region (NWT). Skeletons Missing Their Heads: In 1905 a grisly find in the Nahanni region, Northwest Territories, resulted in "sasquatch implications." Although the Nahanni region is not in BC (it is about 100 miles north of the BC border), I have considered it appropriate for this work.

Whether a sasquatch was actually involved is a matter of speculation. Nevertheless, that the creature could have been involved is a natural outcome.

Officers of the Royal Canadian Mountain Police (RCMP) were led to the skeletons of 2 missing prospectors, Frank and Willie McLeod (brothers), in a remote Nahanni valley. Both skeletons were missing their heads (skulls) and according to a pathologist, they appeared to have been torn off by brute force.

The gruesome find resulted in the valley being given the name "Headless Valley." Subsequently, there was more "foul play" in the same area. At least 13 other men have apparently been murdered in the valley. Some were burned to death in their sleeping bags, some died of gunshot wounds, others just disappeared. In one case, once again a headless skeleton was found. The remains were those of Martin Jorgensen, found beside the burnt remains of his cabin.

Frank McLeod

An expedition to explore the valley was undertaken in 1971, led by British explorer Sir Ranulph Fiennes. He later wrote a book with the title, *Headless Valley*.

In the 1940s or early 1950s a man by the name of R.M. Patterson built a cabin in the valley and lived there for some years. He contended that the so called "mountain men" in the area are not sasquatch. It appears to me that there might be some connection here with the "woodsmen," a being quite different from sasquatch (possibly the Russian almasty), which is said to inhabit parts of Alaska.

In defense of the sasquatch, there are very few reports of serious aggres-

Willie McLeod

sion associated with the creature. It has been known to throw rocks at people, but to my knowledge no one has been hit. As a rule, the creature generally looks at people and then simply walks or runs away. I personally don't believe the sasquatch was involved in the Headless Valley murders, however...

Source: Zander Hollander, 1971. "British Explores Will Probe.'Headless Valley.'" The *Statesman,* Boise, Idaho, USA, June 28, 1971. Also the *Bigfoot Bulletin, No. 26,* April-May-June, 1971, from R.M. Patterson, 1954. *The Dangerous River.* George Allen & Unwin Ltd., London, England, and other genera resources. (Photos: Public Domain.)

1905/YR/00 – Kitimat. Probable Sasquatch Killing: Billy Hall claims he shot and probably killed a very unusual creature (sasquatch?) near Kitimat at the head of Gardner Canal in 1905. Hall was hunting in the area with a friend when they saw what they thought were 2 bears on the mountainside. Hall shot one of them and hastened to collect his kill, his friend remaining in their beached canoe. Arriving close to the spot where the creature had fallen, Hall came upon some sort of creature tending to the creature he had shot (evidently the other "bear") apparently trying to revive it. Terrified, Hall turned and ran back to the canoe, with the surviving creature in angry pursuit. He found his partner asleep inside the canoe, but with strength that only great fear can provide, he whisked the canoe into the water, partner and all, to make his escape.

Source: John Green, 1981. *Sasquatch, The Apes Among Us.* Hancock House Publishers, Surrey, BC, Canada, pp. 367, 368, from a story provided by William Freeman who was told the story by his father, a resident of Kemano at the time.

1906/08/05 – Alberni (VI). Wild Man No Figment of the Imagination: On August 5, 1906 a prospector reported that he saw a "wild man" near Alberni at Horne Lake a few days earlier. The following is what he stated:

A few days ago myself and another prospector dropped right onto the wild man on the shores of Horn Lake, Alberni. The mowgli was clothed in sunshine and a smile except that his body was covered with a growth of hair much like the salmon berry-eating bears that infest the region. The wild man ran with astonishing agility as soon as he saw us.

We found the wickieup in which he had been sheltering and also many traces of where he had been gathering roots along the lake bank for sustenance. That wild man is no figment of the imagination. You can take my word for that.

The Horne Lake area was the scene of a previous multiple witness sighting in 1904. The region has spectacular caves. Although probably known to First Nations people at the time, the caves did not appear in a geological report until 1912. There has always been speculation, and some marginal evidence, that sasquatch might live in caves.

Source: "Alberni Has a Wild Man: Vancouver Island Mowgli Said to Be No Myth – Seen by a Prospector Recently." *Yukon World,* (probably Whitehorse), August 24, 1906.

1906/09/00 – Vancouver. Wild Man—Beard Like Rip Van Winkle: In September 1906 a wild man was reported to be living about 30 miles from Vancouver on tiny Paisley (also Pasley) Island. There are numerous islands along the southern coast of BC, and this is one of the smallest. An article published in a Fairbanks, Alaska, newspaper told the story:

> Another wild man has arisen to contest claims to notoriety with the Vancouver Island Mowgil. The second man lives on Paisley [or Pasley] Island, a small domain, approximately 240 acres in extent, and about thirty miles up the coast from Vancouver. He has the whole island to himself and makes his home in an old hut that used to be occupied by a fisherman.
>
> He has been chased several times by the crew of the local steamer *Era,* who wanted to find out how he managed to live without working for there is no sign of cultivation on the island. The wild man, however, never allowed the sailor men to corner him. He took to the tall timber every time and hid. He has a beard like Rip Van Winkle, almost down to his waist and presents a most uncouth appearance.

For certain, the "wild man" was simply that—a man gone wild. He evidently was able to subsist without "cultivation," so there should be no question as to the ability of sasquatch to provide for themselves.

Source: Scott McClean, 2005. *Big News Prints.* Self-published, p. 47, from "Another Wild Man in British Columbia: Lives Alone on Paisley [Pasley] Island and Flees From Sailors Who Try to Investigate." The *Daily Times,* Fairbanks, Alaska, USA, September 6, 1906.

1907/03/00 – Bishop Cove. Something Monkey-like: In March 1907 something "monkey-like" so terrified the First Nations people in a village at Bishop Cove (now Bishop Bay) that they all frantically scrambled aboard a steamer and left. The cove is located 380 air miles north of Vancouver, and is probably more remote now than it was in 1907.

This incident is one of the most startling in the records of sasquatch research. The details are provided in the following newspaper article:

The location of the incident is now called the Bishop Bay – Monkey Beach Conservancy. It appears the event in 1907 resulted in this name.

Villagers Flee Terrifying Monkey

A monkey-like wild man who appears on the beach at night, who howls in an unearthly fashion between intervals of exertion at clam digging, has been the cause of depopulating an Indian village, according to reports by officers of the steamer *Capilano,* which reached port last night from the north. The *Capilano* on her trip north put into Bishop's Cove [official name is "Bishop"] where there is a small Indian settlement.

As soon as the steamer appeared in sight the inhabitants put off from the shore in canoes and clambered onboard the *Capilano* in a state of terror over what they called a monkey covered with long hair and standing about five feet high which came out on the beach at night to dig clams and howl. The Indians say that they had tried to shoot it but failed, which added to their superstitious fears. The officers of the vessel heard some animals howling along the shore at night but are not prepared to swear that it was the voice of the midnight visitor who has so frightened the Indians.

Source: John Green, 1981. *Sasquatch, The Apes Among Us.* Hancock House Publishers, Surrey, BC, p. 48, from "Villagers Flee Terrifying Monkey." The *Province,* Vancouver, BC, March 8, 1907. Also John Green, Sasquatch Incidents Data Base, Incident Number 1000969. (Photo: Image from Google Earth; Image IBCAO; Image © 2011 TerraMetrics; Image © 2011 DigitalGlobe; © 2011 Cnes/Spot Image.)

1909/05/00 – Harrison Mills (Chehalis Reservation). He Was In a Rage: The ordeal of Peter Williams and his family provides a rare account of a violent sasquatch encounter that occurred near Harrison Mills on the Chehalis Reservation in May 1909. Williams said the fearsome creature chased him home where he barricaded himself inside with his wife and children. The angry giant shook the building to the point of collapse. Williams' frightening experience is presented here in his own words.

Peter Williams, c. 1950.

I was walking along the foot of the mountain about a mile from the Chehalis Reserve. I thought I heard a noise something like a grunt nearby. Looking in the direction in which it came, I was startled to see what I took at first sight to be a huge bear crouched upon a boulder twenty or thirty feet away. I raised my rifle to shoot it, but, as I did, the creature stood up and let out a piercing yell. It was a man—a giant, no less than six and one-half feet in height, and covered with hair. He was in a rage and jumped from the boulder to

The ruins of the Williams' family home.

the ground. I fled, but not before I felt his breath upon my cheek. I never ran so fast before or since—through brush and undergrowth towards the Statloo, or Chehalis River, where my dugout was moored. From time to time, I looked back over my shoulder. The giant was fast overtaking me—a hundred feet separated us; another look and the distance measured less than fifty—the Chehalis [came in view] and in a moment [I was in] the dugout [and] shot across the stream to the opposite bank.

The swift river, however, did not in the least daunt the giant, for he began to wade it immediately. I arrived home almost worn out from running and I felt sick. Taking an anxious look around the house, I was relieved to find the wife and children inside. I bolted the door and barricaded it with everything at hand. Then with my rifle ready, I stood near the door and awaited his coming. If I had not been so excited, I could have easily shot the giant when he began to wade the river. After an anxious waiting of twenty minutes, I heard a noise approaching like the trampling of a horse. I looked through a crack in the old wall. It was the giant. Darkness had not yet set in and I had a good look at him. Except that he was covered with hair and twice the bulk of the average man, there was nothing to distinguish him from the rest of us. He pushed against the wall of the old house with such force that it shook back and forth. The old cedar shook and the timbers creaked and groaned so much under the strain that I was afraid it would fall down and kill us. I whispered to the old woman to take the children under the bed. After prowling and grunting like an animal around the house, he went away. We were glad, for the children and the wife were uncomfortable under the old bedstead. Next morning I found his tracks in the mud around the house, the biggest of either man or beast I had ever seen. The tracks measured twenty-two inches in length, but narrow in proportion to their length.

Source: John Green, 1973. *The Sasquatch File*. Cheam Publishing, Agassiz, BC, Canada, p. 10, 11, from J.W. Burns, 1929. "Introducing B.C.'s Hairy Giants." *MacLean's* magazine, April 1, 1929. The story provided is one of several in Burns' article. Also John Green, Sasquatch Incidents Data Base, Incident Number 1000071. (Photos: Author's file.)

1910/YR/00 – Harrison Mills (Chehalis Reservation). Same Type of Creature Seen: Peter Williams, who had a "hairy giant" encounter the year before, reported that he saw the same type of creature in 1910 while out duck hunting near Harrison Mills, Chehalis Reservation. He was in an area local people call the "prairie." It is on the north side of the Harrison River, about 2 miles from the reservation village. Williams again ran away from the creature and was again pursued. Fortunately, the creature gave up the chase after about 300 or 400 yards.

Later that same day, another man by the name of Paul was chased away from his fishing spot by the same hairy individual. As he did not have his gun, he ran in utter terror, and collapsed from sheer exhaustion near his shack. The creature suddenly stopped a short distance from the shack, and then walked into the forest. Paul had to be carried home by his mother and others of his family.

Source: John Green, 1973. *The Sasquatch File*. Cheam Publishing, Agassiz, BC, p. 10, 11, from J.W. Burns, 1929, "Introducing B.C.'s Hairy Giants," *MacLean's* magazine, Toronto, Ontario, April 1, 1929. The story provided is one of several in Burns' article.

1911/YR/00 – Harrison Mills (Chehalis Reservation). Creatures Were a Man & a Woman: Chehalis resident Peter Williams who reported 2 encounters with hairy giants over the previous 2 years in the Harrison Mills area, Chehalis Reservation, had yet 2 more experiences in 1911 with these frightening creatures.

Early in 1911, Williams and a friend were hunting in the area of Williams first encounter—about a mile from the Chehalis reserve. They saw what appeared to be 2 old tree stumps, but as they approached within 50 feet or so, the stumps suddenly stood up. They were close enough to note that the creatures were a man and a woman. Williams remarked that neither were as big or fierce-looking as the giant he previously met. The hunters ran from the scene, but the creatures did not give chase as happened in Williams past encounters.

A few weeks later, Williams and his wife were fishing in a canoe on the Harrison River, near Harrison Bay. As they paddled around a neck of land, they saw the terrifying giant of past experiences in 1909 and 1910 on the beach about 100 feet away. They looked at the creature for a long time, but it took no notice of them.

Source: John Green, 1973. *The Sasquatch File*. Cheam Publishing, Agassiz, BC, p. 10, 11, from J.W. Burns, 1929. "Introducing B.C.'s Hairy Giants," *MacLean's* magazine, Toronto, Ontario, April 1, 1929. The story provided is one of several in Burns' article. Also John Green, Sasquatch Incidents Data Base, Incident Number 1000072 and 1000074.

1914/YR/00 – Hatzic. She Spoke the Douglas Tongue: Charlie Victor, who reported finding a strange cave in 1898, provided the following story (his own words) of a frightening experience while out hunting near Hatzic in 1914.

I was hunting in the mountains near Hatzic. I had my dog with me. I came out on a plateau where there were several big cedar trees. The dog stood before one of the trees and began to growl and bark at it. On looking up to see what excited him, I noticed a large hole in the tree seven feet from the ground. The dog pawed and leaped upon the trunk, and looked at me to raise him up, which I did, and he went into the hole. The next moment a muffled cry came from the hole. I said to myself: The dog is tearing into a bear, and with my rifle ready, I urged the dog to drive him out, and out came something I took for a bear. I shoot and it fell with a thud to the ground. 'Murder! Oh my!' I spoke to myself in surprise and alarm, for the thing I had shot looked to me like a white boy. He was nude. He was about twelve or fourteen years of age. His hair was black and woolly. Wounded and bleeding, the poor fellow sprawled upon the ground, but when I drew close to examine the extent of his injury, he let out a wild yell, or rather a call, as if he were appealing for help. From across the mountain a long way off rolled a

66

booming voice. Near and more near came the voice and every now and again the boy would return an answer as if directing the owner of the voice. Less than a half-hour, out from the depths of the forest came the strangest and wildest creature one could possibly see.

I raised my rifle, not to shoot, but in case I would have to defend myself. The hairy creature, for that was what it was, walked toward me without the slightest fear. The wild person was a woman. Her face was almost Negro black and her long straight hair fell to her waist. In height she would be about six feet, but her chest and shoulders were well above the average in breadth.

In my time, and this is no boast, I have in more than one emergency strangled bear with my hands, but I'm sure if that wild woman laid hands on me, she'd break every bone in my body. She cast a hasty glance at the boy. Her face took on a demoniacal expression when she saw he was bleeding. She turned upon me savagely, and in the Douglas tongue said: "You have shot my friend."

I explained in the same language—for I'm part Douglas myself—that I had mistaken the boy for a bear and that I was sorry. She did not reply, but began a sort of wild frisk or dance around the boy, chanting in a loud voice for a minute or two, and, as if in answer to her, from the distant woods came the same sort of chanting troll. In her hand she carried something like a snake, about six feet in length, but thinking over the matter since, I believe it was the intestine of some animal. But whatever it was, she constantly struck the ground with it. She picked up the boy with one hairy hand, with as much ease as if he had been a wax doll.

There was a challenge of defiance in her black eyes and dark looks, as she faced and spoke to me a second time and the dreadful words she used set me shaking. She pointed the snake-like thing at me and said: "Siwash, you'll never kill another bear." Her words, expression, and the savage avenging glint in her dark, fiery eyes filled me with fear, and I felt so exhausted from her unwavering gaze that I was no longer able to keep her covered with my rifle. I let it drop.

After relating his story, Charley added that his brave dog that never turned from any bear or cougar, lay whimpering and shivering at his feet while the woman was speaking, "just, as if he understood the meaning of her words." She spoke the words, "Yahoo, yahoo" frequently in a loud voice, and always received a similar reply from the mountain. Charlie felt sure that the woman looked somewhat like the wild man he had seen at Yale many years before, although the woman was the darker of the two. He did not think the boy belonged to the hairy giant [sasquatch] people, "because he was white and she called him her friend." He commented:

> They must have stolen him or run across him in some other way. Indians have always known that wild men lived in the distant mountains, within sixty and one hundred miles east of Vancouver, and of course they may live in other places throughout the province, but I have never heard of it.

Charlie provided this account to John W. Burns in 1929. At that time he said it was his opinion, since he met that wild woman, that because she spoke the Douglas tongue, these beings must be related to First Nations people. Charley had been paralyzed for the last eight years, and he was inclined to think that the words of the wild woman had something to do with this. The old hunter's eyes moistened when he admitted that he had not shot a bear or anything else since that fatal day.

Source: John Green, 1973. *The Sasquatch File.* Cheam Publishing, Agassiz, BC, p. 10, 11, from J.W. Burns, 1929. "Introducing B.C.'s Hairy Giants," *MacLean's* magazine, Toronto, Ontario, April 1, 1929. The story provided is one of several in Burns' article.

1915/YR/00 – Hope. Looked More Like a Human Being: Charles Flood, a prospector, emerged from the woods in 1915 and stated that he and friends saw strange creatures in an unexplored forest region near Hope. He provided the following account:

> Donald McRae, Green Hicks, and myself, were prospecting at Green Drop Lake twenty-five miles south of Hope, and explored an area over an unknown divide on the way back to Hope, near the Holy Cross Mountains. Hicks, who is part

The town of Hope is seen in the foreground here with a marker at about 25 miles south of the town and a circle showing where Flood and his companions had gone. One can immediately see they were among the very few who have gone into this area.

native Indian, told us he had seen alligators at what he called Alligator Lake, and wild humans at what he called Cougar Lake, in the area. Out of curiosity, McRae and I had him take us there. Hicks had been there a week previous looking for a fur trap line. Sure enough, we saw his alligators, but they were black, twice the size of lizards in a small mud lake. A mile further up was Cougar Lake. Several years ago a fire swept over many square miles of mountains which resulted in large areas of mountain huckleberry growth. While we were traveling through the dense berry growth, Hicks suddenly stopped us and drew our attention to a large, light brown creature about eight feet high, standing on its hind legs [standing upright] pulling the berry bushes with one hand or paw toward him and putting berries in his mouth with the other hand, or paw. I stood still wondering while McRae and Hicks were arguing. Hicks said "it's a wild man" and McRae said, "it's a bear." The creature heard us and suddenly disappeared in the brush around 200 yards away. As far as I am concerned the strange creature looked more like a human being, we seen several black and brown bears on the trip, but that "thing" looked altogether different. Huge brown bear are known to be in Alaska, but have never been seen in southern British Columbia. I have never seen anything like this creature, before

Source: John Green, 1981. *Sasquatch, The Apes Among Us.* Hancock House Publishers, Surrey, BC, p 58, from an account provided by Charles Flood in about 1957. Also John Green, Sasquatch Incidents Data Base, Incident Number 1000037. (Photo: Image from Google Earth; © 2011 Cnes/Sport Image; Image U.S. Geological Survey; Image © 2011 IMTCAN; Image © 2011 Province of British Columbia.

1919/YR/00 – Yale. Huge Nude Hairy Man: Charlie Victor, who related a sasquatch encounter in 1914 near Hatzic, stated he had another encounter in 1919 near Yale. He had been bathing with friends in a lake, and while dressing a huge nude hairy man stepped out from behind a rock. Victor stated, "He looked at me for a moment, his eyes were so kind looking that I was about to speak to him, when he turned about and walked into the forest."

Source: John Green, 1973. *The Sasquatch File.* Cheam Publishing, Agassiz, BC, p. 10, 11, from J.W. Burns, 1929, "Introducing B.C.'s Hairy Giants," *MacLean's* magazine, Toronto, Ontario, April 1, 1929. The story provided is one of several in Burns' article. Also John Green, Sasquatch Incidents Data Base, Incident Number 241.

1922/05/00 – Prince George. Primitive Cave or Tree-top Dweller: The following newspaper article tells of a wild man being seen in the Prince George area in May 1922.

Another "Wild Man" is at Large in Northern B.C.

How near does a human being resemble a bear, or how near does a bear resemble a human being? [This] is a busy question just at present in the Prince George district of British Columbia, owing to rumors of a "wild-man" running at large in the upper country. His description—given by those who have seen him—is such that he is easily mistaken for a bear at first sight: narrowly escaping the attentions of the rifle every homesteader in these lonely regions packs along wherever he goes at this time when the she bear is out with her cubs in search for food; hungry and desperate protectors if a man happens between her and her cubs.

This "wild-man" has been seen in different localities lately. Those who got close enough to see it was a human and not a marauding animal, describe him as a type of the primitive cave or tree-top dweller: huge in stature; his few rags streaming out as he flees as a deer when seen. He is evidently a white man, according to them.

This part of BC has had its "wild-men" running at large, more or less, ever since the advent of the G.T.P., supposed to have been working on the railroad construction, afterwards squatting on the wild lands abounding in this district, until they in turn become "wild" themselves, according to the remoteness from supplies or from other human companions.

Another article that appeared in a different newspaper (Prince George paper) on June 2, 1922, shows that the location where the wild man was sighted as the "Chief Lake District."

Source: Magonia Exchange, from Another "Wild Man" Is at Large in Northern B.C." The *Sun*, Vancouver, BC, May 28, 1922. Also John Bindernagel, 2010. *The Discovery of the Sasquatch*. Beachcomber Books, Courtenay, BC, p. 154, from, "Another Wild Man reported Herabouts: Type of Primitive Man, Said to Resemble a Bear: Flees at Sight of Fellow Man." The *Leader*, Prince George, BC, June 2, 1922.

1924/SU/00 – Toba Inlet. The Ostman Abduction: Albert Ostman reported that he was abducted by a sasquatch near Toba Inlet in the summer of 1924. The original documentation of this story is probably the largest account of a sasquatch encounter on record. The entire story is about eleven pages in length. As far as we know, Ostman did not tell many people of his experience directly after the incident. However, he apparently told some people, as a Toba Inlet trapper-friend of John Green said that he had first heard of the sasquatch in the early 1930s—he knew of a young Swede who had been carried off by one.

Albert Ostman

Ostman did not make his story widely known until 1957. At that time, information was appearing in the news about sasquatch research, so Ostman contacted John Green.

Green interviewed Ostman in 1957, and was given a scribbler in which Ostman had handwritten his entire experience. Green then had Ostman appear before a Justice of the Peace and swear that the story he told was true. In Canada, the courts do not take it lightly if one falsifies such a declaration, so there is a measure of credibility here. The following is a summary of the story.

The Ostman Abduction

During this summer of 1924, Albert Ostman, a construction worker, went to look for gold at the head of Toba Inlet.

After a 2-day trek, he set up his camp. When he awoke the next morning, he found that his things had been disturbed, although nothing was missing. He was a heavy sleeper, so was not surprised that he slept through the intrusion.

The next morning he awoke to find the same sort of disturbance, but this time his packsack had been emptied out and some food was missing. After yet another "visit," Ostman determined to stay awake all night to catch the intruder. He climbed into his sleeping bag fully clothed, save his boots, with his rifle by his side in the bag. He placed his boots at the bottom of the bag.

Despite his resolve, he fell asleep, but was then awakened by something picking him up and carrying him, sleeping bag and all. He was bundled up in such a way that he could not move. Whatever was carrying him, also had his packsack, as Ostman could feel food cans touching his back. Having heard of "mountain giants," Ostman reasoned that it was one of these creatures carrying him.

After a three-hour journey, the creature unloaded its cargo onto the ground. Upon emerging from his cocoon and getting himself together, Ostman discovered he was in the company of four sasquatch, 2 adults (male and a female) and 2 children (boy and girl).

Ostman spent six days with his captors, and was able to observe firsthand (and later recount in considerable detail) how the creatures looked and lived. He then made his escape by tricking the adult male into eating a box of snuff (tobacco), making him violently ill. Ostman grabbed his belongings and ran, eventually finding his way back to civilization.

—0—

The entire story has been raked through and analyzed for over 50 years. Any "red flags," as it were, in the story, save one in my opinion, can be reasonably rationalized. The exception deals with the fact that at one point Ostman made a fire and the sasquatch in his company did not react to it. The sasquatch themselves did not use fire (unless Ostman failed to mention this). If, as is generally believed, sasquatch do not use fire, I would think that seeing a man create such would have been a bit of a surprise, or at least a concern. Nevertheless, in one report from the US a sasquatch was very curious with a camper's fire, and actually went to it and pulled out a burning

This drawing of the male sasquatch Ostman encountered was made by Ivan Sanderson under Ostman's direction. Ostman later said that the upswept bang was only on the female.

stick and played with it. Can we assume that sasquatch are simply accustomed to humans making and using fire?

With regard to the details of the creature's appearances that Ostman provides in his story, they generally comply with other sightings, although more details are given. As to the way the creatures lived and provided for themselves, such was about what one would expect.

Although I have now "red-flagged" the entire Ostman story, in some ways it is easier to believe than disregard. The information to which he

Albert Ostman's sworn statement as to the truth of his story.

CANADA }
Province of British Columbia }   IN THE MATTER OF "THE SASQUATCH"
        TO WIT: }

I, Albert Ostman, of Langley Municipality in the Province of British Columbia, retired, do solemnly declare:

That the attached article, signed by me and marked Exhibit "A" is a true copy of the events which happened as set forth therein.

AND I make this solemn Declaration conscientiously believing it to be true, and knowing that it is of the same force and effect as if made under oath and by virtue of The Canada Evidence Act.

DECLARED before me at Langley }
Municipality in the Province }
of British Columbia, the }
Twentieth day of August, A.D. }
1957 }
                              }   *Albert Ostman*
A. M. Fairsmith }
A Justice of the Peace in and
for the Province of British
Columbia.

Albert Ostman, left being interviewed by John Green in 1957. Ostman is holding the scribbler in which he wrote his story. It is not known when Ostman wrote the story. However, I remember scribblers of that nature when I was in grade 1 (1947). This might indicate that he wrote it long before he met John Green.

had access in and before 1957 was very limited, yet his descriptions of the creatures stand up to "current knowledge" (things revealed in other sightings that occurred much later). I also have to wonder if a man like Ostman could have simply made up a story of this nature. Nevertheless, Ostman was a "storyteller"and considered himself a bit of a "writer." He had written in scribblers a number of "romance" stories. Nothing, to my knowledge, however, was about sasquatch encounters or related "adventures."

John Green comments on the Ostman sasquatch encounter are as follows:

Ostman portrays a much more humanlike lifestyle for the sasquatch than is indicated in other reports. For instance, they lived in one place as a family group, with the mother bringing in food for all, and with a sleeping place apparently equipped with crude blankets. The story would be unacceptable if told at a later time, but is included [in Green's data base] because of the physical descriptions of the individuals, which have stood the test of time, and of cross-examination by experts in ape anatomy, but were quite different from the public image of the sasquatch in 1957 when he first wrote the story.

Source: John Green, 2004. *The Best of Sasquatch-Bigfoot.* Hancock House Publishers, Surrey, BC, p. 27, from a handwritten account provided by Albert Ostman. Also Thomas Steenburg, Sasquatch Incidents File, and John Green, Sasquatch Incidents Data Base, Incident Number 1000052. (Photos: Albert Ostman head and shoulders, sworn statement, and Albert Ostman with John Green: J. Green; Drawing of creature, author's file.)

1924/CI/00 – Bella Coola. Scientist Writes About the "Boqs": In 1924 T.W. McIlwraith, an anthropologist, wrote an article about a dwarfish and mischievous creature called a "Boqs" that is said to inhabit the Bella Coola area. His article "Certain Beliefs of the Bella Coola Indians Concerning Animals," was published in the Ontario Archaeological Report of 1924–25. McIlwraith stated:

The beast sometimes resembles a man, its hands especially, and the regions around the eyes being distinctly human. It walks on its hind legs, in a stooping posture, its long arms swinging below its knees; in height it is rather less than the average man. With the exception of its face, the entire body is covered with long hair, the growth being especially profuse on the chest, which is large, corresponding to the great strength of the animal.

First Nations people in the region state that the boqs make life difficult by menacing them in the woods. When angered, the boqs may be heard to shriek and whistle.

Source: John Robert Colombo, 1989. *Mysterious Canada,* Doubleday Canada Ltd., Toronto, Ontario, p. 339.

1926/CI/00 – Harrison Mills (Chehalis Reservation). The Word "Sasquatch" is Developed: The word "sasquatch" was essentially developed by John W. Burns, a teacher at Harrison Mills on the Chehalis Reservation in or about 1926. Burns anglicized the First Nations word "Saskehavis," which literally means "wild man." Burns expressed an interest in the creature with the First Nations people and went on to write and have published sasquatch encounters. In his own words Burns stated:

John W. Burns in 1946.

Because they [First Nations people] knew I would not taunt them, my Chehalis neighbors revealed to me the secrets of the Sasquatch—details never confided to any white man before. The older Indians called the tribe "Saskehavis," literally, "wild men." I named them "Sasquatch," which can be translated freely into English as "hairy giants."

Burns worked on the reservation from 1925 to about 1945. The sasquatch encounters he wrote about are shown in this work under their applicable dates.

Source: John W. Burns, 1954. "My Search for B.C.'s Giant Indians" (as told to Charles V. Tench, *Liberty* magazine (now defunct). (Photo: Ralph Burns.)

1927/09/00 – Agassiz. It Was Covered With Hair Like an Animal: William Point reported that he and Adaline August had a sasquatch encounter after they attended an annual hop-pickers picnic near Agassiz in September 1927. The following is the report he provided on this unusual incident:

Adaline August and myself walked to her father's orchard, which is about four miles from the hop fields. We were walking on the railroad track and within a short distance of the orchard, when the girl noticed something walking along the track coming toward us. I looked up but paid no attention to it, as I thought it was some person on his way to Agassiz. But as he came closer we noticed that his appearance was

Looking south from Agassiz. Mount Baker, Washington, is seen in the distance. Sasquatch have also been sighted in that region.

very odd, and on coming still closer we stood still and were astonished—seeing that the creature was naked and covered with hair like an animal. We were almost paralyzed from fear. I picked up two stones with which I intended to hit him if he attempted to molest us, but within fifty feet or so he stood up [still] and looked at us. He was twice as big as the average man, with arms so long that its hands almost touched the ground. It seemed to me that his eyes were very large and the lower part of his nose was wide and spread over the greater part of his face, which gave the creature such a frightful appearance that I ran away as fast as I could. After a minute or two I looked back and saw that he resumed his journey. The girl had fled before I left, and she ran so fast that I did not overtake her until I was close to Agassiz, where we told the story of our adventure to the Indians who were still enjoying themselves. Old Indians who were present said: the wild man was no doubt a "Sasquatch," a tribe of hairy people whom they claim have always lived in the mountains—in tunnels and caves.

Source: John Green, 1973. *The Sasquatch File*. Cheam Publishing, Agassiz, BC, pp. 10, 11, from J.W. Burns, 1929. "Introducing B.C.'s Hairy Giants," *MacLean's* magazine, Toronto, Ontario, April 1, 1929. The story provided is one of several in Burns' article. (Photo: Image from Google Earth; Image U.S. Geological Survey; Image State of Oregon; Image © 2011 Digital Globe; Image © 2011 IMTCAN.)

1928/YR/00 – Duncan (VI). Henry Napoleon Story: In 1928 historian Jason Ovid Allard, who was investigating sasquatch legends, interviewed Henry Napoleon, a member of the Clallam, Washington tribe, while Napoleon was in Duncan. Napoleon told him of an encounter he had as follows. There is no location provided for the incident, however, it appears it was on Vancouver Island:

> I had been warned not to go too far into the wilderness, but in following a buck I had wounded, I went in further than I expected. It was about twilight when I came across an animal that I believed to be a big bear. The creature looked up and spoke to me in my own tongue. It was a man about seven feet, and his body was very hairy.

Jason Ovid Allard

Napoleon went on to tell that that the creature invited him to meet the rest of his tribe. He followed his towering guide along an underground trail to meet the others. The strange tribe, he said, hibernated in the winter, and during the summer kidnapped women and children from coastal villages. Allard noted that the Puntledge tribe had been virtually wiped out by "strange natives" from Forbidden Plateau years before.

Source: Fuhrmann Files – magazine article (no details) and Wikipedia. (Photo: Public Domain.)

1928/YR/00 – Lavington. An Odd Visitor: In 1928, something strange was seen outside a home in Lavington. A woman ill in bed, asked her daughter to send away a man standing by the fence of their property. The daughter looked out and saw a tall, bulky, furry creature standing behind a fence post, with its hands resting on top of the post. The creature left shortly, whereupon the mother said that it had been farther up the fence line when she first saw it and it walked upright to the position last seen.

Source: John Green, Sasquatch Sighting File (not published).

1928/YR/00 – Bella Coola. Ape Seen & Shot: George Talleo stated he saw and shot at an "ape" near Bella Coola during 1928. He was trapping on a hillside near Bella Coola, east of the South Bentinck Arm, and saw the creature stand up from behind a fallen tree. He shot at it with a small caliber rifle and it fell. Talleo ran from the scene immediately.

Prior to this incident, he noticed that a lot of moss had been stripped from a rock face, and he then found that it was being used to cover a large pile of excrement.

Source: John Green, 1973. *The Sasquatch File.* Cheam Publishing, Agassiz, BC, p. 14, from personal communication. Also John Green Sasquatch Incidents Data Base, Incident Number 1000067.

1929/DE/00 – Windermere. Tall Human Skeletons: In the late 1920s work crews cutting a trail to a lake near Windermere were reported to have uncovered four very tall human skeletons. We are told they ranged in height from 6 feet, 9 inches to 9 feet.

I am sure a skeleton 9 feet tall would have received some scientific attention, but nothing else I have run across mentions this finding. In current times, I believe that if very old skeletons are uncovered, First Nations people demand that they be given such remains for re-burial. Indeed, actual normal height skeletons that were displayed in museums have now been returned to the respective First Nations people.

Source: T.W. Peterson, 1967. "Wanted Dead or Alive: Sasquatch," *Real West* magazine (appears to be defunct), July 1967.

1930/YR/00 – Swindle Island: Wading Sasquatch: A group of adults and children reported seeing a sasquatch on Swindle Island near Kwaka on Kitasu Bay during 1930. Tom Brown (a child at the time), who was with the group , said that they were walking at night when a sasquatch was heard and seen wading in the shallows of the bay. Adults in the group shot at the creature, using rifles, shot guns, and a revolver and they heard it scream. The area was checked in the morning but nothing was found.

Source: John Green, 1973. *The Sasquatch File.* Cheam Publishing, Agassiz, BC, p. 14, from information provided by Tom Brown. Also John Green, Sasquatch Incidents Data Base, Incident Number 1000135.

1934/03/23 – Harrison Mills (Chehalis Reservation). Rocks From Above: Men fishing near Harrison Mills (Chehalis Reservation) at Morris Creek, reported they encountered a sasquatch on March 23, 1934. They said rocks fell around them from a cliff. They looked up and saw a gigantic, hairy man preparing to roll a huge bolder over the edge of the cliff. The men quickly scattered.

Source: John Green, Sasquatch Incidents Data Base, Incident Number 1000138. Green's material is used in, John A. Cherrington, 1992. *The Fraser Valley: A History.* Harbour Publishing, Madeira Park, BC, p. 295.

1934/03/00 – Harrison Mills (Chehalis Reservation). Rock Bombardment: Tom Cedar reported a frightening encounter near Harrison Mills, Chehalis Reservation, in March 1934. He was fishing on the Harrison River when his boat was bombarded with huge rocks from cliffs above . He narrowly escaped after one rock missed his boat by mere inches. Looking up, he saw a huge, hairy, man-like creature waving its arms wildly and stamping its feet, clearly agitated by his presence. The creature frightened Cedar to such a degree that he cut his fishing line and paddled away to safety. The distraught fisherman also stated that salmon have twice been stolen from outside his house where they were tied well above the reach of his dogs.

Source: Scott McClean, 2005. *Big News Prints.* Self-published, p. 49, from "Reports Tell of Canadian Monster Men; Settlers Fifty Miles from Vancouver Describe Hairy Giants." The *Times,* Hammond, Indiana, USA, October 25, 1935. (This reference shows the incident date as "last March," however another reference written in 1934 shows it as happening in that year.)

1934/03/00 – Harrison Mills (Chehalis Reservation). Night Encounter: Frank Dan reported a close encounter with a sasquatch at Harrison Mills, Chehalis Reservation, in March 1934. Dan lived close to the Harrison River. He went outside at night to see why his dog wouldn't stop barking and saw the "hairy giant" nearby. It was naked and covered with fluffy hair from its head to its feet. There was a small hairless area around its eyes. It was extremely tall and had a very muscular build. Dan raced back into his house and bolted the door. The unwanted visitor strolled back into the underbrush and disappeared.

The press had a bit of a field day with this report stating that fear had gripped the people and men were never far from a rifle. However, it appears this was just the case of a curious sasquatch.

Source: Don Hunter with René Dahinden, 1993. *Sasquatch-Bigfoot: The Search for North America's Incredible Creature.* McClelland & Stewart Inc., Toronto, Ontario, pp. 34-35, from "Sasquatch is Again on Rampage," The *Province, Vancouver,* BC, March 3.1934. Also Scott McClean, 2005. *Big News Prints,* Self-published, p.47 from "Terrible Sasquatch Abroad in the Land." The *Oshkosh Northwestern,* Wisconsin, USA, March 3, 1934.

1934/04/09 – Vancouver. First Official Sasquatch Hunt: What might be deemed as the first official sasquatch hunt was undertaken in April 1934 by 2 Americans, J.F. Blakeney and C.K. Blakeney, brothers and medical students at the University of California. They left Vancouver BC for the mountain on April 9, 1934.

I have no record of what, if anything, the brothers found. It appears they had been keeping their eye on newspaper reports and probably saw a magazine article on the sasquatch by John W. Burns that appeared in *MacLean's* magazine in April 1929 (contents of which are presented in this work in chronological order). However, not all of the incidents I have presented so far appeared in print until much later than 1934. Whatever the case, it's "hats off" to our American cousins for taking this initiative.

Source: "Californians Out to Bag Legendary 'Sasquatch.'"*The Fresno Bee,* Fresno California, USA, April 9, 1934.

1934/05/00 – Harrison Mills: A Buzzing Sound: Mrs. James Caufield reported seeing a sasquatch on her farm in Harrison Mills in May 1934. Mrs. Caufield relates that she was washing clothes in a river when she heard a buzzing sound similar to that made by a humming bird:

> I turned my head, but instead of a bird there stood the most terrible thing I ever saw in my life. I thought I'd die for the thing that made the funny noise was a big man covered with hair from head to foot. He was looking at me and I couldn't help looking at him. I guess he was a Sasquatch so I covered my eyes with my hands, for the Indians say that if a Sasquatch catches your eye you are in his power. They hypnotize you. I felt faint and as I backed away to get to the house I tripped and fell. As he came nearer I screamed and fainted.

Mrs. Caufield's scream brought her husband running out of the house just in time to see the creature run off into the bush.

Source: The *Star*, Toronto, Ontario, May 28, 1934 (no details). Also Scott McClean collection, "Hairy Tribe of Wild Men in Vancouver." The *Daily Dispatch,* Brainerd, Minnesota, USA, July 10, 1934 (used for the entry). Also John Green, Sasquatch Incidents Data Base, Incident Number 1000136.

1934/07/29 – Lincoln, Nebraska, USA. The Dickie Article: By 1934 news of the sasquatch in BC evidently trickled through to a noted US newspaperman by the name of Francis Dickie. He wrote an article that appeared in the *Sunday Journal and Star* Lincoln, Nebraska, on July 29, 1934. For the most part, Dickie rewrote the incidents that had been covered by John W. Burns in his *MacLean's* magazine article (1929)—all provided separately in this work. However, Dickie did have some new material for the time. What made the Dickie article noteworthy was the startling heading, map of sightings, artwork and photographs.

I believe Francis Dickie is shown in this photograph taken in 1927.

I have shown a partial image of the article below so the reader can appreciate its impact. I have also provided a reprint of the article (next page). I think the article is important from the standpoint that it reflects the "state of the sasquatch issue," as it were, in 1934 (77 years ago). References are shown as to "misinformation."

Are they the Last Cavemen?

British Columbia Startled by the Appearance of "Sasquatch," a Strange Race of Hairy Giants
By Francis Dickie

It is peculiarly in keeping with the topsy-turvy year of violently varying weather, universal human unrest, droughts,
grasshopper plagues and other phenomena that there now
comes from various eyewitnesses the report of seeing
some of the "Sasquatch," those weird hairy men reported
for twenty years to dwell in the tremendous and unexplored
mountain region of British Columbia, Canada.

Their reported return is particularly in keeping with this
unusual year, as remarkable for the number of appearances
of various startling monsters sighted from Scotland to the
Caribbean, from the Pacific to the Mediterranean, the reality of which is affirmed by scores of eyewitnesses. Moreover,
the statements of some of these people, in-so-far as curious
denizens of the oceans are concerned, have been borne
out, for within a short time of each other, at a dozen places
on the European coast, the remains of incredible monsters
of the deep have been cast up.

Of all these mysterious earthly visitants, perhaps the
"Sasquatch" is the least known, by reason of the rarity of
their appearance and the reluctance of those who have
seen them to talk.

The existence of a troglodyte race inhabiting the mountains of British Columbia in many of the vast caves is a tribal legend among the Chehalis Indians and those of the
Skwah Reservation, near Chilliwack, in the Harrison Lake
district, about a hundred miles east of Vancouver. Among
the Indians the race has been known for centuries by the
name "Sasquatch," **[Note 1]** or hairy men.

But reports of these creatures being seen frequently at
various times over a period of the last twenty years, and
more frequently in recent weeks, have caused a number of
people to raise the question if these strange creatures may
not be more than an Indian legend of the past, and that

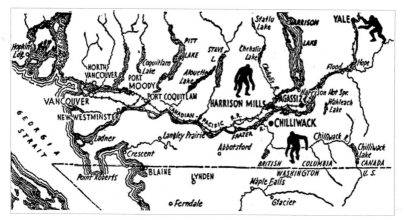

This map which appears in the Dickie article appears to be the first published "sasquatch map," at least the first for BC.

some of this race of cavern dwellers are still living in the unexplored vastness of British Columbia.

The Sasquatch have been seen, according to the statements from both white men and Indians. The wild, hairy men have mostly been reported in the Harrison Lake district, but also as far east as the mountainous region of Yale, on the main line of the Canadian Pacific Railway.

The repeated reports of eyewitnesses of seeing one or more of the huge hairy men in recent years, and more particularly in the last month, and the mounting number of the reports of eyewitnesses now seem to point strongly that the old tribal legend, long contemptuously flouted by the white man, is true, and that at least a few of this mysterious race may still inhabit the solitudes nearby where once they were numerous. The possibility of this is further borne out when it is recalled that the remains of a giant race of men recently have been unearthed in the mountainous region of Mexico. **[Note 2]**

The chief difficulty, in fact the whole task of an investigator, in matters of such phenomena as Sasquatch or sea serpents, is, of course, the credibility of the witnesses. If untruthful, what motive lies behind their story? In the case of the Sasquatch, the element of credence is heightened because in most cases the witnesses have been reluctant ones, some of them not revealing their stories for years.

From a careful comparison of all eyewitness statements to date, all are closely in agreement as to the following facts: The Sasquatch are gigantic men, varying from six and one-half to seven feet in height. One, and only one, witness states the nose of them to be very broad, and the arms long, reaching below the knee. All but one are agreed as to the hideousness of the face.

However, as in most instances the Sasquatch were not seen close up, it is natural the descriptions remain very general. Those people who have been close were so terror-stricken that their accounts are vague. Yet, aside from one of the most recent happenings, in only two other cases have the Sasquatch shown themselves hostile.

The fact that some of these strange people have just been reported close to civilization at this time accurately compares with dates noted by the Chehalis Indians. The Indians have oral records covering three generations. According to them, members of the tribe have seen in the springtime every fourth year the light of a great fire on one f the highest peaks in the Chehalis Range. The fire burns for four nights, riding in a very high, thin column. Sometimes it is suddenly extinguished, to rise again a little later. That this is some periodic mark of a return to a certain place of worship at some ancient shrine, or a communication with members in some remote mountain fastness, are possible conjectures. **[Note 3]** These periodic returns to some ancient gathering place do bring these people close to what are now civilized areas.

A few days ago, a middle-aged Indian, Tom Cedar, was trout fishing from his canoe on Morris Creek , a tributary of the Harrison. He was near a rocky terraced bank. Suddenly a large rock struck the water so close to his canoe that he was drenched by the splash. Looking up, he saw with amazement a huge hairy man above him just as he threw another rock. This also barely missed the canoe. Cedar paddled rapidly upstream to the settlement.

By way of noting an odd coincidence, this particular

stream, now called Morris Creek, was known as Saskakau when the white man first arrived, and is so called on old maps. Nearby are caverns which were investigated by Captain Warde, forty years a resident in the district. He states they bear evidence of habitation. Upon the walls are some crude drawings. **[Note 4]** In this region, according to the Indians, two large bands of Sasquatch fought a long time ago until both were brought almost to extinction.

The other evidence of hostile intention of some of these creatures dates back twenty years and consists of the statements of two Indians, Peter and Paul Williams, of Chehalis. The following is very much a condensed resume:

"On an evening in May," states Peter, "I was about a mile from the reserve, near the foot of the mountain, when what I at first took to be a bear rose up in the underbrush. It was between six and seven feet tall, covered with hair. I turned and ran through the underbrush to my dugout. The hairy man came after me. I paddled across the stream, which is not very deep, and the man waded after. I reached the house where my wife and child were inside. I bolted the door. Presently the hairy man arrived. It was growing dark. He prowled around, grunting and growling, but after a little while went away."

About the same time, Paul was chased from a creek where he was fishing. But the giant did not run after him very far, and apparently the action was only to drive the man away to get the fish he had taken.

On another occasion in the next year, Peter and another man came upon two giants so close as to distinguish a man and a woman. Though the Indians ran, they were not pursued.

Charley Victor, now living at Chilliwack, relates that he and a little group of companions, while bathing in a mountain lake near Yale, suddenly looked up to see a huge man, naked and hairy, looking down upon them from among the trees.

"His big eyes looked very kind, and I was about to speak to him when he drew back into the trees," related Charley.

Here we have the only witness who give a favorable reaction to sight of the mysterious race.

This took place many years ago and at a point about a hundred miles from where the majority of the Sasquatch have been reported seen in recent times.

The next account of which any fully recorded evidence is now to be seen deals with September 1927, near the little mountain town of Agassiz, which is very near the points at which all the other Sasquatch have been reported. A party of hop-pickers were picnicking here. On their way to this, a man, named Herbert Point, **[Note 5]** and a girl, Adeline August, were walking when they saw a strange creature approaching. "He was twice as big as the average man, with hands [arms?] so long they nearly touched the ground, and his nose seemed spread all over his face. His body was covered with hair like an animal. He stopped within fifty feet of us. We ran away as fast as we could." The lines in quotes are excerpts from a letter written by the man in answer to a query of what he had seen.

Emma Paul, seen here, who is mentioned in this article, but not shown, is said to have told missionaries that she met sasquatch three times. Unfortunately there are no details of these incidents. This image was obtained from an article by John W. Burns, 1954. It is likely Emma was a relative of Chief Alexander Paul, 1990s. (See 1995/8/03 – Harrison Mills (Chehalis Reservation). Footprints Found and Later a Sighting.)

Within recent weeks Emma Paul and Millie Saul, two other members of the Chehalis Reserve, saw one of the Sasquatch near their home on the fringe of the woods. Several nights later he was heard prowling around the home of Millie Saul, and one rubbed his hand over the window frame.

To date, the last report was from Harrison Mills, a small hamlet on the Harrison River.

The woman, on hearing a humming noise, looked up to see a big man covered with hair on the edge of the clearing. She was frightened. Taking a backward step, she fell into one of the half-full laundry tubs at which she had been working. When she had extricated herself and looked again, the man had disappeared.

Such, in brief, are the legendary and eyewitness stories regarding the Sasquatch.

The scientific board connected with the Museum of Vancouver is skeptical regarding the existence of any such remnant of a race that once might have roamed the forested regions.

An objection that the climate is too rigorous for a naked race, no matter how hairy, might be answered by pointing to the Feugians [should be Fuegians] – **[Note 6]**, who live in a much more inhospitable one.

The eyewitness reports have always been reluctantly given. There may be many more. The chief objection among the natives to telling white inquirers is fear of ridicule. This sensitiveness is much stronger among natives than whites.

Here, for the present, the matter must rest. Perhaps further witnesses may be heard in the future. Remembering, however, in judging the possibilities of the existence of the Sasquatch, how many people have seen sea serpents and that remains of strange creatures have been recently washed on various shores. **[Note 7]** It is quite within the bounds of probability that just as there are unknown forms of life in the boundless depths of the ocean, equally so may there be in the enormous wilderness stretches of British Columbia wild hairy men roaming.

— 0 —

The captions for the photographs shown in the article are as follows (left to right, ending with the map.)

—*Millie Saul of the Chehalis Reserve, had a harrowing experience with the sasquatch when one approached her home during the afternoon. Several nights later she heard a prowler, and glancing up, saw him rubbing his hand over the window pane. She screamed and the giant visitor disappeared.*

—*Entrance to one of the great caves thought to be the home of some of the giant Sasquatch.*

—*Tom Cedar, an Indian, was fishing in Morris Creek. Suddenly a huge rock struck the water near his canoe. Looking up, he saw a hairy-figured giant, poised to hurl another stone at him.*

—*It was near this Indian house where the first Sasquatch were reported seen years ago. The house was abandoned.*

—*Every four years great columns of fire are seen on certain mountain tops, thought to be signals from the Sasquatch. The arrow shows one of these signal fires. (See Note 3)*

—*Every four years a strange race of giants is reported at various places in British Columbia. They are reported this year as shown on the map.*

Note 1: The term "sasquatch" was not created until the 1920s. The original term was "saskehavis" (wild men). John W. Burns, a teacher on the Chehalis reservation anglicized the term making it the current "sasquatch."

Note 2: There is no proof of this finding.

Note 3: There is absolutely no proof of "sasquatch gatherings" and no credibility as to sasquatch making fires. The idea that sasquatch can and do use fire has not been proven. Naturally, if it could be proven, then we would have a good indication of the creature's nature—humans are the only beings that use fire.

Note 4: I have no evidence of "crude drawings" in any caves in the Chehalis area. There are some pictographs in the Harrison region done by early First Nations people, but they are not in caves. One is said to depict a sasquatch. It appears that this was the source for the information provided.

Note 5: My reference for the first name shows William Point.

Note 6: Fuegians are the indigenous inhabitants of Tierra del Fuego, at the southern tip of South America. The Fuegian population was devastated by diseases brought by "outsiders" and their numbers were reduced from several thousand in the 19th century to hundreds in the 20th century. There are no full-blooded native Fuegians today; the last died in 1999. (From Wikipedia).

Note 7: *No "strange creatures" that have washed onto any shores have been proven to be something other than a recognized species. In other words, sea serpents are still in the realm of cryptozoology (notwithstanding **possible** photographic evidence of an 11-foot Cadborosaurus).*

As can be seen, there are some issues with the article. Nevertheless, it was a major piece of work in the 1930s, and as such forms an important part of BC's sasquatch history.

Research done by Francis Dickie and John W. Burns resulted in the entry on sasquatch in the *Encyclopedia Canadiana* (1970) which reads as follows:

Encyclopedia Canadiana (1970 edition)

Sasquatch or Saskehavas, legendary tribe of aboriginal giants: Indian folklore places their habitat mainly in the vicinity of Harrison Lake, B.C., some 60 miles from Vancouver, though they have been reported as far inland as Kamloops. Known originally to the Indians—most of whom firmly believe in the existence of this mysterious race—as Saskehavas (wild men), they are called by the more skeptical whites Sasquatch (hairy men). They are described by Indians who claim to have seen them as hairy monsters between 7 and 9 feet tall, of subhuman appearance, with wide flat noses and abnormally long arms. They are believed by the Chehalis Indians of the Harrison Lake area to be descendants of two bands of giants who were almost exterminated in battle many years ago. They are said to inhabit remote mountain caves and meet periodically near the top of Morris Mountain, upon which fires have been observed at regular intervals for many years. The earliest known written record of a belief in Sasquatch is that of Alexander Caulfield Anderson of the Hudson's Bay Co., who established a post near Harrison Lake in 1846; in his reports he frequently mentioned the wild giants of the mountains. The finding in 1932 of remains of a long-extinct race of giants in Mexico gave some impetus to the belief that the remnants of a prehistoric race of troglodytes may have survived in BC. Of re-

cent years several small search expeditions have explored the Harrison Lake area without success.

References. Dickie, F., "Cave Men in B.C.?" *Toronto Star Weekly,* July 21, 1934; Burns, J.W., "My Search for B.C.'s Giant Indians," *Liberty,* Dec. 1954.

Francis Dickie

We see that the erroneous references in Dickie's article are shown here (fires, sasquatch gatherings, race of giants in Mexico). Furthermore, the reference to Alexander Caulfield Anderson is without any foundation as previously provided in this work.

Source: Scott McClean collection and Magonia Exchange, from Francis Dickie, 1934. "Are they the Last Cavemen? British Columbia Startled by the Appearance of 'Sasquatch,' a Strange Race of Hairy Giants." *Sunday Journal and Star,* Lincoln, Nebraska, July 29, 1934. (Photos: Author's collection – articles as stated, and *Encyclopedia Canadiana,* now defunct.)

1934/08/00 – Haney. Human Face On a Fur-clad Body: Francis Hann and his companion saw what they described as a creature with a "human face on a fur-clad body" near Haney at the head of Pitt Lake in August 1934. They used binoculars to observe what they thought was a bear feeding on berries and were astounded at what they saw. After the creature left, they went to where it had been to examine any tracks it may have left. Large human-like prints were found that did not show any claw marks. The men came to the conclusion that they had seen a sasquatch.

The foregoing report is essentially based on a letter dated December 15, 1968, to the *Sun.* The following is a full reprint of that letter:

Mr. John Rodgers C-O
Vancouver Sun
Vancouver, B.C.

Dear John: Very interested in your article this date on the Sasquatch and I will certainly get a copy of John Green's booklet now published. Here is the story. Thirty-five years ago there was a stockbroker office on Dunsmuir operated by Cartwright and Crickmere. Cartie had a cabin cruiser,

Pitt Lake in 2010. It is still a remote area, and sasquatch sightings in the region have continued.

was a bachelor, hard as nails in business and with a heart of gold for those he liked, and an experienced outdoorsman. One weekend he asked my wife and I along on a party of eight to go to the head of Pitt Lake. I was an ardent rock hound, and this was virgin territory for me to possibly add to my collection, and Cartie wanted to run down a clue he had as to some lost mining prospect. In the morning we left the rest of the party to amuse themselves for the day and Cartie and I climbed some fifteen hundred feet and rested at the edge of a small plateau to eat our lunch. We had our haversacks and small hammers but were otherwise unarmed. A movement behind a thicket some quarter mile away caught my eye and I said "Cartie, there is something down there." He looked then asked for the field glasses. We both thought it was a black bear feeding on berries, then he exclaimed, "here, look at its face!"

Through the glasses it was quite plain—a human face on a fur clad body. "What the hell," I said, "he must be a hermit or something of the kind but look at the size of him." Cartie replied, "Wait until he leaves and let us go down and look at the tracks." So we waited. I don't think the creature saw us, though he or she may have sensed us, as presently it went way across the plateau and vanished among the rocks. We went down after a suitable interval and examined the tracks which were quite distinct. Cartie looked pretty grim and said, "Let's go back." What you have just seen is a Sasquatch; you...you will just be laughed at and you will have a miserable time of it! Just forget the whole thing and keep quiet." So I did, and I have, until now.

(Francis Hann)

Source: John Green. 1970. *Year of the Sasquatch*. Cheam Publishing, Agassiz, BC, p. 62, from a letter by Francis Hann sent to John Rodgers, the *Sun*, Vancouver, BC, December 15, 1968. Also John Green, Sasquatch Incidents Data Base, Incident Number 1000020. (Photo: C. Murphy.)

1934/YR/00 – 100 Mile House. Quarter Mile of Tracks: A 14-year-old girl (last name Goffin) reported finding a long line of unusual tracks near 100 Mile House in 1934. The girl, who lived at Hawkins Lake, was out looking for her family's cows and found a quarter mile of tracks that looked human, had prominent toes, and showed no sign of front feet. There was a strong smell in the area, and as she followed the tracks she saw signs indicating something had bedded down. She also found feces similar to that of humans.

Source: John Green, Sasquatch Incidents Data Base, Incident Number 1000081.

1935/10/00 – Harrison Hot Springs. Straddled a Log & Paddled With Hands & Feet: Several people reported they saw 2 sasquatch near Harrison Hot Springs, Morris Valley/Creek area in October 1935. They watched as the two creatures scrambled out of the forest and proceeded across the marshy bogs along the Harrison River.

The creatures demonstrated incredible agility. They leapt down the rocky slope to the river with the "agility and lightness of mountain goats." They walked without breaking stride across the swamp flats in the shadow of Little Mystery Mountain. They then crossed Morris Creek by straddling a log and paddling with their hands and feet. Once across, they climbed hand over hand up the nearly perpendicular cliff to Gibraltar Point. According to the eyewitnesses, snatches of their unusual "language" could be heard during this time.

When at the top of the cliff, they were observed to be carrying clubs. We are also told (assumed by other witnesses) that they stayed away from trails used by people and had gone out of their way to avoid a herd of cattle in their path.

How the creatures came to have clubs at the top of the cliff is a mystery. There is no mention of them having such when they climbed the cliff. It appears some fiction has crept into the story.

Source: "Reports Tell of Canadian Monster Men; Settlers Fifty Miles from Vancouver Describe Hairy Giants," *The Times,* Hammond, Indiana, USA, October 25, 1935. Also: John Green, Sasquatch Incidents Data Base, Incident Number 1000137.

1937/YR/00 – Lillooet. Eight-foot Skeleton: Floyd Dillon claims he uncovered a nearly 8-foot-tall skeleton with an 8-inch tail while digging a trench behind his home along the Fraser River near Lillooet in 1937. His wife, Francis, said the skull was twice the size of a man's skull.

It appears Dillon just left the bones where he found them, although probably not deeply buried. By the time he decided to send them to the Provincial Museum in Victoria, BC, most had disappeared. Only the damaged skull and a few bones were found. We are told these were sent to the museum. However, the museum claims they never arrived.

The reference to an 8-inch tail is odd, however, Dillon likely mistook something for a "tail." As to the bones disappearing, animals would take them if they were not fully (deeply) buried again.

Source: The *Sun*, Vancouver, BC, April 5, 1957.

1937/YR/00 – Osoyoos. Looked Just Like a Monkey: Mrs. Jane Patterson told of seeing a monkey-like animal on a ranch near Osoyoos, Anarchist Mountain area, where she was staying in 1937. She saw the creature sitting on the ground "like you'd sit on the floor," with its hands on its knees and its back against a tree. She said it was taller than she was, had light brown hair all over its body, and eyes like slits. In her own words it, "looked just like a monkey to me."

Patterson had wandered over to an old abandoned house on the ranch property to look for rhubarb, which she had been told was growing in the garden.

She ducked under a tree branch and, upon straightening up, came face to face with the unusual creature about ten feet away. She inadvertently said to the creature, "Oh, there you are," but does not know why she said this. The creature just blinked its eyes. Patterson backed away and headed for home, hastening her pace when beyond the other side of the old house. She told her husband of her experience, but he would not go with her back to the old house, stating he did not wish to see a monkey. He finally agreed to go three days later, but nothing was seen.

John Green, 1981. *Sasquatch, the Apes Among Us.* Hancock House Publishers, Surrey, BC, pp. 63-64, from personal communication. Also John Green, Sasquatch Incidents Data Base, Incident Number 1000139.

1938/YR/00 – Harrison Mills (Chehalis Reservation). Bear & Sasquatch Fight:

James Cranebrook and three companions said they witnessed a fight between a bear and a sasquatch near Harrison Mills, Chehalis Reservation, in 1938. The incident took place on Morris Valley Road "north of Chehalis" (I assume they mean the reservation center.) They said the sasquatch won the fight.

Source: Daphne Sleigh,1990. *The People of the Harrison.* Abbotsford Printing, Abbotsford, BC, (self published), p. 251.

1939/YR/00 – North Bend. Playful Sasquatch:

Burns Yeomans of Deroche stated that he and a friend saw strange creatures "wrestling" near North Bend in 1939. The men had climbed to the ridge between Harrison Lake and the Fraser Canyon. They looked down on the east side into a small valley and saw a group of dark colored creatures running about on 2 legs and wrestling. They frequently threw each other to the ground, but always got back up on 2 legs. The never went on "all fours."

Source: John Green, 1973. *The Sasquatch File.* Cheam Publishing, Agassiz, BC, p.14, from personal communication. Also John Green, Sasquatch Incidents Data Base, Incident Number 1000054.

1939/DE/00 – Klemtu. Playing Around As They Walked.

Joe Robinson reported on a sighting of strange creatures his father had about 25 miles east of Klemtu in the early 1930s. While hunting near a lake the elder Mr. Robinson said he saw two 5-foot-tall, hair-covered black creatures playing around as they walked downhill towards the lake. It was dusk, so Robinson shone his flashlight on

The Klemtu area measures significantly in sasquatch sightings. The current (2007) people population is 505 souls.

them and they shaded their eyes. He then shot at them, but we don't know what happened after that because Robinson blacked out (I have no explanation for this).

Source: John Green, Sasquatch Incidents Data Base, Incident Number 1000134. Photo: Image from Google Earth. Image ©2011 Digital Globe; Image © 2011 Terra-Metrics; © 2011 Cnes/Spot Image; Image © 2011 Province of British Columbia.)

1939/DE/00 – Harrison Mills. Life-like Mask: Ambrose Point, a Chehalis Reservation artist, unveiled an unusual sasquatch mask in the 1930s. Unlike the usual First Nations carvings that draw on mythology, Point's creation smacks of reality. It is very large, probably reflecting the size of the creature it depicts, and is not decorated with colorful lines. Instead, we see a simple lifelike face. The wide nose and tall upper lip convey the feeling of an ape-like creature.

The Chehalis area has long been known for sasquatch sightings. Is it likely Point actually saw one of the creatures and created his mask from memory?

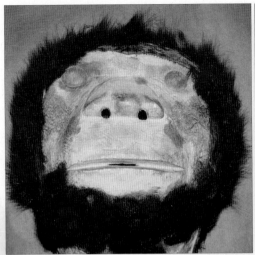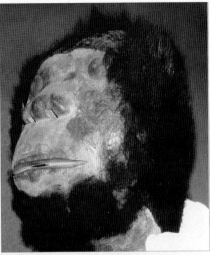

Sasquatch mask carved by Ambrose Point. It was donated to the Vancouver Museum in 1938.

Source: Christopher Murphy, 2010. *Know the Sasquatch: Sequel and Update to Meet the Sasquatch.* Hancock House Publishers Ltd., Surrey, BC, p. 22. (Photos: C.Murphy.)

1939/DE/00 – Alert Bay. D'sonoqua Seen: In the 1930s, Ellen Neal, a young girl who lived in Alert Bay, reported that she saw a hair-covered D'sonoqua (sasquatch) walking up a beach and into the trees. Later she and others went to the spot where the creature had been and they found huge tracks in the sand.

I believe the young lady referenced here went on to become a noted artist and is mentioned in connection with Chief August Jack Khahtsalano and the Jacko story (see 1884/00/00 – Vancouver. Apeboy Displayed?). She is also mentioned in another incident (see 1964/04/00 – Turnour Island: House Moved). However, I note that references to Ellen (the artist) spell her last name as "Neal," Neel, and "Neil." Wikipedia shows "Neel." It appears they are all referencing the same person.

Source: John Green, 1973. *The Sasquatch File.* Cheam Publishing, Agassiz, BC, p. 14. Also John Green, Sasquatch Incidents Data Base, Incident Number 1000075. Also Wikipedia.

1941/10/00 – Agassiz. Something Coming Out of the Woods.
Jeannie Chapman reported a frightening experience she had at her home near Agassiz, Ruby Creek area, in October 1941. She lived with her husband, George, and their three children (8, 7, and 5 years old) in a small isolated house on the banks of the Fraser River. George Chapman was a railroad maintenance worker at Ruby Creek. During the daytime in October 1941 he was surprised to see his family running down the tracks towards him. Jeannie excitedly told her husband that a sasquatch was after them.

As it happened, one of the children had been playing in the front yard of the house. The child came running into the house shouting that a "big cow is coming out of the woods." Jeannie looked out and saw an ape-like creature, 7.5 feet tall, covered in dark hair, approaching the house. Terrified at the sight, she gathered her children and fled. George and other men went to the house and found 16-inch footprints that led to a shed, where a heavy barrel of fish had been dumped out. The prints then led across a field and into the mountains. Footprints on each side of a wire fence, between 4 and 5 feet high, gave a clue as to the height of the creature. It apparently had taken the fence in stride.

The Chapmans returned to their home, but were continually bothered by unusual howling noises and their agitated dogs (which appeared to sense an unusual "presence"). The family abandoned the house within a week. The house was later occupied by another family, but was again abandoned after a short stay and left to the elements.

The incident was thoroughly investigated by Joe Dunn, a deputy sheriff in Whatcom County, Washington, George Cousins (who made a deposition as to the size of the tracks found), and Gustav Tyfting (who worked with George Chapman) and his wife

The Chapman house some years after it had been abandoned.

Esse. A cast was made of one of the creature's footprints which measured 17 inches. (Additional photographs relative to this event are in the color section.)

Source: John Green, 1981. *Sasquatch, the Apes Among Us.* Hancock House Publishers, Surrey, BC, pp. 51-52. Also, from information provided by Joe Dunn, George Cousins and Gustav and Esse Tyfting as provided by Stephen Franklin, 1959. "The Trail of the Sasquatch." *The Standard,* Weekend Magazine, Montreal, Quebec, April 11, 1959. Also John Green, Sasquatch Incidents Data Base, Incident Number 1000009, and research by Thomas Steenburg. (Photos: Chapman house, J. Green; Tyfting portrait, author's file.)

Gustav Tyfting

1941/11/00 – Harrison Hot Springs. At Least 14 Feet Tall: Three canoes full of First Nations people reported fleeing from a sasquatch at Port Douglas (head of Harrison Lake) in late November 1941. Jimmy Douglas, who fled with his family, said the creature was at least 14 feet tall. The witnesses said they were very sure the oddity was not a bear. They said it walked on 2 legs like a man.

Port Douglas on Harrison Lake. During the gold rush in the mid and late 1800s, it was a major settlement. It is now home to just a small number of First Nations people.

Source: "Sasquatch Return Frightens Indians in British Columbia." The *Independent*, Long Beach, Washington, USA, November 28, 1941. Photo: Image from Google Earth; Image © 2011 DigitalGlobe; Image © Province of British Columbia; © 2011 Cnes/Spot Image.)

1943/CI/00 – Hope. Man Attacked By a Sasquatch: A man reported an altercation with a sasquatch near Hope, about a mile from the community of Katz, around 1943. While the man was out berry picking with his wife and others, he strayed away from the group. A sasquatch, over 6 feet tall, came running from behind some rocks and hit the man—"hit him on the head, and side, and arms." The man yelled and the others in his group came running to his assistance. The sasquatch ran off.

The man's wife reasoned that because the sasquatch was small by sasquatch standards, others had treated it badly, so it took out its frustrations on First Nations people.

The man was said to have had a "crooked arm" until he died in 1955.

Source: Don Hunter with René Dahinden, 1993. *Sasquatch/Bigfoot: The Search for North America's Incredible Creature.* McClelland & Stewart Inc., Toronto, Ontario, pp. 81, 82, from a letter sent to the Harrison Project (sasquatch hunt) Council, 1957, by Mary Joe, the daughter of the man who was attacked.

1944/08/00 – Bella Coola. Bending Down In Water: George Robson, Clayton Mack and his wife Clara, reported they saw a sasquatch in the Bella Coola area at Jacobsen Bay in August 1944. The 3 were in a boat at the time, and Mack said the brown-colored creature was bending down in the water near the shore. He thought it was a grizzly checking drift material for food. When it heard the boat motor, it stood up, turned and looked back (towards the 3 people), then turned again and walked on 2 legs through drifting logs up to the forest. On the way, it looked back at the three. Upon reaching the forest, it spread trees with its hands and disappeared.

George Robson in 1991.

Mack, a game guide specializing in grizzlies, said that he had never seen a grizzly do these things. He decided what was seen could not have been a bear. In an interview, much later, Mack provided the following account:

> I saw this thing walking on the breach; a light brown color, standing about eight to nine feet tall. I thought it was a big bear—but he didn't go down on his four legs. I was nosing right towards

Clayton Mack in the 1960s or 70s.

> him but I was about four hundred yards away so I didn't get a good look at his face. I kept wondering because he didn't go down on his four legs. He was right on the edge of the water; he stood up straight and looked at me, then turned and walked up to the timber. Then halfway up to the driftwood logs he stopped and turned and looked at me, twisting his head round. I never seen a grizzly bear on its hind feet stand and twist round with just its head; they always turn the whole body, and they go down on four feet to do it. Then he went on and got to the logs and walked on top of them. He got to the timber—it's second-growth—and it looked as though he just reached out and spread the trees apart as he walked through them—young spruce and hemlock trees.

We are also told that George Olson, who worked at the Tallheo Cannery at Jacobsen Bay also saw what he believes was a sasquatch in the same area. However, there is no date for this incident.

Source: "Sasquatch reported in Valley." *Bella Coola News*, January 27, 1977. Also Don Hunter with René Dahinden, 1993. *Sasquatch-Bigfoot, The Search for North America's Incredible Creature*. McClelland & Stewart Inc. Toronto, Ontario, p. 32, from an interview with Clayton Mack. Also John Green, Sasquatch Incidents Data Base, Incident Number 1000065. (Photos: Robson portrait, Andy Robson [grandson]; Clayton Mack portrait, J. Green.)

1945/04/00 – Swindle Island. Where the Apes Stay: Chief Tom Brown reported that he found huge human-like tracks only minutes old on the beach at Meyers Pass on Swindle Island in April 1945. This area has a First Nations name that in English translates to "where the apes stay."

Source: John Green, 1973. *The Sasquatch File.* Cheam Publishing, Agassiz, BC, 14, from personal communication.

1945/CI/00 – Bella Coola. As Big as a Moose: The captain of a fishing boat reported that he saw an erect hairy creature "as big as a moose and about the same color" near Bella Coola—shoreline at Jacobson Bay, in the early 1940s. It ran on 2 legs into the forest, and as it ran, hair swinging beneath its forearms could be seen.

Source: John Green, 1973. *The Sasquatch File.* Cheam Publishing, Agassiz, BC, p. 14, from personal communication. Also "Sasquatch Reported in Valley," a newspaper article, 1973, no details. Also John Green, Sasquatch Incidents Data Base, Incident Number 1000059.

1945/CI/00 – Coombs (VI). Hair Streamed Out: Alex Oakes stated that what he believes was a sasquatch ran in front of his car not far from his home at Coombs in the early 1940s. He said it was running very fast, and hurdled the fences on both sides of the road. What particularly struck him was the way its hair streamed out behind its shoulders as it ran and leaped. He said it was the color of a brown bear and estimated its height at about seven feet. In describing it, he said it looked "damn near human."

Source: John Green, 1981. *Sasquatch, the Apes Among Us.* Hancock House Publishers, Surrey, BC, pp. 57, 58, from personal communication. Also John Green, Sasquatch Incidents Data Base, Incident Number 1000140.

1945/CI/00 – Osoyoos. Faster Than a Man Could Run: Mrs. Hazel Smith reported that her brother-in-law, Don Dainard, saw a strange hair-covered animal near Osoyoos in about 1945.

While horseback riding southwest of the Patterson ranch at Bridesville, Dainard saw an animal sitting on a log. The creature (we assume upon realizing it was being seen) took off running faster on its "hind legs" than a man could run.

Source: John Green, Sasquatch Incidents Data Base, Incident Number 1000142.

1947/YR/00 – Bella Coola. Not Bounding Like a Bear: Bill Mounce and the skipper of their tug reported seeing a odd animal in the Bella Coola area in 1947. The men were between Kwatna Inlet and the South Bentinck River in Burke Channel. The creature was about 200 yards from the south shore. They said it was the color of a grizzly bear and it was walking upright on large rocks on the shore. It went up a rock slide using all fours, but as a human would, not bounding like a bear. Mounce argued it was a bear, the skipper insisted that it was a sasquatch.

Source: John Green, Sasquatch Incidents Data Base, Incident Number 1000143.

1947/YR/00 – Vancouver. Skin-clad Sasquatch? Mr. and Mrs. Werner reported a strange encounter near Vancouver on Grouse Mountain in 1947. While the couple were traveling in their jeep along a logging road, they saw 2 unusual creatures that were naked except had "a skin wrapped around them." Both had shoulder-length hair, but no mention is made of body hair. Both were barefoot and had "huge feet." One was about 8 feet tall, the other 6 feet tall. They had very bushy eyebrows over very small eyes, and flat, wide noses. The taller was carrying a stick over its shoulder with what may have been a bag tied to the end. It appeared to be leading the other, which possibly had its hands tied together. The taller one was quite slim, the smaller, very broad.

Grouse Mountain, as it presently appears. The lower reaches of the mountain have undergone excessive housing development to the point where back yards are virtual cliffs.

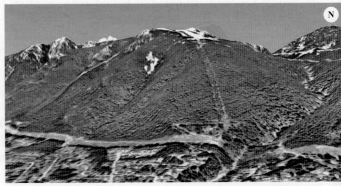

Source: Don Hunter with René Dahinden, 1993. *Sasquatch-Bigfoot, The Search for North America's Incredible Creature.* McClelland & Steward Inc., Toronto, Ontario, pp 101, 102. (Photo: Image from Google Earth; © Cnes/Spot Image; Image © 2011 Province of British Columbia; Image © 2011 GeoEye; Image © 2011 DigitalGlobe.)

1948/YR/00 – Harrison Mills (Chehalis Reservation). Bicycle Chase: Henry Charlie reported a strange incident near Harrison Mills on the Chehalis Reservation in 1948. Charlie was cycling towards Harrison Mills. He saw 2 unusual creatures come off the hill opposite the Fenn Pretty property. One of the creatures chased him for a distance of more than a mile down Morris Valley Road. He described it as between 7 and 8 feet tall, and covered in dark hair, except for its face. Charlie said that the creature had no problem keeping pace as he pedaled as fast as he could. He saw it only when he glanced behind himself, and eventually it wasn't there anymore.

Source: John Green, 1973. *The Sasquatch File.* Cheam Publishing, Agassiz, BC, p. 14, from personal communication. Also John Green, Sasquatch Incidents Data Base, Incident Number 1000015.

1949/07/00 – Harrison Hot Springs. Horse Balked at Going Farther: Mrs. G. Mason reported seeing an odd animal near Harrison Hot Springs in July 1949. She was horseback riding on a bridal trail when her horse balked at going farther. She led it forward and it balked more, so she tied it up and walked up the trail. She came upon a creature about 50 feet away she thought was a bear and approached it, "to feed it sugar." It shuffled off on all fours, then stood erect and walked away. She described it as being 7 to 8 feet tall with unkempt brown hair. It was heavily built with medium length arms. She noted that it did not have a snout. Mrs. Mason stated that her horse refused to continue down the bridle trail before the creature was completely gone (evidently until the horse no longer sensed its presence).

Source: John Green, Sasquatch Incidents Data Base, Incident Number 1000029.

1949/10/00 – New Hazelton. Like a Bear: Frank Luxton, a highway foreman who was working in the New Hazelton area reported seeing an unusual creature in October 1949. Luxton was in his car at dusk and saw a large creature like a bear walk across the road on its hind legs. He stated that it was about 8 feet tall, covered in dark brown hair, and had a flat face and low forehead. Luxton said the creature looked at him and then ran down towards the river.

Source: John Green, 1973. *The Sasquatch File.* Cheam Publishing, Agassiz, BC, p. 14, from information provided to Bob Titmus by Frank Luxton. Also John Green, Sasquatch Incidents Data Base, Incident Number 1000055.

1949/DE/00 – Agassiz. Huge Barefoot Tracks: An Agassiz farmer reported finding huge barefoot tracks in snow in one of his fields during the 1940s. He followed them for some distance beyond his property and up a slope of a nearby mountain. He started to feel uneasy so decided to turn back and go home.

Source: John Green, 1973. *The Sasquatch File*. Cheam Publishing, Agassiz, BC, p. 14, from personal communication.

1949/CI/00. Vanderhoof. Thought It Was a Gorilla: Mr. Hentges reported seeing an odd animal in the Vanderhoof area in about 1949. Hentges was driving from Cluculz Lake to Vanderhoof on a highway under construction and saw what he thought was a gorilla walk across the road about a quarter mile away on a straight stretch. It was on its hind legs, with arms hanging down. When Hentges reached the point where it crossed the road he could not see it, so assumed it was hiding behind one of the piles of brush cleared from the right of way. He reported the incident to the police.

Source: John Green, Sasquatch Incidents Data Base, Incident Number 1000961.

The Period from 1950 to Date

The 1950s marked the beginning of the modern era in sasquatch history. New roads were rapidly pushed into wilderness areas and vastly improved automobiles encouraged people to explore the great outdoors. Television became available to middle income people and, along with other improved media, they became more aware of newsworthy events. By the late 1950s, the possible existence of sasquatch became common knowledge. By the end of the century, substantial evidence had been collected and the creature was now being studied more intently by professional people. (Photo: R. Patterson, public domain.)

1950/12/00 – Roderick Island. Tracks in Frost: Paul Hopkins reported finding large "ape tracks" in frost on a disuse wharf on Roderick Island in December 1950. Hopkins had landed at an abandoned abalone cannery (several buildings) on the wharf to get a barrel. He walked by a building, and when he returned there were ape tracks leading out of the building (presumably to the interior of the island) in half-inch-deep frost.

Source: John Green, 1973. *The Sasquatch File.* Cheam Publishing, Agassiz, BC, p. 14, from personal communication. Also John Green, Sasquatch Incidents Data Base, Incident Number 1000084.

1950/CI/00 – Smithers. Trees Stripped: Two loggers exploring the woods in the Smithers area reported that they found large human-like tracks on a game trail around 1950. The distance between the steps was about 5 feet on level ground, and 4 feet when going up a steep hill. They followed the tracks and as they proceeded they saw that alder trees on each side of the trail had their branches stripped bare up to 20 feet. The trees had 4-inch diameter trunks at ground level. They were evidently bent over by something to access the leaves and buds at such a high level. One of the loggers tried to bend a tree, but was unable to do so. Small twigs were found on the ground as though they had been spit out.

Source: BFRO website, from "The Sasquatch in Northern British Columbia, Canada," HBCC UFO Research, April 07, 2001.

1951/11/00 – McBride. Lean-to Slept In By Something: Macky and Slim Gudell reported an unusual occurrence in the McBride area during November 1951. The men were hunting and had left half a goat hanging in a tree while away overnight. When they returned to their campsite, they found 15 to 16-inch barefoot, human-like footprints in 2 inches of snow, estimated to be just a few minutes old. Whatever had made the tracks appeared to have slept in their lean-to; also, the half goat was gone.

The actual location of the incident was given as "Horse Creek." It is believed this was actually Horsey Creek, which is in the McBride area.

Source: John Green, 1973. *The Sasquatch File.* Cheam Publishing, Agassiz, BC, p.18, from a report provided to René Dahinden. Also John Green, Sasquatch Incidents Data Base, Incident Number 1000082.

1951/YR/00 – Swindle Island. Tracks Entered Lake Kitasu. Residents from Klemtu reported that they saw unusual barefoot human-like tracks on Swindle Island in 1951. The tracks crossed a sand bar and entered Lake Kitasu. Most of the tracks were small (not very long), however some were fairly large, with the largest being about 14 inches long. Swindle Island is considerably north of Vancouver Island. The area and lake in May would be too cold for human swimmers.

Source: John Green, 1973. *The Sasquatch File.* Cheam Publishing, Agassiz, BC, p. 18, from a personal report provided by Bob Titmus.

1952/YR/00 – Terrace. Watched the Man For a While: A man reported that while driving he saw an odd creature standing erect beside the the main road near Terrace in 1952. The man slowed down and stopped his car. The creature watched him for a while, then turned and walked away. It stopped in the underbrush and turned around and looked at the man again. As it was dusk its facial features could not be seen. The man said there were two other people who claimed to have seen the same thing in the same general area that year.

Source: John Green, Sasquatch Incidents Data Base, Incident Number 1000145.

1953/09/00 – Courtenay. "Thought It Was a Friend": Jack Twist reported that he saw a strange animal near Courtenay along the Oyster River (20 miles northwest of Courtenay) in September 1953. He was on a camping trip and stated that while walking back to camp on an old logging road, he saw a dark figure about 200 to 300 yards away. He thought it was one of his friends and called out, but did not get an answer. He continued walking straight towards the figure. When he got close enough to see what it was, he saw a huge creature covered in dark hair. It turned and faced him, then turned again and, walking upright, wandered into the forest.

Twist heard it moving through the heavy brush. Judging from overhanging trees where it had been, he estimated the creature's height as at least 8 feet.

Source: John Green, 1973. *The Sasquatch File.* Cheam Publishing, Agassiz, BC, p. 18, from personal communication. Also John Green, Sasquatch Incidents Data Base, Incident Number 1000147.

1953/CI/00 – Esquimalt (VI). Massive Bone: A Vancouver couple reported unearthing a massive bone, shaped like a human limb and foot, on the Esquimalt First Nations Reservation in about 1953. One of the Esquimalt tribal legends tells of its braves having killed a giant many years ago.

Source: T.W. Peterson, 1967. "Wanted Dead or Alive: Sasquatch." *Real West* magazine (now believed to be defunct), July 1967.

1954/YR/00 – Chilco Lake. Sasquatch Pass Named: Dr. George V.B. Cochran, president of the New York-based Explorers Club, and his team found large human-like footprints in snow above Chilco Lake in 1954. The explorers were in search of a route to the Homathko ice fields at the time. As a result of the finding, the club named the area Sasquatch Pass. This name now has official status.

Sasquatch prints in snow at high elevations have been found before and since, leading researchers to wonder why the creature goes

Sasquatch Pass, so named for large footprints found there in 1954.

up above the snow line. For certain, there is not much vegetation up there and few animals, so "searching for food" does not appear to be the reason. But with all wild animals just about everything they do is food related.

A theory developed by Frank Beebe and the late Don Abbot, makes surprising sense. They point out that wolverines take their kills up to the snow and bury them where they naturally freeze and are preserved. When the going gets tough at lower elevations, the wolverine goes up and retrieves its frozen stash. It is taken to a lower elevation where it thaws and becomes as good as fresh meat. It is reasoned that sasquatch would be far more intelligent that wolverines, so likely know about this process.

Source: Christopher L. Murphy, 2010. *Know the Sasquatch.* Hancock House Publishers, Surrey, BC., p. 213, also John Napier, 1972. *Bigfoot: Startling Evidence of Another Form of Life on Earth Now.* Berkley Publishing Corporation, New York, New York, USA, p 172. (Photo: Russ Kinne.)

1955/10/00 – Tete Jaune Cache. Definitely Not a Bear: William Roe reported that he had a long and up-close look at what definitely appears to have been a sasquatch near Tete Jaune Cache, on Mica Mountain, in October 1955.

Roe, a highway worker and experienced hunter and trapper, decided to hike up to a deserted mine on Mica Mountain for something to do. Just as he came within sight of the mine, he spotted what he thought was a grizzly bear half hidden in the bush about 75 yards away. He had his rifle with him, but did not wish to shoot the animal, as he had no way of getting it out. He therefore calmly sat down on a rock behind a bush and observed the scene. A few moments later, the beast rose up and stepped into the open on 2 legs. He now saw that it definitely was not a bear, but what appeared to be a man-like creature, about 6 feet tall, covered in dark brown, silver-tipped hair.

The creature, unaware of Roe's presence, walked directly towards him. Roe then observed by its breasts that it was female. It proceeded to the edge of the bush where Roe was hiding—within 20 feet of his position. Here it crouched and began eating leaves from a bush. Roe was able to observe many important details as to how the creature walked, its physical makeup, and its habit of eating by drawing branches through its teeth.

As to its appearance, Roe tells us:

The the head was higher at the back than at the front. The nose was broad and flat. The lips and chin protruded further than its nose. But the hair that covered it—leaving bare only the parts of its face around the mouth, nose

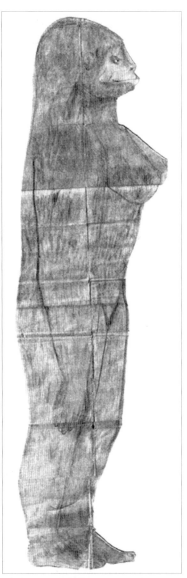

A drawing of the creature seen by William Roe created by his daughter, Myrtle, under his direction. Some aspects do not tally with Roe's description.

The long sloping mountain in the background of this photograph is Mica Mountain.

and ears—made it resemble an animal as much as a human. None of the hair, even on the back of its head was longer than an inch, and that on its face much shorter. Its ears were shaped like a human's ears. But its eyes were small and black like a bear's. And its neck also was unhuman, thicker and shorter than any man's I have ever seen.

In subsequent correspondence with Roe, he said that the creature's fingernails were not like a bear's, but short and heavy like a man's fingernails. Also, there were no bulging muscles, but the animal was as deep as it was wide.

When the creature noticed Roe, a look of amazement crossed its face, which Roe found comical and he chuckled to himself. Remaining crouched, the creature backed away 3 or 4 steps. It thereupon straightened up, turned, and rapidly walked away in the direction whence it had come, glancing back twice at Roe over its shoulder.

Realizing he had stumbled on something of great scientific interest, Roe leveled his rifle at the creature to kill it. However, he changed his mind because he felt it was human. In the distance, the creature threw its head back on 2 occasions and emitted a peculiar noise that Roe described as "half laugh and half language." Roe's examination of feces in the area, which he believed was from the creature, convinced him that it was strictly a vegetarian.

What I have provided here is just a summary of Roe's report, which is highly detailed. Also, I need to mention that he provided a sworn statement that the account he provided was true to the "best of his powers of observation and recollection."

Unfortunately, all research with William Roe was done by telephone or letter. No researcher, to my knowledge, personally met him. However, his daughter, Myrtle (married name is now Myrtle Walton), who prepared the drawing of the creature under her father's direction (see previous page), in a television documentary (1970s) talked about creating the drawing . As William was not seen in the documentary, I presume it was made after his death.

Source: John Green, 1981, *Sasquatch: The Apes Among Us.* Hancock House Publishers, Surrey, BC, pp. 52–57 (from a deposition and letter to John Green). Also Don Lory, 1955. "Hunter Spots Sasquatch Girl, Just couldn't shoot her." The *Province,* Vancouver, BC. Also "Sasquatch Seen at Close Range Two Years Ago," the *Agassiz-Harrison Advance,* BC, September 5, 1957. Also John Green, Sasquatch Incidents Data Base, Incident Number 1000011. (Photos: Drawing, scanned from original drawing owned by J. Green; Mica Mountain, C. Murphy.)

1956/03/30 – Flood (or Floods). Sasquatch Couple: Stan Hunt reported that while driving the Trans-Canada Highway to Vancouver on March 30, 1956, he saw a 7-foot, upright creature, covered in gray hair cross the highway near Flood. As he drove past the point of the crossing, he saw a second creature of similar appearance standing beside the road. He recounted the incident as follows:

It was just near Flood, the other side of Hope. The light was just coming up. I almost ran into this gray horse on the road. That's the reason I remember the color. A little further on I was going about 45 to 50 miles an hour when I saw these weird things. One crossed the road from the river side. The other was already in the bushes. I could just see the top of its head. The creature was walking upright. I thought at first it was a great big bear only it wasn't that big around. The hair was thinner—not matted like an animal. The whole thing was kind of eerie. There isn't much traffic at that time in the morning. I didn't bother to stop.

Hunt apparently later recalled getting a better look at the creature in the bushes as he stated that it was, "gangly, not stocky like a bear."

Source: "Hairy Giant Tale Popping Up Again." The *News, Vernon, BC,* May 3, 1956. Also, John Green, 1973. *The Sasquatch File.* Cheam Publishing, Agassiz, BC, p. 18, from personal communication; also, Don Hunter with René Dahinden, 1993. *Sasquatch/Bigfoot: The Search for North America's Incredible Creature.* McClelland & Stewart Inc., Toronto, Ontario, p. 102. Also John Green, Sasquatch Incidents Data Base, Incident Number 1000040.

1956/07/00 – Harrison Hot Springs. Sasquatch Hunt Planned. In July 1956 several newspapers reported that sasquatch researcher René Dahinden and his partner, Anton Ruesch, planed to look for the sasquatch in the Harrison Lake district. We are told that the men apparently chose this time as it coincided with the legendary meeting of sasquatch creatures on Morris Mountain which is on the Chehalis reservation. The meeting was said to occur every 4 years and 1956 was one of those years.

Apparently at this time some credibility was given to the stories about Morris Mountain. Nevertheless, a couple of pieces of new information were provided in the articles.

We are told that a devout missionary in New Mexico tried to convert the creatures, which he referred to as *Karen Kowahs* (meaning they walked in streams and caught fish in their hands). After white people moved to the area the creatures occupied, there was a dispute which resulted in the New Mexico sasquatch tribe being placed in box cars and shipped to the central US. From that region they drifted to other areas. This is the only information I have run across on this incident, so for certain it is just legend.

Furthermore, it was stated that 3 or 4 years before (i.e., 1952 or 1953) a Vancouver couple was reputed to have found the skeleton of a man that was more than 7 feet tall in the BC interior mountains. Again, there is nothing further on this alleged finding. The last report I have on a large skeleton being found was in 1937 (see 1937/YR/00 – Lillooet. Eight Foot Skeleton).

Source: "New Hunt on for Indian Giants of Fraser Valley." Vernon newspaper or The *Sun* or The *Province*, Vancouver, BC, June or July 1956. A similar article, "To Track Down Legendary Tribe" was featured in The *Herald,* Lethbridge, Alberta, June 16, 1956.

1956/08/00 – Sechelt. Creature Standing In Doorway: A man who said he was one of 3 people staying in a cabin near Sechelt reported a frightening experience that occurred in August 1956. One night one of the men opened the door to go out, screamed and came back into the room. After he calmed down he said he had seen a huge creature standing in a doorway leading to a shed.

Source: John Green, Sasquatch Incidents Data Base, Incident Number 1000148.

1956/08/00 – Sechelt. Rocks Thrown at US Visitor: A visitor from Erie Pennsylvania had a frightening experience while visiting Sechelt in August 1956. He was walking up a long driveway to his cabin when a huge dark figure on a bank above hurled rocks at him. The visitor said that a friend had told him a sasquatch had been seen in the area earlier in the month. The friend was apparently one of the men in the previous Sechelt incident.

Source: John Green, Sasquatch Incidents Data Base, Incident Number 1000148.

1956/CI/00 – Bella Coola. Eating Blueberries: A hunter and 2 friends reported that they saw a sasquatch inland from Bella Coola c. 1956 during August. They watched the creature as it ate blueberries at the edge of a highway. They said it had a human-like face except for its protruding mouth.

Source: John Bindernagel, 1998. *North America's Great Ape: The Sasquatch.* Beachcomber Books, Courtenay, BC, pp.38, 210, from a report provided by a Port Hardy resident.

1957/03/00 – Harrison Hot Springs. Dahinden Speaks Out: Sasquatch researcher René Dahinden spoke out to the press at Harrison Hot Springs in March 1957 with regard to his belief in the sasquatch. He said said he is convinced that Indian legends about the creatures are not myths and that the creatures are probably "closely al-

René Dahinden in 1959.

lied" with the abominable snowman of the Nepal mountains. He added that he would ask the BC Centennial Committee to help him organize an expedition to find one of the creatures in connection with the upcoming 1958 celebrations.

Source: Scott McClean collection, "Missing Link." The *Herald,* Lethbridge, Alberta, March 9, 1957. (Photo: Cropped from a photo by J. Green.)

1957/04/10 – Harrison Hot Springs. A Search For the Sasquatch: On April 10, 1957 the Harrison Centennial Committee voted "A Search for the Sasquatch"as its 1958 project, to be led by the noted researcher René Dahinden, who summed up his position as follows:

I don't want to catch a Sasquatch, I don't believe much in the catching business. But we should get movies of them, and try to tame them. Then we could bring scientists in to study them in their natural surroundings. It would be a tremendous find for science.

I have spoken to three people who saw Sasquatch during the last year. They are reliable people, and one, Stan Hunt of Vernon, saw two by the roadside at Flood, near Hope.

Nobody can say there is no Sasquatch. Anybody fair would have to say there is too much evidence about them to dismiss.

In May 1938, the Hon. Wells Gray said at a Harrison Hot Springs festival that Sasquatch are merely legendary monsters, and they do not exist today. That was all he could say because the Indians made an angry protest. The Indian Chief, Flying Eagle, went to the mike and thundered. "The speaker is wrong. I, Flying Eagle, say some white men have seen the Sasquatch. Many Indians have seen them, and spoken to them. Sasquatch still live all around here. Indians do not lie."

In 1846, Alexander C. Anderson, Hudson's Bay inspector, wrote in his official report of the "wild giants of the mountains." He also wrote that his party was met by a bombardment of rocks hurled by Sasquatch.

Indian Legend in the Okanagan Valley says "big men from the mountains" existed around 1850. They have been seen at Merritt, and big bones were found in caves around Shuswap.

Information we have tells us they are about seven feet high, make footprints about 22 inches long, have long hair over their bodies, and long arms.

They have, so to say, the shape of a stone-age man. The Indians say they do not have any kind of weapons and they live off game, such as birds and deer, which they are supposed to hypnotize.

I am sure we can find a Sasquatch, if they are still alive. I don't want to say just where the spot is, but I have three men ready to go with me to hunt for it. They are Charles Flood of Burnaby, who saw a Sasquatch near Greendrop Lake, in the Chilliwack Lake area, in 1915; Klaus Wittenborg of New Westminster; and Mike Hosy of Hope, now working at Banff.

I am disappointed at Canadians. In the Himalayas, many thousands of pounds were spent by English expeditions seeking the Abominable Snowman. Here it is even better. We have more evidence about the Sasquatch in a place within 100 miles of Vancouver, but nobody wants to do anything about it.

Source: "René Dahinden, 1957. "René Believes in Sasquatch: Once Made Indians Angry." The *Province*, Vancouver, BC, April 10, 1957 (the article shows, "by René Dahinden").

1957/04/21 – Victoria. Provincial Archives Searched: In April 1957, René Dahinden and John Green, both avid sasquatch researchers, searched BC's archives in Victoria for sighting reports of North America's elusive hairy giant. They hoped to find more evidence to bolster their claim that the creatures do indeed exist, and thereby get government funded research to confirm its reality. Dahinden, whose bid for direct financial support to search for the creature was rejected by the BC Centennial Board, pointed out that sasquatch lore brings the province worldwide publicity, "which no money can buy." He contended that this fact justifies government funding. He wanted to lead a sasquatch-hunting expedition through the mountains and forests around Harrison Lake, the alleged habitat of the legendary wild man.

Nevertheless, while his direct appeal failed, the board still pondered an application from the village of Harrison Hot Springs for leave to spend its centennial grant on a sasquatch hunt. Dahinden stated:

> There is more evidence for the existence of the Sasquatch than there is for the Abominable Snowman, and yet they spent between $50,000 and $100,000 on an expedition to the Himalayas in search of the snowman. With $10,000 we could conduct an all-out search for the sasquatch—and I believe it would be only a matter of time before we came up with something.

The 2 searchers traveled to Cobble Hill on April 20 and saw Bruce A. McKelvie, noted BC historian and author, who believed that the sasquatch may exist. McKelvie mentioned the report that a sasquatch had once been captured in BC, discovery of giant bones in several parts of the province, and persistent belief in wild giants among Indians all over BC and as far away as Fort St. James, where the First Nations people believe not only is there a race of giants, but also a race of little people.

The foundation for such beliefs, McKelvie stated, might be the fact that a race of giants conquerors are said to have driven out an aboriginal race of small people who took up residence in the wilderness. From that time forward, word might have passed around the tribe: "Watch your step in the woods, or the little people will get you."

Should a race of tall people have been driven out by a more warlike people of smaller stature, the saying would have been: "Watch out for the giants; they're hiding in the mountains."

"If they found pygmies in Africa," McKelvie said, "why shouldn't they find giants in British Columbia?"

Source: G.E. Mortimore, 1957. "Priceless Publicity: Sasquatch Boosts B.C." The Daily Colonist, Victoria, BC, April 21, 1957.

1957/04/24 – Harrison Hot Springs. Plans for Search Dampened: On April 24, 1957, sasquatch seeker René Dahinden told the Harrison Centennial Committee members that even if they can't financially support his proposed expedition, he might be able to get private sponsors. Reports that Victoria [government] people were not looking favorably at the funding dampened hopes of financial aid from that source. The Harrison people sided with the determined researcher and together they tentatively made plans for a search that fall.

Dahinden stated that the hairy giants migrate from one area to another each spring and fall, so it's too late to watch them on the move this spring, which makes fall the earliest time for the search.

Tony Burger, chairman of both the local centennial committee and the board of village commissioners, said that the expedition would definitely be a search, not a hunt. "We don't want to shoot one or capture one—we just want to watch them, photograph them and secure positive evidence they exist," he stated.

It would only be a small expedition. Besides Dahinden, licensed guide Jack Kirkman would go along and two other members, maybe a photographer and a scientist, Burger stated. It was expected that the group would be flown into sasquatch land by Cascade Air Services. Kirkman was to establish several base camps in the area. Dahinden was staying at Harrison, at least until word was received from the BC Centennial Committee.

Burger added that his committee had received a sworn affidavit from a person [William Roe] who claimed to have seen a sasquatch from a distance of fewer than 10 feet [actually within 20 feet]. A plea was broadcast for sketches or photographs by people who said they had seen the creature.

Source: "Plans for a Search." The *Progress,* Chilliwack BC, April 24, 1957. Also "Tallyho Sasquatch! Head Hunter Picked." The *Province,* Vancouver BC, April 23, 1957.

1957/05/08 – Lillooet. Tired of Sasquatch Politics: In May 1957 a group of citizens in Lillooet said they were tired of "sasquatch politics" and would take the matter into their own hands. They claimed the sasquatch lived only in the Lillooet region and that the only skeleton resembling the creature was found there. An expedition sponsored by and financed by civic-minded citizens was planned for May 10, 1957. A recent sasquatch sighting further spurred the group to action—a man reported that he saw a tall, hairy man fishing in the nearby Stein River.

The expedition members were Dr. Ted Oliver [no specifics];

Noel Baker, a merchant who was to supply guns and camping equipment; Tony West, a civil engineer; and an unidentified Pacific Great Eastern Railway official (who hoped to dissuade the creature from rolling rocks onto trains).

The expedition planned to set out from the mouth of the Stein River with the call, "Bring back a sasquatch by fair means or foul." Although the sasquatch was now their target, Dr Oliver stated that the foursome had originally planned a grizzly bear hunt, but then put the sasquatch on the top of the list.

Source: Alex MacGillivray, 1957, "Lillooet Hunters Jump Gun; Search for Sasquatch May 10." The *Province,*Vancouver, BC, May 8, 1957.

1957/08/25: Vernon. Dahinden Returns After Month-long Search. Sasquatch searcher René Dahinden arrived in Vernon on August 25, 1957 after a month long search for North America's elusive mountain giant. Dahinden left Lumby, BC in mid-summer to find evidence of the creature. He searched various BC areas, by land, water and air.

His aerial surveillance was compliments of a survey crew equipped with a Dehavilland Beaver float plane. The crew was mapping country and invited Dahinden along. They flew from Harrison Hot Springs to Fort Douglas, and then further west and north to Boot Lake, which is on the border of Garibaldi National Park. Dahinden stated:

> We think we were about the first people up in that country. We saw a lot of signs, but it might have been bears. There was very little game. We were too high up. I think the sasquatch are lower down, probably around Ruby Creek.

Dahinden accompanied the survey group on a climb up Mount Brackenridge, which is near Fort Douglas. It took them 6 hours to reach the 6,000-foot level. He was now biting his nails thinking about his future plans. Although nothing was found, Dahinden was still hopeful, "I'm going again if I get a sponsor," he firmly stated.

Source: Jim Peters, 1957. "Sasquatch Man Still on the Trail." The *News,*Vernon BC, August 25, 1957.

1957/YR/00 East Sooke (VI). Large Bone With Foot: During 1957 a First Nations man presented a large leg bone to a newspaper reporter doing research in East Sooke. The bone, which was shaped like a a large human limb and foot, was claimed to be from a sasquatch, and was offered as proof of the creature's existence. The relic had been uncovered by a man operating a bulldozer, excavating for a building foundation. The size of the bone indicated that it came from a creature that was at least 15 feet tall. The man who had the bone said he wanted to have it placed on display in the Empress Hotel in Victoria.

Source: T.W. Patterson, 1994. "Sasquatch." The *Times Colonist,* Victoria, BC, July 10, 1994.

1957/YR/00 – Agassiz. Green Takes Up the Search: In 1957 John Green, owner/publisher of the *Agassiz-Harrison Advance* newspaper, took up the search for North America's elusive sasquatch. Green, a professional journalist, worked for newspapers in Toronto, Victoria, and Vancouver, before buying the *Agassiz-Harrison Advance* in 1954.

During the previous three years or so, Green heard stories of sasquatch from people in the Harrison area, which has long had a history of sasquatch-related incidents. "I took little interest in the subject," he said, "until I learned that some people in the community I had come to respect had been witnesses to an incident at nearby Ruby Creek in 1941."

This incident involved a Mrs. Chapman who was at home with her three children when a sasquatch came onto their property. Mrs. Chapman grabbed her children and fled to get her husband who worked nearby on the railroad. The incident was thoroughly investigated and huge human-like footprints and other evidence found on the property definitely verified Mrs. Chapman's frightening experience. (See entry:1941/10/00 – Agassiz. Something Coming out of the Woods, for full details.)

At this writing, John Green has spent 52 years researching the sasquatch. He has written several books on the subject, the most noteworthy being *Sasquatch: The Apes Among Us,* which is the most detailed and most authoritative work on the subject ever written. All of John's books have been used extensively for the material provided in this book.

Source: Christopher L. Murphy, 2009. *Know the Sasquatch: Sequel and Update to Meet the Sasquatch.* Hancock House Publishers, Surrey, BC, pp. 37, 227. Green's work, *Sasquatch: the Apes Among Us,* was originally published by Cheam Publishing Ltd, Agassiz, BC, in 1979. It was republished by Hancock House Publishers in 1981 and 2006.

1957/CI/00 – East Sooke (VI). Tore Trees Out By Roots: Jimmy Fraser of the Songhees tribe recalled to a reporter in 1972 his encounter with a sasquatch near East Sooke at Matheson Lake in about 1957. While hunting in this area he heard an ear-splitting roar and saw a gigantic hairy man—"maybe 18 feet tall." Terror stricken, Fraser ran as the creature tore trees out by their roots and hurled them at him.

Jimmy Fraser in 1972.

Source: T. W. Patterson, 1972. "Horne Lake's Wild Man," The *Daily Colonist,* Victoria, BC, Canada, September 24, 1972. (Photo: Author's file.)

1958/SU/00 – Harrison Hot Springs. A Hairy Monster: Rick Hedberg aged 9, and his younger brother Gary, aged 7, were frightened by a strange creature they saw at Harrison Hot Springs in the summer of 1958. The boys' parents (local residents) were highly concerned with the story their boys told them.

They liked to hike in the wooded hills around the hot springs at Harrison (near the Harrison Hot Springs Hotel) and while up in this area said they encountered a "hairy monster." It was standing upright by a large cedar tree, about 75 feet away, facing half away from them, with its face hidden by the tree. It was about 7 feet tall and covered from head to foot with dark gray hair. It had an arm and elbow (parts they could see) like a human, not like a bear's leg. The boys remarked that it smelled, "worse than a bear."

Upon spotting the creature, Rick managed to stay calm and slowly backed away, while Gerry yelled, panicked, and ran back the way they had come. After Gary yelled, the creature ran on 2 legs straight down the steep slope beyond the tree (towards Harrison Lake). Rick lost sight of it in about 3 seconds. Having no idea which way it might have turned, or whether it was going to come after him, Rick decided to follow his younger brother out of the area. He did not see the creature again.

When Gary came out of the woods just behind the hotel, he realized his brother was not with him. In a state of shock he yelled for help, attracting the attention of several adults nearby. Rick, in the meantime, slowly walked out of the woods, looking over his shoulder now and then to make sure the monster was not following him.

He emerged from the woods to see the adults trying to calm his brother, and attempting to get the facts about a hairy monster that might be eating Rick. When Gary saw Rick, he shouted with joy at seeing him alive.

A drawing by Rick Hedberg of the creature he and his brother saw. Rick was sure that the creature had a hand, not a paw.

At this point the adults, save one kindly lady, lost interest and dismissed the whole incident as young boys trying to get attention. The lady told the boys to go home and tell their parents what had happened. The boys told their parents, who listened and then dismissed the whole matter, telling them never to speak of the incident again, and to stay out of the woods behind the hotel.

Rick took his parents advice and stayed silent for 37 years. Then, as an adult living in Calgary, Alberta, Rick took an interest in sasquatch research and became friends with researcher Thomas Steenburg in about 1993. It was at least 2 years, however, before Rick decided to tell Steenburg of his and his brother's sasquatch encounter at Harrison.

In relating the story to Steenburg, Rick reasoned that the creature was seemingly hiding behind the large cedar tree. It was not hiding from them, but from the human voices that came from the hotel hot springs structure (downhill and just out of sight from its position). The creature had its right arm around the tree as though it

The Harrison Hot Springs Hotel with the hot spring structure indicated. It was behind this structure that the brothers had their sasquatch encounter. This photograph was taken in 2010. The forest down the mountain side is the same as it would have been in 1958.

were holding itself steady. Rick is sure it had a large hand at the end of its arm, not a paw. He said that at first he thought it was a large grizzly bear.

Source: Thomas Steenburg, 2000. *In Search of Giants.* Hancock House Publishers, Surrey, BC, pp 69–75. Also John Green, Sasquatch Incidents Data Base, Incident Number 1001334. (Photos: Drawing, R. Hedberg; Harrison view, C. Murphy.)

1958/08/00 – New Hazelton. Ladies Report Sighting. Rose Hibden and a friend reported that they saw a hair-covered erect creature in the New Hazelton area during August 1958. The creature was seen on the road in daylight.

Source: John Green, Sasquatch Incidents Data Base, Incident Number 1000051.

1958/08/00 – New Hazelton. Crossed the Road: A man reported seeing a huge, erect, hair-covered creature cross the road in front of his car about 13 miles south of New Hazelton during August 1958. Another sighting, (earlier or later time in the same month) of the same or similar creature appears to have been in the same area (see previous entry).

Source: John Green, 1973. *The Sasquatch File.* Cheam Publishing, Agassiz, BC, p.18, from information provided by Bob Titmus. Also John Green, Sasquatch Incidents Data Base, Incident Number 1000045.

1958/10/29 – Yale. Residents Interviewed: An article that appeared in the July 3, 1884 edition of the *Daily Colonist* (Victoria, BC) brought noted sasquatch researchers John Green and René Dahinden, together with their wives, to Yale on October 29, 1958. The old article tells of the capture near the area of what is now considered to be a sasquatch boy who was taken to the outskirts of Yale (see 1884/06/30 – Yale: Jacko the Ape-boy, for details).

Among others, Green interviewed August Castle who was a young boy at the time. He did not actually see the creature, but said he recalled hearing later that such had been caught. Annie York of Spuzzum, who is noted for her wealth of historical knowledge on the region, was also interviewed. Although she could not provide any information on this incident, she related many stories of sasquatch sightings at different points in the Fraser Canyon area.

The researchers also examined a cave in the vicinity that was reported to have been the home of a sasquatch. There were indications that some form of life had inhabited the cave at some time, but nothing was found to indicate exactly what it was.

Source: A. Milliken, 1958. "Sasquatch Hunt Moves from Harrison to Fraser Canyon." The *Sun,* Vancouver, BC, October 29, 1958.

1958/12/08 – Bella Coola: Gaze Returned & Creature Runs Away: Hunters George Robson and Bert Solhjell reported a "big ape" sighting in the Bella Coola area at Burnt Bridge on December 8, 1958.

The men were seated by their fire when they noticed a huge, dark, erect creature with long arms—like a big ape—standing in waist-high brush watching them about 50 yards away. As soon as they returned its gaze, the creature ran off quickly, disappearing over a rise. The men went immediately over to watch it run down the hill, but it had already disappeared. They found only one heel print in a patch of snow, and concluded that the creature had been deliberately dodging the snow patches. To do so they felt it would have to leap much farther than a human could, and also a human could not have moved fast enough to get out of sight so quickly. They reasoned that their fire might have attracted the oddity. Two other hunters, seated at the other side of the fire, did not see the creature. Later loud screams were heard.

Source: John Green, 1973. *The Sasquatch File.* Cheam Publishing, Agassiz, BC, p. 18, from personal communication. Also John Green, Sasquatch Incidents Data Base, Incident Number 1000038.

1958/YR/00 – Vancouver.

Oilman Needed: The *Sun* newspaper drew attention to the need for funding to find the sasquatch in an article published in 1958. The newspaper headed the article "Harrison Needs a Texas Oilman." This was in reference to the funding Tom Slick, a Texas oil millionaire, had provided to find the yeti. Although Slick is not mentioned by name, the article outlines his accomplishments and states that he will be undertaking another costly yeti expedition. It then states:

Tom Slick, second from the left, with members c his Pacific Northwest Expedition (PNE) in 195 Slick went on to finance explorations in BC undertaken by Bob Titmus.

The mountains back of Harrison are considerably less rugged than the Himalayas. Surely a sasquatch hunt might be an attractive undertaking for even a junior-grade (or two-Cadillac) Texan.

Tom Slick did, in fact, take up the search for the sasquatch in the US the following year with his Pacific Northwest Expedition, and later he funded Bob Titmus for research in BC.

Source: "Harrison Needs a Texas Oilman." The *Sun* Vancouver, BC, 1958 (actual date not known). (Photo: J. Green.)

1959/03/00 – Aristazabal Island. Strong Odor Sensed: Lawrence Hopkins stated he saw an "ape" while waiting on the shore of Aristazabal Island to be picked up by friends during March 1959. He had just shot a deer and had evidently dragged it to the shoreline. Hopkins says he sensed a strong animal odor and when he turned around saw the creature emerging from the underbrush directly behind him. Fearful that the animal might attack, Hopkins ran to the water and swam to a small islet just offshore. When he felt the creature had left, he found a drift log and paddled back to the island to await his friends.

Source: John Green, 1973. *The Sasquatch File.* Cheam Publishing, Agassiz, BC, p. 18, from information provided by Bob Titmus. Also John Green, Sasquatch Incidents Data Base, Incident Number 1000056

1959/08/00 – Kamloops. Queen Elizabeth II & Prince Philip Given Sasquatch Presentation: While on a visit to Canada in 1959, Queen Elizabeth II and Prince Philip vacationed in the Kamloops area (late August). They stayed at a remote mountain lake and were given a presentation on the sasquatch, which included a general overview of the creature and incidents involving Serephine Long (1871), Jacko (1884), Bishop Cove (1907), Albert Ostman (1924), William Roe (1955), and Stan Hunt (1956).

John Green and René Dahinden were responsible for the publicity for this royal event.

Canadian postage stamp showing Queen Elizabeth II and Prince Philip. The stamp commemorated the 1957 royal visit to Canada.

Source: "Snowman Legend Heard by Queen: British Columbians Told Royal Visitors of Own Mysterious Giant." The *New York Times,* New York, USA., August 30, 1959. (Photo: Postage stamp, © Canada Post Corp., 1957.)

1959/09/00 – Enderby. Nose Just a Flat Area With 2 Holes. Mrs. Bellvue reported seeing a sasquatch while camped with her husband near Enderby at Hidden Lake on Labor Day weekend 1959 (September 5, 6 and 7). The couple had been there for 5 days. Mrs. Bellvue left their tent area to collect some campfire kindling at about 7:30 p.m. She suddenly became intensely aware that she was being watched. She looked up and off to a small knoll with a pine tree about 50 feet away. Beside the tree, and partially concealed by lower branches, stood a tall, heavy, human-like figure. It appeared to be well over 6 feet tall and was covered with rust-colored hair which lightened around the chest area. Its forehead sloped back, and parts of its face were without hair. The nose was, "just a flat area with 2 holes." The mouth appeared to be little more than a slit. The oddity just stood there motionless studying Mrs. Bellvue. She slowly backed away and returned to their campsite.

A little view of a section of Hidden Lake. It's now a popular spot for camping, but continues to be highly remote, surrounded by a vast forest.

She did not immediately tell her husband what she had seen, "When all of a sudden I knew we had to get out of there. That's when I told him about it."

Her husband did not show the surprise she expected. He said, "The night before I'd been fixing my fishing tackle when suddenly something told me it was okay to spend that night there, but we'd better leave tomorrow." His wife's story confirmed his feeling.

The next day the couple packed up their camping equipment and as they dismantled the last tent poles they heard the sound of running feet moving gradually away from them through the bush. The sounds faded and disappeared. The Bellvues then left the area.

Source: Don Hunter with René Dahinden, 1993. *Sasquatch/Bigfoot: The Search for North America's Incredible Creature.* McClelland & Stewart Inc., Toronto, Ontario, pp. 99, 100, from a radio station interview and later personal contact by René Dahinden. Also, John Green, 1973. *The Sasquatch File.* Cheam Publishing, Agassiz, BC, p. 18, from personal communication. Also John Green, Sasquatch Incidents Data Base, Incident Number 1000014. (Photo: C. Murphy.)

1959/AU/00 – Clinton. Massive Legs. H. Dudeck reported that he saw something highly unusual in the Clinton–Lillooet area during the fall of 1959. Dudeck was hunting when he saw some massive legs covered with yellowish-brown hair protruding from behind a stump. He froze in his tracks for a few moments, then retreated without seeing the torso or head of whatever was behind the stump. After a couple of hours he returned to the spot and followed huge foot tracks in the snow to a ravine.

Source: John Green, Sasquatch Incidents Data Base, Incident Number 1000227.

1959/DE/00 – Armstrong area. Seven-foot Skeleton Uncovered: A road gang working in the Armstrong area in the 1950s reported they uncovered a skeleton that was nearly 7 feet tall. Coroner Arthur Phair of Lillooet was informed of the find and he notified government official in Victoria, BC.

This information was actually gathered from the coroner's report to the government officials. In all likelihood, the coroner determined that the skeleton was very old and simply that of a tall First Nations man. He likely had the bones reinterred at a cemetery or turned them over to the First Nations people in the area for burial.

I can reason that if the skull was highly unusual (unlike a normal human skull, save for size) Phair would have sent the skeleton to the Provincial Museum in Victoria. If this was the case, nothing to my knowledge was ever released by the museum on the finding.

Incidents of this nature lead us to muse as to what is really in the museum's storage.

Source: The *Daily Colonist,* Victoria, BC, May 5, 1957.

1959/DE/00 – Rocky Mountains. Moved Like a Monkey: Roy Miller reported that he had a possible sasquatch encounter in the Rocky Mountains in the 1950s. He was following a sheep trail on a high plateau and when he stooped over to avoid branches he encountered a sasquatch-like creature coming the other way. It immediately leaped off into the brush. Miller said it was as tall as he was and had reddish brown hair on its face. He added that it did not have enough hair on its body and its "arms" were too long to be a bear. He said its movements were like those of a monkey.

Source: John Green, Sasquatch Sighting-Incident File, from *The Herald,* Calgary, Alberta, January 31, 1981. Also John Green, Sasquatch Incidents Data Base, Incident Number 247.

1959/DE/00 – Hope. Sasquatch Caves: Sometime after the word "sasquatch" came into common use, a number of caves near Hope were given the name, "Sasquatch Caves." The word "sasquatch" was essentially developed by John W. Burns, in or about 1926 (see 1926/CI/00 – Harrison Mills (Chehalis Reservation). The Word "Sasquatch" is Developed), I believe the caves would haves been given the name sometime in the 1950s, but such could have been earlier. I have not been able to find any history on them.

The "caves" are not really caves in the true sense of the word. At some remote point in history, it appears the entire side of a mountain collapsed and enormous boulders crashed down. They landed one atop the other and in this process caused "openings" or "caves." From those I looked at, the caves do not go in very far, but certainly far enough to provide effective protection against the elements.

That sasquatch might have used the caves at some time is possible, but there is no proof or evidence supporting this. It simply appears BC's reputation for sasquatch encounters resulted in the caves being given the name.

I have been told that for a time, the caves were on a tourist attraction bus route. I believe there was a gift shop with refreshments,

A view of the "caves" area (top) and the opening of a cave, which was between 5 and 6 feet high. I don't believe any of the caves would be very stable. The slightest earth tremor would likely cause the rocks to shift and collapse on each other. There is no proof of any sasquatch connection with the caves.

and tourists could wander up the trail and peer into the dark openings. At this time, there is only a motel at the site. The "caves" are part of the motel's property, but have been left to nature and teenagers with paint spray cans—nothing is sacred.

Source: Christopher L. Murphy, 2010. *Know the Sasquatch: Sequel and Update to Meet the Sasquatch.* Hancock House Publishers, Surrey, BC, p. 33, and general research. (Photos: C. Murphy.)

1959/DE/00 – Mainland Bay. Sasquatch Scares Off Fisherman: Arthur Boundsound Sr., an experienced fisherman and trapper, was frightened away from his anchorage in a remote mainland bay during the 1950s. The man reported that he had anchored his boat to brew a pot of tea. A sasquatch appeared on the shore and threw a small log towards the man's boat. The man immediately pulled anchor and moved to another spot.

Source: John Bindernagel, 1998. *North America's Great Ape: The Sasquatch.* Beachcomber Books, Courtenay, BC, pp. 118, 225, from a report provided by Arthur Boundsound's nephew.

1959/DE/00 – Gilford Island. Digging for Clams: Edwin James, along with a group of friends and family members, reported that they surprised a sasquatch digging for clams on Gilford Island in the 1950s.

Source: John Bindernagel, 2000. "Sasquatches in Our Woods." *Beautiful BC* magazine, summer, 2000.

1959/DE/00. Peachland. Cook Sees Sasquatch: A cook named Sophie, working at a logging camp between Peachland and Princeton, reported that she saw a sasquatch in the early 1950s. Sophie said that all the men were out falling trees at the time.

Source: *Peachland Memories.* Peachland Historical Society, 1984.

1960/02/00 – Price Island. Size Of a Small Man: Joe Hopkins of Klemtu reported that he saw an upright ape-like creature on the beach of Price Island at Higgins Pass in February 1960. Hopkins was digging clams at the time and saw the oddity walk up the beach and into the trees. He stated it was about the size of a small man. It had apparently been down by the water, and as Hopkins came around a point he noticed it walking away.

John Green, 1973. *The Sasquatch File.* Cheam Publishing, Agassiz, BC, p. 18, from personal communication. Also John Green, Sasquatch Incidents Data Base, Incident Number 1000028.

1960/02/00 – Roderick Island. Blood Found in Snow: Fishermen Timothy Robinson and Samson Duncan stated that they saw and shot at a small ape-like creature on the beach of Roderick Island at Watson Bay in the winter of 1960. The men were on their fishing boat at the time, and after the creature ran away they went ashore and found blood on the snow. They stated that they were afraid to follow it.

Source: John Green, 1973. *The Sasquatch File.* Cheam Publishing, Agassiz, BC, p. 20, from information provided by Bob Titmus. Also John Green, Sasquatch Incidents Data Base, Incident Number 1000008

1960/02/00 – Langley. Four-and-a-half Feet Tall: Mrs. D. Mott wrote that her younger son was frightened by something strange at their farm in Langley in February 1960. Her boy went out at about

7:00 a.m. to clear the barn before milking. He came back into the house very agitated, saying he had seen a strange creature outside through a wide opening in the boards of a shed. He said the creature was about four and one half feet high, covered with shaggy hair, and walking like a human, but slightly crouched. Mrs. Mott (and probably her husband) investigated but nothing was found. They said they did not look for tracks because children had trampled the snow down while playing the evening before.

Source: John Green, Sasquatch Incidents Data Base, Incident Number 1000150.

1960/08/00 – Nelson. A Great Beast: Woodsman, hunter, and fisherman John Bringsli reported that he came face-to-face with a "great beast" while picking huckleberries near Nelson (Kokanee Glacier Park, Six Mile area) at Lemmon Creek in early August 1960. The encounter took place at the head of the creek, on a deserted logging road. His frightening experience is presented here in his own words:

I had just stopped my 1931 coupe on a deserted logging road and walked about 100 yards into the bush. I was picking huckleberries.

I had just started to pick berries and was moving slowly through the bush. I had only been there about 15 minutes.

For no particular reason, I glance up and that's when I saw this great beast. It was standing about 50 feet away on a slight rise in the ground, staring at me.

The sight of this animal paralyzed me. It was seven to nine feet tall with long legs and short powerful arms with hair covering its body. The first thing I thought was...what a strange looking bear.

It had very wide shoulders and a flat face with ears flat against the side of its head. It looked more like a big hairy ape.

It just stood there staring at me. Arms of the animal were bent slightly, and most astounding was that it had hands ... not claws. It was about 8 a.m., and I could see it very clearly.

The most peculiar thing about it was the strange bluish-

133

gray tinge of color of its long hair. It had no neck. Its ape-like head appeared to be fastened directly to its wide shoulders.

When the creature started to move towards him, he fled to his car and quickly left the area. He returned the next day with friends and upon inspection the group found a 16- to 17-inch footprint.

Bringsli's observation of short arms is a little surprising. Generally, very long arms are seen.

Face to face with a "great beast." John Bringsli is seen here in a drawing by Duncan Hopkins.

Source "Berry Picker Met Hairy Giant Near Nelson, B.C., Last Summer." The *News,* Nelson, BC, October 1960. Also John Green, Sasquatch Incidents Data Base, Incident Number 1000039. (Photo: Artwork, D. Hopkins.)

1960/08/00 – Chilliwack. Stones Rain For About 2 Hours: A woman reported an unusual experience she had near Chilliwack at Cultus Lake (provincial campground) in August 1960. She and four family members had arrived late and decided not to pitch their tents, but to sleep out in the open. The site they chose was a flat gravel area, far from the other camp sites.

Sometime after midnight, the woman was awakened by the sound of small stones falling around her and her sleeping family. The woman alone was awakened, and she lay in fear for about 2 hours as the stones continued to fall. In the morning, 40 to 50 small stones were counted laying on the gravel surface near and between the sleeping bags. The stones, although larger than the small, even-sized pea gravel (ground covering), were less than one-and-a-half inches in diameter. The woman did not detect any odor, and heard nothing else other than the stones landing. Upon inspecting the area, an obvious trail was seen (new since the previous evening) which indicated a large animal had walked through a nearby stand of tall stinging nettle.

A view of Cultus Lake looking south. Mount Baker, Washington, is seen in the distance. The border between Canada and the US is close to the lake—running horizontally about at the arrow shown on the right.

A border marker near Cultus Lake (invariably covered in graffiti) showing the boundary between Canada and the US. The photo was taken from the Canadian side, as seen by raised lettering. On the other side of the marker, "United States" is shown, and on the side the information seen on the left. I am sure the bor-der must be patrolled, but have never heard of a sasquatch sighting report by either US or Canadian border offi-cials. I find this a little odd.

135

Source: John Bindernagel, 1998. *North America's Great Ape: The Sasquatch.* Beachcomber Books, Courtenay, BC, p. 117, 225, from information provided in 1997 by a campsite government worker. (Photos: Cultus area, Image from Google Earth; Image © 2011 DigitalGlobe; Image © 2011 IMTCAN; Image U.S. Geological Survey; Image State of Oregon; Border marker and side detail, C. Murphy.)

1961/04/19 – Harrison Hot Springs. Puzzling Print: Mrs. Gerry Stary reported that she found a large human-like footprint near here home, about 6 miles from the Harrison Hot Springs Hotel, on April 19, 1961. She found it while walking her little girl to the school bus. The print was on the gravel road leading to her home. It measured about 12 inches long and 10 inches wide at the toes.

Mrs. Stary stated that she had heard unusual noises outside her home on the previous night. She also said that a broken tea cup was taken from her garbage can over night and carried some distance down the road.

There was only one clear footprint impression; however, other indentations that might have been footprints were observed. No print indications were found leading to the road, which might indicate a hoax. A photograph was taken of the print.

Source: "Abominable Snowman Headed for Chilliwack." The *Enterprise,* Chilliwack BC, April 26, 1961.

1961/09/00 – Cumberland. Seemed To Be Checking Where They Camped. Bill Winnig reported seeing a strange creature near Cumberland in September 1961. Winnig and another youth had camped for several days on a sand bar at the head of Forbush Lake in the Puntledge Valley. As they were leaving, Winnig looked back and saw an animal walk out of the bush on 2 legs onto the sand bar. It squatted, stood again, and then walked into the forest. It was dark colored and had no neck. Winnig remarked that it seemed to be checking the sand bar at the spot where the 2 had camped.

Source: John Green, Sasquatch Incidents Data Base, Incident Number 1001356.

1961/10/00 – Nelson. Silhouetted In the Moonlight. John Bringsli, who reported seeing a sasquatch near Nelson in Kokanee Glacier Park in 1960 (see 1960/08/00 – Nelson. A "Great Beast") had another sighting here in October 1961. He had gone to the area with another man hoping to see another sasquatch. At midnight they saw the upper body of one the creatures silhouetted in the moonlight. It was in the trees just short of where the trail went out onto open rock. Bringsli estimated the creature was 7 to 8 feet tall.

Source: John Green, Sasquatch Incidents Data Base, Incident Number 1000036.

1961/10/00 – Swindle Island. Long String of Large Footprints: Bob Titmus, who was researching sasquatch with funding from oil millionaire Tom Slick, reported finding a long string of large footprints on a small island offshore from Swindle Island in October 1961. He saw the tracks with binoculars from his boat. As there was no way to get ashore other than to swim, Titmus stripped off his clothing and did just that.

The tracks were about 13 to 13.5 inches long and approximately 6 inches wide at the ball of the foot. The stride [pace] was about 4 feet. Some of the impressions were quite deep, although he could see that the creature was only walking. The tracks came out of the water and angled toward the timber and undergrowth. They paralleled the growth line for about 125 feet and then entered it. He stated that the tracks appeared very similar to 14-inch tracks that have been found in California.

He was unable to spend a great deal of time on the island due to the cold weather and for fear of his boat breaking anchor.

Source: John Green, 1973. *The Sasquatch File.* Cheam Publishing, Agassiz, BC, p. 25, from information provided by Bob Titmus in a report to Tom Slick.

1961/YR/00 – Moricetown. Prints Were Awe Inspiring: A couple sitting down to breakfast at their residence about 4 miles south of Moricetown reported that they watched in amazement as a sasquatch strolled by during 1961. The creature, which they described as about 8 feet tall, black, hair-covered, very heavy and with a flat face, walked erect across a field and then across the highway. After it left sight, they inspected its path and found 300 yards of 5-toed, flat, 16-inch long footprints that sank 4 to 5 inches deep in the field. The depth of the prints, compared to their own, was "awe inspiring."

Source: John Green, 1973. *The Sasquatch File.* Cheam Publishing, Agassiz, BC, p. 26, from information provided by the witnesses to Bob Titmus. Also John Green, Sasquatch Incidents Data Base, Incident Number 100061.

1962/04/00 – Bella Coola. Four Sasquatch Seen Together: Harry Squiness reported that while camped with his family near Bella Coola at Anahim Lake, Goose Point area, in April 1962, he observed 4 sasquatch. Squiness said that the couple's baby was crying and he saw a head and forearm come in thorough his tent flap. The creature had a monkey face, but its head was bigger than a human's. Also the arm was covered with long, dark brown hair. Squiness grabbed for his flashlight and gun, but the flashlight was not working. At the same time the creature wandered off. After leaving his tent to investigate, Squiness threw gasoline on the campfire and in the bright flare he saw 4 of the creatures. They raised up after laying face down on the ground at the edge of his camp area, about 14 feet from his tent. They all walked slowly away, "walking like men." All were about 8 feet tall.

Squiness called to them, "Hey, what you doing out there? Hey, come back!" However, the creatures just kept walking away. He thought that the reason one of the creatures peeked into his tent was because the baby was crying.

The next day, Squiness found big finger marks in the dust on a poplar tree with its bark skinned about 8 feet up.

This is an interesting account from the standpoint that a crying baby was involved. There are a number of incidents in which sasquatch are apparently attracted to children—the creatures simply stand and watch the children at play. In Russia, there is an account of a mother who left a baby to fetch water. The baby apparently started to cry, and when the mother returned she found a female almasty cuddling her baby. Perhaps something like this could happen with a sasquatch. I have mused that a "set up" along the lines of the Squiness experience, using a recording, just might attract one of the creatures.

Source: John Green, 1973, *The Sasquatch File*. Cheam Publishing Ltd., Agassiz, BC, p. 26. Also Don Hunter with René Dahinden, 1993. *Sasquatch/Bigfoot: The Search for North America's Incredible Creature*. McClelland & Stewart Inc., Toronto, Ontario, p. 33. Also John Green, Sasquatch Incidents Data Base, Incident Number 1000066.

1962/04/00 – Bella Coola. Mother & Child: A young woman reported that she spotted a sasquatch mother holding a child by the hand on the banks of the Bella Coola River (right in Bella Coola) in

April 1962. The woman was out walking with her own 2 children when she saw the creatures. Word was also circulating in the little town at about the same time that other people had seen the "mom" at night.

Sasquatch sightings of a lone creature are by far the norm. However, it stands to reason that those solitary "big guys" were once toddlers—and this is a little proof.

The Bella Coola River (and valley), facing east towards the town. The wilderness in this area is very primitive, and has long been considered a very likely area for sasquatch sightings.

Source: John Green, 1973. *The Sasquatch File.* Cheam Publishing, Agassiz, BC, p. 26, from personal communication. Also John Green, Sasquatch Incidents Data Base, Incident Number 1000152. (Photo: C. Murphy.)

1962/06/08 – Kitimat. About 10 "Apes": A commercial fisherman reported that he saw about 10 "apes"near Kitimat on June 8, 1962. He was cleaning fish with his boat tied to the shore. Several creatures like "apes" came out on an overhanging tree, he guessed about 10 of them. Thoroughly shocked, the fisherman started his boat motor and fled, breaking his shorelines.

He at first said they were "apes," but later changed his story to "bears." However, bears don't usually come in such large numbers, and we have to wonder if bears would have greatly shocked the man (although I think 10 would have been at least a surprise).

Source: John Green, Sasquatch Incidents Data Base, Incident Number 1000004.

1962/07/23 – Wells. Light Gray Animal: Alec Lindstrom and friend George Bryant reported a long distance sighting of a strange creature in the Wells area near Stoney Lake on July 23, 1962. The men were in a boat on the lake when they saw an erect light gray animal standing on the shore more than half a mile away. They said it looked to be between 9 and 10 feet tall and very heavy. They went towards it and it eventually turned and ran into the trees. They observed the creature for about 6 to 10 minutes.

Source: John Green, 1973. *The Sasquatch File* . Cheam Publishing, Agassiz, BC, p. 26, from personal communication. Also John Green, Sasquatch Incidents Data Base, Incident Number 1000069.

1962/SU/00 – Aristazabal Island. Many Tracks Found: Bob Titmus reported that during the summer of 1962 he found 1,200 yards of bipedal tracks—much larger than normal human tracks—in deep moss on Aristazabal Island.

Source: John Green, 1973. *The Sasquatch File.* Cheam Publishing, Agassiz, BC, p. 26, from information provided by Bob Titmus

1962/08/00 – Devastation Channel. Pace of 42 Inches. Bob Titmus reported that during August 1962 he found flat, 14-inch tracks with a 42-inch pace in a creek bed on an island in Devastation Channel.

Source: John Green, 1973. *The Sasquatch File.* Cheam Publishing, Agassiz, BC, p. 26, from information provided by Bob Titmus

1962/08/00 – Hixon. Hair-covered Man: A woman stated that she saw a tall, hair-covered man near Hixon in August 1962. She said he was about 7 feet tall and covered in black hair. The oddity came towards her as she was walking along a local creek. When he noticed her, he jumped into the bushes.

Source: John Green, 1973. *The Sasquatch File*. Cheam Publishing, Agassiz, BC, p. 26, from information provided by René Dahinden who talked to the witness.

1962/08/00 – Quesnel. Small, Black Eyes: Mrs. Calhoun reported seeing a man-like creature between Quesnel and Prince George (Canyon Creek area – not official name) in August 1962. She was on a prospecting and fishing trip and while waiting by the creek for her daughter to return with lunch, she heard a noise behind her. Thinking her daughter had returned, she turned to speak to her. Instead of her daughter, about 10 yards away there stood a hairy man-like creature. She had a hunting rifle and quickly positioned it for self-protection. Fortunately, the creature simply stared and then retreated to the bushes. Reflecting on the incident, Mrs. Calhoun stated the following:

> My first, fleeting impression was that it was a human with very long arms. But it took me weeks to get out of my mind the look it was giving me from its small, black eyes as it stood there. It was like an ape—but like a human as well. It had blond-brown hair on it chest and long, loose, matted hair on its head. It had high cheekbones, a wide flat nose, a forehead that sloped back, and a mouth that stuck out. It opened its mouth, but didn't make a sound; just stood there looking at me. I started moving away and the thing jumped into the bush and disappeared. All the time as I was backing off down the creek, I knew I was being watched.

Mrs. Calhoun later said that the creature had something around its waist—animal skin or garment, she couldn't recall.

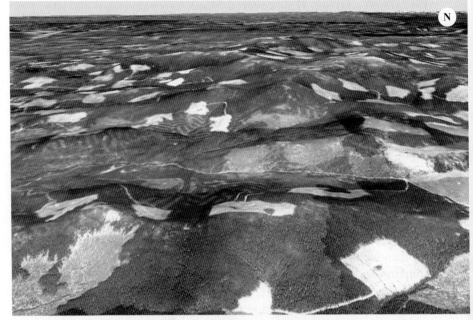

A section of the region between Quesnel and Prince George. The relatively low rolling hills provide easy access for logging operations, as can be seen by the clear-cut areas. I was taken aback when I first saw clear-cuts from a plane. Logging companies do not essentially log areas that can be seen from a main road to avoid public outcry.

Source: Don Hunter with René Dahinden, 1993. *Sasquatch/Bigfoot: The Search for North America's Incredible Creature.* McClelland & Stewart Inc., Toronto, Ontario, pp. 100, 101. Dahinden interviewed Mrs. Calhoun. Photo: Image from Google Earth; Image © 2011 DigitalGlobe; © Cnes/Spot Image; Image © Province of British Columbia; Image © 2011 TerraMetrics.)

1962/AU/00 – Swindle Island. Three Sets of Tracks. Bob Titmus reported that during the fall of 1962 he found 3 sets of tracks, about 14, 13, and 12 inches long, on a Swindle Island beach.

John Green, 1973.*The Sasquatch File.* Cheam Publishing, Agassiz, BC, p. 26, from information provided by Bob Titmus

1962/11/00 – Chilliwack. Eyes Glared: Joe Grigg, a Pacific Stage Lines bus driver, reported that he did a "good turn" and saw something very unusual near Chilliwack, Vedder Crossing area, in November 1962.

He had returned to his depot late at night with 2 soldiers on his bus, and there was no one there to meet them. Grigg was going off shift, so gave them a lift in his own car to the army camp at Vedder Crossing (C.F.B Chilliwack), which was not far out of his way. As he drove home towards Yarrow in heavy rain, he saw what he thought was a very big man in a fur coat standing on the edge of Vedder Mountain Road. In Grigg's own words:

And then as I got close to him, I realized it wasn't a man or anything I had ever seen that looked like one, and when he looked at me my headlights hit his eyes and they glared. Then he slowly kind of half walked and half loped across the road, and I kind of turned my car and followed him with my headlights. Then he stopped along the edge of the road and looked at me for a second and then jumped up onto the bank.

Grigg, who during this time was 20 to 25 feet from the creature, said it was about 6 feet tall, and did not have clothes—just quite long hair all over its body with a reddish glow. He estimated the creature's weight to be well over 400 pounds. When asked about facial features he stated:

Thomas Steenburg is seen here standing in the spot where the creature jumped up onto the bank.

I saw part of his face when I was right close to him...very ape-like, flat-nosed, some of it quite hairy, and a very thin-lined mouth. The one outstanding feature is, I didn't notice any ears. The only other thing was the lack of a neck. He had little or no neck at all.

143

After the creature was out of sight, Grigg marked the road shoulder with his foot and returned the next day. He noted that the oddity had risen over 6 feet (later measured at 7 feet, 10 inches) in its jump, and landed on a large rock at the bottom of the bank.

Source: John Green, 1981. *Sasquatch, The Apes Among Us.* Hancock House Publishers, Surrey, BC, p. 398. Also Don Hunter with René Dahinden, 1993. *Sasquatch-Bigfoot, The Search for North America's Incredible Creature.* McClelland & Stewart Inc., Toronto, Ontario, pp. 66, 67. Also John Green, Sasquatch Incidents Data Base, Incident Number 1000033. (Photo: B. Blount.)

1962/12/00 – Bella Coola. Four Sets of Footprints: Bob Titmus reported that while driving across the high plateau above the Bella Coola Valley (Tweedsmuir Park) in December 1962, he found 4 sets of adult sasquatch footprints that crossed the road. The tracks were in very deep snow, and the weather was very cold.

This finding appears to be a bit of a record for sasquatch prints, which are usually the result of one or 2 individuals traveling together.

Source: John Green, 1973. *The Sasquatch File,* Cheam Publishing Ltd. Agassiz, BC, p.26, from information provided by Bob Titmus.

1962/CI/00 – Bella Coola. Million Dollar Reward: In 1962 or earlier Tom Slick declared he was willing to pay a $1 million reward for a sasquatch—dead or alive. It appears he made the announcement in Bella Coola. Slick had been earnestly seeking the creature since 1959, and had a number of men in his employ scouring forests in the Pacific Northwest.

Although it is common knowledge that *Life* magazine offered a $100,000 reward for a sasquatch

Cliff Kopas

(dead or alive) in the late 1960s, Slick had previously offered $1 million, according to Cliff Kopas of Bella Coola. Unfortunately, Tom Slick died on October 6, 1962, so it is doubtful the offer was valid for very long. Nevertheless, it was an indication of what one could possibly get for a sasquatch.

It appears, however, that Slick's offer was not made public until 1967. The 1967 *Real West* magazine article on the offer refers to Kopas as mentioning it in a "recent edition"of the *British Columbia Digest* magazine. Oddly, in the *Real West* article the offer is talked about as still being valid, which I am sure was not the case.

Source: T.W. Paterson, 1967. "Wanted, Dead or Alive Sasquatch". *Real West* magazine (appears to be defunct), July 1967. (Photos: Cliff Kopas, J. Green; .Poster, Y. Leclerc.)

1963/07/00 – Kemano. Titmus Sees Unusual Figures: Bob Titmus reported that in July 1963 he observed 3 brown figures as they climbed a cliff south of Kemano. Titmus was about 2 miles away but could see arm and leg movements like those of humans.

Source: John Green, 1973, *The Sasquatch File*. Cheam Publishing Ltd., Agassiz, BC, p. 27, from information provide by Bob Titmus. Also John Green, Sasquatch Incidents Data Base, Incident Number 1000041.

1963/SU/00 – Price Island. Sasquatch Roofed Nest: In the summer of 1963, Ray Roberts and his wife found what appeared to be a sasquatch nest on Price Island.

The couple were walking the shore beachcombing for glass fish net floats when they noticed the strong smell of "something rotting" at the forest edge. They continued walking up the beach and found a large nest-like structure of grass and other vegetation within a crude shelter of poles and other driftwood, just above the high tide line. The nest was at least 8 feet across and the low entrance of the structure faced the forest rather than the beach. There was a substantial roof of poles and logs. Humans were eliminated as a possible explanation because of the remoteness of the area and the strong smell.

Source: John Bindernagel, 1998. *North America's Great Ape: The Sasquatch.* Beachcomber Books, Courtenay, BC, p.69, 70.

1963/YR/00 – Bella Bella. Seen On Shore: Jack Wilson reported that he saw a big, erect, hair-covered animal on the shore of an island near Bella Bella in 1963. Wilson related this information to Bob Titmus.

Source: John Green, 1973, *The Sasquatch File.* Cheam Publishing Ltd., Agassiz, BC, p. 27, from information provided by Bob Titmus. Also John Green, Sasquatch Incidents Data Base, Incident Number 1000156.

1963/YR/00 – Minstrel Island. Monstrous Thing: A man reported that he saw a "monstrous thing" on a shore near Minstrel Island during 1963. He was in a rowboat traveling to the flats at the head of Thompson Sound and apparently saw the creature in a stooped position. He said he saw the creature get up on its "hind legs" and move off erect. The man was so astounded with what he saw that he reported the incident to the BC Provincial Museum (now Royal Museum) in Victoria.

Source: John Green, 1973, *The Sasquatch File.* Cheam Publishing Ltd., Agassiz, BC, p. 26. Also John Green, Sasquatch Incidents Data Base, Incident Number 1000155.

1964/WI/00 – Squamish. Very Large Tracks: A man and his wife reported that they found a long line of large "naked foot tracks" in snow north of Squamish in the winter of 1964. The tracks were apparently made by a bipedal animal.

The man said that the tracks were half as long again as his size 12 rubber boots. The couple followed the tracks, but apparently did not see what had made them.

Source: John Green, 1973. *The Sasquatch File.* Cheam Publishing, Agassiz, BC, p. 27 from information provided by René Dahinden who interviewed the male witness.

1964/04/00 – Turnour Island: House Moved: Ellen Neal reported that people in a village on Turnour Island said that a sasquatch had approached a house there in April 1964 and shoved it partially off its foundations. It is said the creature pushed at a corner of the house, apparently with enough force to move the structure.

The man who owned the house said that he had seen just a bear in the area, however, it was noted later that he cut down all the trees around his house.

Source: John Green, 1973, *The Sasquatch File*. Cheam Publishing Ltd., Agassiz, BC, p. 27. Also John Green, Sasquatch Incidents Data Base, Incident Number 1000044.

1964/07/00 – Chilliwack. Four White Dots: Five people reported that they saw 2 unusual creatures near Chilliwack at Cultus Lake on July 3, 1964. The group was driving on a dirt road around 2:30 a.m., looking for a place to camp. They saw 4 white dots up ahead, and as they passed, the dots turned out to be the eyes of two 7-foot-tall creatures, one heavier than the other, both covered with shaggy dark hair except on their feet. The creatures stayed on the road and kept walking as the group went by, but turned their faces away. The driver turned around for another look and the creatures stepped into the woods.

After this experience, the campers were so shaken that they drove to Chilliwack and parked in a school ground for the night.

Source: John Green, 1973. *The Sasquatch File*. Cheam Publishing, Agassiz, BC, 27, from information provided directly by the campers. Also John Green, Sasquatch Incidents Data Base, Incident Number 1000157. (The campers were: Mr. and Mrs. Rodney Doherty, Mike Dika, George Swencera, Linda Dumas and Wolfgang Schultz.)

1964/CI/00 – Agassiz. Tracks With Short Toes: Mr. and Mrs Robertson found large unusual footprints near Agassiz on Herrling Island while out rock hunting in about 1964. The island is located across the Fraser River from Agassiz. The prints were 14 inches long, 7 inches wide (widest point), and had "short toes." They were found in sand along the island's shoreline.

Source: John Green, 1973. *The Sasquatch File*. Cheam Publishing, Agassiz, BC, p. 27, from information provided directly by Mrs. Jean Robertson.

1965/05/31 – Mission. Cows Stared At It: Mrs. Seraphine Jasper reported that on May 31, 1965 she saw a sasquatch near her home in the Mission area, Nicomen Island. The creature was in a pasture with cows. She stated that it was large and covered in black hair. It had appeared from nearby bush land. She said that the creature kept moving around and the cows tended to wander over and stare at it. She became frightened and left, so did not see the creature leave the area.

Mrs. Jasper was an elderly lady and had a superstitious fear of bad luck on seeing a sasquatch. As a result, she did not watch it, just glanced towards it from time to time.

Source: John Green, 1973. *The Sasquatch File,* Cheam Publishing Ltd. Agassiz, BC, p.33. Also,"Beware! Sasquatch Roaming Once More," the *Sun,*Vancouver BC, June 1, 1968. Also John Green, Sasquatch Incidents Data Base, Incident Number 1000030.

1965/06/28 – Squamish. Something Dragged To an Ice Hole. Two brothers, professional prospectors, reported that on June 28, 1965 while traveling on foot at about the 4,000-foot level near Squamish in the Garibaldi Park area (between Pitt Lake and Squamish), they found some fairly fresh and very unusual tracks in the snow. The prints were enormous, 22 to 24 inches long, and 10 to 12 inches wide. They were perfectly flat and showed 4 clear toe impressions, with the big toe on the inside of the foot like a man's. The stride [pace] was double a man's stride. Snow in the bottom of the prints was tinted pink.

Parallel to the tracks were 3 grooves in the snow. The prints were widely spaced from side to side, and there was a wide but shallow drag mark running along between them. Outside the line of prints on either side, but close to them, were narrower, deeper grooves. The snow was old and hard, but both the prints and the outer drag marks sank in about 2 inches. One of the men sketched all this in his notebook on the spot. They followed the trail up the valley until they came to a small lake that was still frozen over. The tracks led out on the ice to a place where a large hole had been made, with the broken pieces of ice lifted out and piled around. There were footprints but no drag mark leading away from the hole. It was later reasoned that humans would not have had the weight or strength to make the tracks or drag anything heavy enough to make the drag marks in the snow.

The Garibaldi Park area (between Pitt Lake and Squamish).

Baffled, the men went on around the side of the lake, until they noticed in the trees on the other side a human-like figure standing and watching them. It was auburn in color except for its hands, where the color lightened gradually almost to yellow. As sketched on the spot, it had a human-shaped head set directly on very square shoulders, and its forearms and hands bulged like canoe paddles. It was swaying slightly as if shifting from foot to foot, and its hanging arms swayed as well. They couldn't make out its face because of the distance, but the features seemed flat. It was just noon, and they sat down and had a cigarette and chocolate bar while they watched the figure and tried to estimate its size. Counting the sets of branches on the evergreens where it stood, and comparing them with those on the side of the lake, they decided it was between 10 and 15 feet tall. It just continued standing there, so finally they went on. When they came back later, there were more tracks around, but the oddity was gone.

The following day, after climbing over the ridge onto a plateau, they came to some very small lakes where there were a lot of smaller tracks, so old that all that was left of them was compacted snow sticking up above the level of the melting snow around them. Some of the prints, however were reasonably clear. They were 18 inches long, and they led out to a place on one of the tiny lakes where the snow had been pushed back and a hole more than 5 feet wide made in the ice. Other smaller tracks, about 10 inches long, did not go out on the ice.

A few days later one of the brothers went back to the spot in a helicopter with a newspaper reporter. They photographed the big tracks in the valley, which were badly melted out. They then found some fresher-looking tracks on a ridge, but could not land here. These tracks led to the edge of a cliff with no snow down below it.

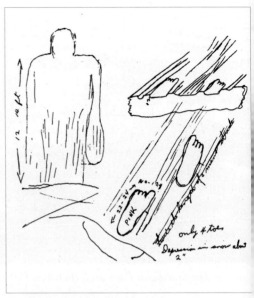

A drawing provided by one of the prospectors.

This is one of the most unusual reports ever submitted on sasquatch experiences. John Green thoroughly investigated this incident and concluded that the men did see what they claimed to have seen. He and René Dahinden even flew over the area to see if they could see any ice holes—checking every lake within 50 miles of the area—however they did not find any.

At the time, the report was not made widely known. However, about 2 years later (September 1967) the brothers met with BC Recreation Minister Ken Kiernan and told him of their experience. It was at this time that the BC Provincial Government expressed interest in the sasquatch issue and appealed for anyone with tangible evidence of the creature to come forward (see 1967/09/19 – Victoria. Government Gets Involved.)

On September 30, 1967 Provincial Anthropologist Don Abbott stated that this sighting was the best account so far received of the sasquatch (see 1967/09/30 – Victoria. Government Impressed). At this time, the brothers said that the tracks they found in snow looked similar to those photographed in Northern California in late August 1967 (see. 1967/09/01 – Victoria. Not a Subject for Mirth).

The fact that the prints showed only 4 toes is unusual, but not unheard of. From my own research, I have concluded that this happens sometimes because the little toe fails to register as deeply as the other toes (i.e., it is not subjected to the same weight).

Source: John Green, 1981. *Sasquatch, The Apes Among Us.* Hancock House Publishers, Surrey, BC, pp. 435-438. Also "Sasquatch Hides." The *Sun,* Vancouver, BC, September 28, 1967 and "Sighting Praised," same newspaper, September 30, 1967. Also John Green, Sasquatch Incidents Data Base, Incident Number 1000017. (Photos: Map, image from Google Earth: © 2011 Cnes/Spot Image; Image © 2011 DigitalGlobe; Image © 2011 IMTCAN; Image © Province of British Columbia; Drawing, one of the witnesses.)

1965/07/00 – Butedale. Three Creatures Seen. Jack Taylor reported seeing 3 strange creatures on and near a small island close to Butedale in July 1965. Taylor was in a boat going to a fishing spot, and when he was within about 75 yards of the island he saw on a rock 2 gigantic bipedal creatures, dark in color and covered in hair. He also noticed something in the water that was surging forward with no apparent arm movements towards his boat and the rock where the other 2 creatures were. Closer examination revealed the swimming creature was the same as those on the rock. Taylor quickly sped off, tearing by a friend in another boat who had arranged to meet him at the fishing spot.

Source: John Green, 1981. *Sasquatch: The Apes Among Us.* Hancock House Publishers, Surrey, BC, p. 432. The witnesses were interviewed by Bob Titmus. Also John Green, Sasquatch Incidents Data Base, Incident Number 1000060.

1965/07/00 – Hope. Lumbered Across Road: Mel Mortimer reported that he saw a strange creature near Hope during July 1965. Mortimer was driving east on the Hope-Princeton Highway, just east of the Hope slide (on the stretch of highway abandoned after the bypass was completed). When he rounded a curve, he saw a creature 8 feet tall and 5 feet wide, erect but slightly hunched, about 200 feet ahead. It lumbered across the road with its arms hanging. It was deep brown in color but lighter below the head. It did not appear to have a neck. It crossed a sloped ditch in one stride [pace] and then crossed about 100 feet of grass before it disappeared in the trees.

Source: John Green, Sasquatch Incidents Data Base, Incident Number 1000963.

1965/08/00 – Kitimat. Creature On the Shore: Several First Nations men and a group of non-First Nations surveyors reported seeing an unusual creature near Kitimat at Clio Bay, during August 1965.

The men were in 2 boats. The surveyors had a separate craft. The entire group of men observed a "huge" 9- to 10-foot-tall hair-covered creature (dark brown hair) on the Clio Bay shore. It was on a rock at the edge of the shore. The men estimated they were about 200 yards away. After a short time the creature walked away on 2 legs.

Source: John Green, 1973. *The Sasquatch File,* Cheam Publishing Ltd. Agassiz, BC, p.33, from information provided by Bob Titmus who talked to several of the witnesses. Also John Green Sasquatch Incidents Data Base, Incident Number 1000007.

1965/10/00 – Fort St. John. "Weetago" Seen: A family reported that they saw a sasquatch cross the highway near Fort St. John at about 6:30 p.m. during October 1965. As they drove around a curve, the creature crossed in front of their pick-up truck. The father, mother, and 2 children saw the oddity. The creature was described as being taller than the cab of the truck, having black/brown "smoothish" hair, and a large odd-shaped head. There was a lingering smell of egg. The father called it a "hairy Indian" and "Weetago" (Cree for "big hairy man").

Source: BFRO website. Incident researched by Cindy Dosen.

1965/AU/00 – Harrison Mills. Wearing Fur Around Its Waist: A man stated that he saw something odd in the bushes near Harrison Mills in the fall of 1965. He went into the bushes on foot and saw a small, dark female creature of some sort with long hair. He said that it appeared to be wearing fur around its waist.

Source: John Green, 1973. *The Sasquatch File*. Cheam Publishing, Agassiz, BC, p. 33, from information provided by René Dahinden who talked with the witness.

1965/11/00 – Bella Coola. White Streak Below Head: James Nelson reported that he saw a strange creature near Bella Coola, Green Bay area, in November 1965. He was hunting at the time and observed a black form that looked like a man. He then noticed that it had a white streak on the front below its head.

Nelson said he was in a logged area (slash) and the creature was about 200 yards away. As to its size, he said that it was quite big because he had to climb over and crawl under logs and trees whereas the creature simply walked over them. He observed that it took the creature only about 10 minutes to cross the slashing. He found this remarkable because it took him 2 hours to cover the same distance.

Source: John Green, 1973. *On the Track of the Sasquatch/Bigfoot.* Ballantine Books, New York, NY, USA, p.60-62. Also John Green, Sasquatch Incidents Data Base, Incident Number 1000006.

1965/YR/00 – Agassiz. Sasquatch Incident Compilation: In 1965 John Green, publisher of the *Advance* at Agassiz, finished compilation of a report listing 120 sasquatch-related incidents, including both sightings and the finding of unusual tracks.

Source: Newspaper article, 1965 (no details).

1965/CI/00 – North Vancouver. White As a Sheet: Three young boys were frightened by a strange creature they saw in the North Vancouver area in about 1965. The boys were hunting frogs in a pond and when they looked up they saw an erect creature covered with long hair across the pond. It broke off a 4-inch tree at shoulder height. The boys said it was a female.

The mother of one of the boys telephoned a radio station in Vancouver and related the incident. She said her son was "as white as a sheet."

Source: John Green, Sasquatch Incidents Data Base, Incident Number 1000158.

1965/CI/00 – Gold Bridge. Seen Through Rifle Scope: Game Guide Chilco Choate stated that in the fall of the early 1960s he and 2 hunters he was guiding saw a strange creature near Gold Bridge in the Bridge River area. Through the telescopic sights of their rifles they saw the head and shoulders of something man-like covered in hair, and might have had a sort of beard. The creature was about 200 yards away from their position. The group was high up at the headwaters of Relay Creek at the time.

Source: John Green, 1973. *The Sasquatch File.* Cheam Publishing Ltd., Agassiz, BC, p. 26, from a letter sent to Dr. Grover Krantz, also: John Green, Sasquatch Incidents Data Base, Incident Number 1000154.

1966/05/00 – Harrison Hot Springs. Sasquatch Not Welcome: In May 1966 the town of Harrison Hot Springs removed its "welcome" sign showing a sasquatch in a top hat. Many residents were apparently not happy with the creature being associated with the town. Harrison has long been known as the "gateway to sasquatch country."

Source: Thomas Steenburg, 2000. *In Search of Giants: Bigfoot/Sasquatch Encounters,* from the *Province,* Vancouver, May 24, 1966. Hancock House Publishers Ltd., Surrey, BC, pp. 48, 49. (Photo: Vancouver Province.)

1966/05/00 – Spillimacheen. Strange Behavior: A hunter believes he saw 3 sasquatch in the Spillimacheen area in May 1966. He glimpsed what he thought were bears at a distance of 35 to 40 feet, but they were upright and "scuffling" with each other, moving further away in this process. There was about 4 inches of fresh snow on the ground.

Afraid of being seen, the hunter moved away to an embankment where he crouched behind a log. The creatures were now about 150 feet from his position. As the hunter observed them, they moved further away to about 300 feet. At this point, one of the creatures appeared to copulate with another, but not in the fashion typical of bears—one creature had laid on its back, and the 2 were face to face. The third creature was standing around nearby.

Although the hunter thought the behavior odd, he still believed that what he was seeing were bears. As the creature standing had stopped moving, the hunter took a shot at it, but missed, and all 3 creatures ran into the woods on 2 legs.

Source: BFRO website. Incident investigated by Kevin Withers.

1966/07/09 – Richmond. Fence Post Torn Out: Bob Gilmore reported that he saw a strange animal near his Richmond farm on July 9, 1966. Gilmore was putting up a fence in a boggy area that had some small trees when he heard his cattle running in the field nearby. He went to see what was happening and observed them being chased by a large hairy creature, "running on all twos." He said he was about 400 yards away from the creature. He went on to state that on that same day he discovered a corner fence post a half-mile from the house that had been torn out and tossed aside.

Sasquatch researchers John Green and René Dahinden investigated the incidents. Green said the fence post was sunk 2 feet into peat and was wired to other posts. He speculated that it might have been removed by the creature.

Source: "Sasquatch Hunt Seeks Footprints." The *Sun,* Vancouver BC, July 23, 1966. Also John Green, Sasquatch Incidents Data Base, Incident Number 1000024.

1966/07/14 – Richmond. Big Wooly Animal: Don Gilmore (brother of Bob Gilmore—previous entry), reported that he and his hired hand Hank Prost saw an odd creature in Richmond on about July 14, 1966. The men were working of the roof of Gilmore's house when they saw a "big woolly animal" stampede 100 head of cattle about one-quarter mile away. He said that the creature went on all fours, like a bear.

The Gilmore house. The roof would have allowed a good view of the entire farm.

Part of the Gilmore farm as it presently (2010) appears. The area marked "New Development" was farmland in 1966.

Gilmore added that the cattle were "range cattle" recently brought in for finishing, and were not accustomed to the barnyard, but they stampeded up to the buildings on several occasions.

Source: "Sasquatch Hunt Seeks Footprints." The *Sun,* Vancouver, BC, July 23, 1966. Also John Green, Sasquatch Incidents Data Base, Incident Number 1000024. (Photos: Gilmore house, C. Murphy; Gilmore farm, Image from Google Earth; Image © 2010 Digital Globe; Image © 2010 Province of British Columbia; © 2010 Google.

1966/07/21 & 22 – Richmond. Three Sightings in 24 Hours: On July 21, 1966 (about a week after the Don Gilmore incident and near his residence in Richmond—previous entry) John Osborne, an artist, saw and later sketched a slim, brown-hair-covered, man-like creature that was standing in a patch of bush about 75 yards away. The artist was sitting on a dike at the foot of No. 3 Road, Richmond, sketching when he saw the creature.

Osborne said it was about 6 feet 8 inches to 7 feet tall, "It wasn't ape-like," he said. "It was like a big hairy man." Osborne observed the creature for 10 to 15 seconds, then it disappeared in the bush.

That same night in the same general area, Darlene Leaf, who was babysitting at a neighbor's farm, heard the horses making a great fuss in the barn at about 10:00 p.m. She looked out and stated that she saw the head and shoulders of something like a man looking over a row of raspberry bushes that were about 6 feet high.

Early the following morning, store owner John McKernan was in his car on a dike road in the same area when he stated that he saw an upright, hairy giant cross a dirt road leading from the dike to a farm. It came out from behind a tree and disappeared in some bushes beside a large drainage ditch.

Sasquatch researchers John Green and René Dahinden investigated this last sighting, but were unable to find any tracks. However, they did note that something large had bedded down in the bushes.

Richmond is a suburb

The Gilmore farm in relation to the Richmond bog.

of Metro Vancouver which is heavily populated. However, these reports came from the shore of Richmond (also called Lulu Island) which is a considerable distance from the city of Vancouver. The Gilmore farm is across the Fraser River from a very large bog where a variety of wild animals reside.

The artist who drew the creature he saw was later reported to have made a joke out of his experience, so there is some concern as to the credibility of this incident.

Source: John Green, 1973. *The Sasquatch File,* Cheam Publishing Ltd., Agassiz, BC, p. 33. The artist incident is from "Hairy Man 6 ft 8 in to 7 ft Tall Seen in Woods." The *Sun,* Vancouver, BC, July 22, 1966, also "Richmond Thing: It was Hairy Sasquatchwise," The *Sun,* Vancouver, BC, July 27, 1966. Also John Green, Sasquatch Incidents Data Base, Incident Number 1000021, 1000022, 1000023. (Photo: Image from Google Earth; Image City of Surrey; Image © 2010 DigitalGlobe; © Google; Image © Province of British Columbia.

1966/07/00 – White Rock. It Fooled Around: Mr. Letoul, reported seeing a sasquatch near his house in White Rock in July 1966. His house was built against a steep bank. A large, light-colored sasquatch came up the trail to his house in the moonlight and "fooled around" in the area for a time, then went up a ladder and over the roof of Letoul's house onto the hill abutting the back of the house. He reported the incident to the RCMP and Constable C.D. Church went to Letoul's home and investigated. Letoul prepared and gave the constable a drawing of the creature. The constable was unable to reach a conclusion on the incident.

Source: John Green, 1973. *The Sasquatch File*. Cheam Publishing Ltd., Agassiz, BC, p. 34. Also John Green, Sasquatch Incidents Data Base, Incident Number 1000160.

1966/08/00 – Penticton. Strange Shadow: A woman with her family parked near Penticton in a tent trailer was greatly frightened during the night in August 1966. They were in a campground and she woke up and saw the shadow of a strange creature on the tent. She was sure it was not a bear. The next morning, it was found that a cooler which had been under the trailer had been removed and some food items taken. The cooler had been crushed, but not strewn about as is normal with animals. There were no claw marks or teeth marks.

Source: BFRO website. Investigated by Blaine McMillan.

1966/CI/00 – Boston Bar. Beautiful Tawny Brown Coat: Dave Mathews and a friend reported that they saw an odd creature while hunting near Boston Bar, Fire Lake area, in about 1966. While scanning a berry patch with their rifle scopes, they saw what they thought was a bear with a beautiful tawny brown coat. As they watched, the animal stood up on 2 legs, looked at them and walked off upright.

Source: John Green, Sasquatch Incidents Data Base, Incident Number 1000159.

1967/WI/00 – Bella Coola. Footprints & Yelling: Clayton Mack reported finding unusual tracks and hearing strange "yelling" near Bella Coola at the head of Gardner Canal in the winter of 1967. He was acting as a guide in the employ of 2 US hunters, a lawyer and a doctor, at the time of the incident. He described the tracks as follows:

> They weren't too big; a little bigger than my footprints. I took my shoes off and stepped alongside them. The fellow with me, Willie Schooner, said, "It's a grizzly bear." I said, "No, its got no front footprints, those are all hind prints." Most of the prints were on top of logs, like he's moving along, but some where in between the logs and I could see real plain. It had

been there just a few minutes before I got there because it was snowing, powder snow, and there was no fresh snow on top of the prints. When I came back into Bella Coola, and told about it, a lot of people believed me; they know I wouldn't say I seen something if I didn't.

When the group were in their cabin, at about 2:00 a.m. something yelled "right at the window." Mack described the sound as making a sharp, harsh exhalation. It then yelled again down on the beach, then on up a hill. Later, when they were leaving the area, the lawyer duplicated the sound the creature made and "got an answer."

Source: Don Hunter with René Dahinden, 1993. *Sasquatch/Bigfoot: The Search for North America's Incredible Creature.* McClelland & Stewart Inc., Toronto, Ontario, pp. 31, 32, from an interview with Clayton Mack.

1967/02/00 – Hartley Bay. Screamed Like a Woman: Two men reported that they saw an ape on an island in Hartley Bay during February 1967. The creature was covered in black hair and about 3 to 4 feet tall. The men said they shot at it and it screamed like a woman. The next day they found blood where the creature had been, but no tracks.

Source: John Green, 1973. *The Sasquatch File.* Cheam Publishing Ltd., Agassiz, BC, p. 34, from information provided by the witnesses to Bob Titmus. Also John Green, Sasquatch Incidents Data Base, Incident Number 1000057.

1967/02/00 – Quatna River. Ape Tracks: Clayton Mack reported that he found unusual tracks in snow along the Quatna River (about 2 miles up the river from the Pacific ocean) in February 1967. He said the tracks were larger than those of a man, and of a different shape. Mack was about 2 miles up the Quatna River from the sea. He was certain they were ape tracks. He said the creature had been walking on logs with its feet turned crossways on the log so that the tracks had no shape, but he followed them to a place where the creature had to step on flat ground. He then observed the shape of the tracks.

Source: John Green, 1973. *The Sasquatch File.* Cheam Publishing Ltd., Agassiz, BC, p. 34, from information provided by Clayton Mack.

1967/07/00 – Prince George. Legs Obscured By Long Hair: Bill Kleinhout, Ian Pointz and a helicopter pilot, Mr. Dugan, reported seeing a strange animal near Prince George at Summit Lake in July 1967. They said they saw an erect, black-hair-covered creature striding uphill with long strides [paces]. It was ascending an erosional draw. Its legs were obscured by long hair. It did not appear to have a neck, yet seemed able to swivel its head 180 degrees left or right. The men followed it for a few minutes, but were low on fuel and had to get back to their base.

A view of Summit Lake. It is located about 25 miles north of Prince George.

Source: John Green, Sasquatch Incidents Data Base, Incident Number 1000164. Photo: Image from Google Earth; © 2011 Cnes/spot Image; Image © 2011 Terra-Metrics; Image © 2011 DigitalGlobe; Image © 2011 Province of British Columbia.

1967/09/01 – Victoria (VI). Not a Subject for Mirth: BC Provincial Museum scientist Don Abbott arrived back in Victoria on September 1, 1967 after examining hundreds of giant, human-like footprints found on Blue Creek Mountain, California. The prints were on, and along, a road that was under construction.

As it happened, sasquatch researchers John Green and René Dahinden went to the footprint site in late August and decided to ask for a BC scientist to come down and look at them. Abbott, an anthropologist with the museum, drove down to the site (arrived there on August 27, 1967).

Upon his return to Victoria, Abbott conceded that what he saw shook him. He stated:

> I was laughing at the whole idea all the way down. Now I don't think it's a subject for mirth anymore. You realize that a scientist could ruin his reputation by going out on a limb and saying the creatures [sasquatch] exist. So I won't say I believe in them, but I am genuinely puzzled.

Many photographs along with plaster casts were taken of the mysterious prints. Commenting further on the finding, Abbott stated:

> Those prints were not made by any known animals. That leaves only two possibilities. Either someone is pulling one of the most complicated stunts ever, or the prints were made by a large bipedal primate that has never been described before. Both alternatives are close to being impossible.

(Top) A line of footprints found on Blue Creek Mountain, California. These prints were about 15 inches long. Other prints were about 13 inches. (Below) Don Abbott attempting to lift out one of the giant footprints that had been treated with glue.

161

Over 590 footprints of 2 sizes were counted on and near the Blue Creek Mountain road. Many prints, however, had been obliterated by road construction workers. Researchers estimated that the original total number of prints was probably well over 1,000.

It was the report of this incident that prompted Roger Patterson and Robert Gimlin to go to northern California and attempt to get footage of footprints for a film documentary on the sasquatch. The 2 men went there in early October 1967, and on October 20 filmed an alleged sasquatch at Bluff Creek, which is in the same area as Blue Creek Mountain.

Source: John Green. 1981. *Sasquatch, The Apes among Us.* Hancock House Publishers, Surrey, BC, pp, 74–79. The quotes are believed to be from a *Sun,* Vancouver, article (no details). (Photos: J. Green.)

1967/09/19 – Victoria (VI). Government Gets Involved: In September 1967 the sasquatch came as close as it ever got to a proper government-sponsored initiative to prove the creature's existence. Through the tireless efforts of John Green and René Dahinden, the matter was placed before Kenneth Kiernan, BC's recreation minister, in Victoria. A Vancouver newspaper article published on September 19, 1967 had the following heading: **Victoria Wants Evidence: Got Any Proof of Sasquatch?** It stated in part:

Kenneth Kiernan in 1954.

People with tangible evidence of the existence of sasquatch were invited today to get in touch with British Columbia's Provincial museum director.

The invitation was announced after Recreation Minister Kenneth Kiernan, museum director Dr. Clifford Carl, and other government scientists met here with sasquatch experts [John Green and René Dahinden].

Kiernan told the *Sun,* people who believe they have seen the hairy monster, or have any evidence in the form of bones, may contact Dr. Carl.

Kiernan, however, said he was unable to make up his mind whether the prints already provided as evidence were made by a prankster or by sasquatch. He said the provincial government does not feel there is enough evidence yet to launch a full-scale hunt for the creature. "But if it is a form of native fauna," he added, "we should find out about it." He went on to add that the government will not laugh at residents of BC who report seeing the sasquatch or who can produce tangible evidence.

Source: "Victoria Wants Evidence: Got Any Proof of Sasquatch?" The *Sun,* Vancouver, BC, September 19, 1967. (Photo: Author's file.)

1967/09/19 – Agassiz. Getting Serious: A Vancouver newspaper article published on September 19, 1967 featured an interview with John Green at Agassiz. It was a follow up article to the news of the BC government calling for sasquatch evidence (see previous entry).

In the article, Green said he used to write about sasquatch in his newspaper, but he stopped for some time because the reaction of the general public was "that I'm a nut."

He added he was glad that somebody was getting serious about the subject.

Source: "Sasquatch Seen, Like a Gorilla." The *Sun,* Vancouver BC, September 19, 1967.

1967/09/28 – Victoria (VI). Keeping an Open Mind: On September 28, 1967 it was reported in a Vancouver newspaper that the BC Provincial Museum, Victoria, was following up on new leads related to the sasquatch. Recreation Minister Kenneth Kiernan said that although the government still has not got enough evidence to undertake a sasquatch hunting expedition it was "keeping an open mind."

Source: "Sasquatch Hides." The *Sun,* Vancouver BC, September 28, 1967.

1967/09/00 – Comox (VI). Teenage Girls Encounter a Sasquatch: Two teenage girls reported that they came face to face with a sasquatch near Comox in the Comox valley area during September 1967. The girls heard the sound of something slapping the water around a nearby point. They walked around the point and found themselves face-to-face with a 7-foot-tall sasquatch, completely covered in long dripping-wet reddish-brown hair. In one hand the creature held three or four ducks; in the other a 4-foot-long stick. After several seconds the sasquatch and the girls fled in opposite directions. The girls assumed that the stick had accounted for the slapping sounds they heard, and it had somehow been used to kill the ducks. One of the girls stated that the creature was definitely not human.

Source: John Bindernagel, 1998. *North America's Great Ape: The Sasquatch*. Beachcomber Books, Courtenay, BC., p. 126, 227, from an interview with one of the girls in the 1990s.

1967/09/30 – Victoria (VI). Government Impressed: On September 30, 1967 a Vancouver newspaper reported that the BC government was impressed with the 1965 sighting in the Garibaldi Park area (see 1965/06/28 – Garibaldi Park area. Something Dragged to Ice Hole). In an article headed "Sighting Praised," we are informed that provincial museum anthropologist Don Abbott said that it was "the best account so far received in the government's quest for sasquatch information." The 2 brothers who had the sighting informed Abbott that the prints they saw looked similar to the prints photographed in Norther California (see 1967/09/01 – Victoria. Not a Subject for Mirth). Recreation Minister Kenneth Kiernan also interviewed the brothers.

Source: "Sighting Praised." The *Sun,* Vancouver, BC, September 30, 1967.

1967/10/23 – Vancouver. Screening Arranged for California Film: On October 23, 1967, arrangements were made for University of British Columbia (UBC) scientists to see a film of an alleged sasquatch taken on October 20, 1967 at Bluff Creek, California. The one-minute 16mm film, taken by Roger Patterson and Bob Gimlin of Yakima County, Washington, shows the creature walking along a gravel sand bar.

John Green and René Dahinden had viewed the film in Yakima on October 22, 1967. They convinced Patterson to arrange with Don Abbott of the BC Provincial Museum to have the film shown to UBC scientists. Abbott arranged a screening for October 26.

Green and Dahinden had naturally concluded that with BC

"primed" on the sasquatch issue, scientists here would be the best to analyze the film.

Source: Christopher L. Murphy, 2008. *Bigfoot Film Journal*. Hancock House Publishers, Surrey, BC, p. 59.

1967/10/26 – Vancouver. The Webster Interview: Jack Webster, a noted Vancouver radio personality, interviewed Roger Patterson and Bob Gimlin (see previous entry) in Vancouver on October 26, 1967.

During the in-depth interview, Webster questioned Gimlin on why he had not shot the creature, given *Life* magazine had a standing $100,000 dollar reward for a bigfoot, dead or alive. Webster stated that, in the same situation, he thinks he would have shot it. Patterson took the question and calmly replied, "I don't think you would have if you had seen the humanness of it. I think it would take a person with a little bit of murder in his heart to shoot something like this." This reply was absolutely perfect for the hard-hearted radio host.

Source: Jack Webster, 1967. Radio Interview, October 26, 1967.

1967/10/26 – Vancouver. UBC Scientists First to See Film of Alleged Sasquatch: On October 26, 1967 scientists at the University of British Columbia, and from the provincial museum in Victoria, were shown the Patterson/Gimlin film of what was purported to be a sasquatch or bigfoot filmed in Northern California. Also shown was a second film showing the creature's footprints. and plaster casts made from footprints. The first screenings were held at a session headed by Dr. Ian McTaggart-Cowan. The main film was later shown to the press at the Hotel Georgia in Vancouver.

The UBC scientists were under orders not to express an opinion one way or the other on the film footage. The museum scientists, however, were not restricted, and 2 of them, Don Abbott and Frank Beebe, provided statements.

Don Abbott stated:

> I think most people assumed that it was a fake, and were content to let it go as that.

...Further information is needed before the provincial government will aid in the search for the animal.

...In the past, scientists have always laughed off reports of this kind, but the mere fact that a group of 50 or 60 scientists are prepared to get together and discuss it at least semi-seriously indicates a change in attitude. Before too long, someone is going to make a major attempt to get to the bottom on this thing somehow.

...It is about as hard to believe the film is a fake as it is to admit that such a creature really lives. If there's a chance to follow-up scientifically, my curiosity is built to the point where I'd want to go along with it. Like most scientists, however, I'm not ready to put my reputation on the line until something concrete shows up—something like bones or a skull.

Frank Beebe stated:

...I'm not convinced, but I think the film is genuine. And if I were out in the mountains and I saw a thing like this one, I wouldn't shoot it. I'd be too afraid of how human it would look under the fur. From a scientific standpoint, one of the hardest facts to go against is that there is no evidence anywhere in the Western Hemisphere of primate (ape, monkey) evolution—and the creature in the film is definitely a primate. So either a large primate got stranded in North America, or the film is a fake.

Frank Beebe (1990s).

...The animal's gait was too masculine, so was its build, yet it was purported to be a female. Also, it went around obstructions and not over them. Animals in the woods usually go over things, not around them.

...The film gave the impression of great size in the creature—there was no accurate scale.

...The background appeared quite genuine and there was no suggestion that the figure had been faked.

...If it was fakery, it was very clever fakery, and it had to be a man in a monkey suit. We can't conclude that because there was nothing in the film to give us its proper proportions. If a series of footprints 1½ inches deep and 14½ inches long shown in a second piece of film by Patterson were, as he claimed, left by the creature, then only one conclusion could be drawn. If that is true and we have no means of proving or disproving this, then I've no doubt certainly what we saw was a Sasquatch.

As to the thoughts of Dr. Ian McTaggart Cowan, there is nothing "official." Nevertheless, from information provided by René Dahinden and John Green, the following provides some insights.

Dr. Ian McTaggart-Cowan attempted to establish a physical height for the creature in the film. The purpose of the exercise was to determine if the creature was within the range of human proportions. McTaggart-Cowan used a yardstick to measure the creature's foot. He then determined that the height of the creature was 5.9 times the size of its foot. He then measured one of the plaster casts, arriving at 14.5 inches. Based on these 2 figures the creature's height is 85.55 inches or 7 feet 1.55 inches. However, the doctor decided to reduce the size of the plaster cast to 14 inches. The reason for this action was to allow for foot slippage. Given the new figures, the creature height equaled 82.5 inches or 6 feet 10.5 inches. He then subtracted another 1.5 inches for an unknown reason to arrive at a final figure of 81 inches or 6 feet 9 inches.

Dr. McTaggart-Cowan stated that the creature filmed was male because it walked like a human male. He was then questioned as to the creature's breasts. On this point he said that he could not see any breasts distinctly enough to identify them as mammary glands.

McTaggart-Cowan then made a profound statement: "The more a thing deviates from the known, the better the proof of its existence must be." With these few words, he tactfully told the audience that when it comes to bigfoot, the film is just not adequate proof.

One of the scientists who attended the press screening, Professor W.J Houck of Humboldt State College, also provided a short statement: "I'm not going to call it a hoax," he said, "yet the alternative is still too fantastic to accept. Where does that leave me? Darned if I know."

Houck said he doubted his college would back any attempt to launch a scientific inquiry, but the situation might vary according to the state of the college budget and the possibility of further evidence.

John Green would later write about the screening:

A frame from the film taken by Roger Patterson and Bob Gimlin showing the sasquatch they encountered. The creature reasonably met the descriptions provided by eye-witnesses going back over 100 years.

None of the scientists at that first screening was a primatologist or physical anthropologist, and their opinions were asked for after a brief look at the film with no time for research or reflection; so their com-

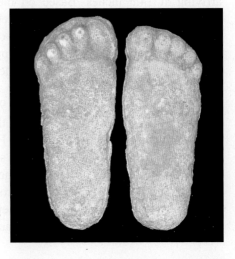

Copies of the footprint casts shown to the scientists.

ments, while damaging, are probably excusable. There is no such excuse for their modern colleagues."

Nevertheless, one would think that there might have been some sort of an "official" statement from the university on the film. All we have in this regard is the following statement provided by a "UBC spokesman."

It's either a mighty big man in a monkey suit or it's a Sasquatch.

There was no request to subject the film to in-depth scientific analysis. In short, the university simply, and very quietly, walked away from the issue.

Dr. Clifford Carl of the provincial museum, however, did provide an official statement, which actually made the situation worse rather than better. He stated:

The Provincial Museum feels it is impossible to determine with any degree of certainty either that the Sasquatch is real or that it is a hoax.

For certain, Carl was referring to the sasquatch seen *in the film,* not the creature in general. The least he could have said was "At this time...," and left the door open. Keep in mind that government involvement in the sasquatch issue had just commenced a little over one month earlier.

Dr. Clifford Carl (1960s).

Don Abbot and Frank Beebe's "official reports" to Kenneth Kiernan were effectively negative and it was subsequently decided by the provincial museum, provincial government and the University of British Columbia that the sasquatch issue did not warrant any further attention.

Remarkably, in latter years, Beebe relented, and unofficially supported the possible existence of the creature. Also, Abbott with Beebe offered an unofficial theory on sasquatch sustenance.

In retrospect, I have reasoned that the submission of the Patterson/Gimlin film was very bad timing. Had the government been involved in sasquatch research without the film for say a year or so, the film would have been viewed from an entirely different standpoint. Odd though it may seem, we have not yet been able to prove beyond a doubt that the film is either authentic or a hoax.

Roger Patterson holding casts of the creature's footprints. The photograph was in the Sun newspaper, October 27, 1967, in an article headed: "Hairy 'Thing' Puzzles Experts: Not Convinced About Sasquatch."

Source: Christopher L. Murphy, 2008. *Bigfoot Film Journal*. Hancock House Publishers, Surrey, BC, pp. 63–65. (Photos: Frank Beebe, D. Hancock; Sasquatch, R. Patterson, pubic domain; Casts, C. Murphy; Dr. C. Carl Portrait, J. Green; Patterson with casts, Brian Kent, *Sun,* Vancouver.)

1967/10/28 – Vancouver. The Killing Question: Although the BC government had "killed" its sasquatch research initiative, the question arose as to the legality of killing one of the creatures given they do exist.

On October 28, 1967 George Curtis, Dean of the University of British Columbia Law School, Vancouver, waded into the issue and provided some advice. It all depends on whether or not the sasquatch is a human being. If it is human, it is undoubtedly illegal to shoot it. If it is not human there is probably no law against shooting it unless it's covered by game regulations. However, a tricky situation arises if one shoots a sasquatch in the belief that it is an animal, but then finds out after it is dead that it was a human. Curtis came to the conclusion that anyone seeing one of the hairy giants would be well advised to hold fire un-

The question as to whether or not one should shoot a sasquatch brought about heated arguments. Dmitri Bayanov detailed the entire issue in this book. His conclusion was that killing a sasquatch for the benefit of science was not justified.

til it's status is established. "It would be wrong to shoot it if it was not known whether it was a human or not," he said.

Provincial museum anthropologist Don Abbott, however, stated that if the sasquatch exists, he doubts it is a human being. Referring to his recent viewing of the Patterson/Gimlin film, he stated:

> "If that thing we saw is really a sasquatch, I don't think it's a human being. It's definitely hominid, but not homo sapiens. It's nearly human, but not in the same sense."

He offered that he's quite sure nobody would be prosecuted for shooting one. "But if it's somebody in an ape suit, he added, "that's a different kettle of fish."

Source: Nat Cole, 1967. "If You Shoot a Sasquatch, You'd Better Hire a Lawyer." The *Sun*, Vancouver, BC, October 28, 1967. (Photo: Book cover, D. Bayanov, artwork by Lydia Bourtseva.)

1967/10/31 – Vancouver. A Crippled Giant: Mike Cramond, the outdoors expert for the *Province* newspaper, Vancouver, reported that he had attended the press screening of the Patterson/Gimlin film on October 26, 1967 in Vancouver. He was reasonably impressed with the film, and this prompted him to "talk about the sasquatch" in an article published on October 31, 1967. The following is one of the stories he related:

> Years ago, when I was on the B.C. police force we had a very intelligent young West Coast Indian incarcerated for a minor misdemeanor. We whiled away many hours encouraging him to talk of his Indian way of life. I guided him into discussion of native legends and beliefs.
>
> He startled me one day by admitting to an encounter with what he called *Sotsochtah,* a crippled giant of the forest having one large foot, and one small withered one. He had been with a companion who became alarmed and fled when he wanted to shoot it.
>
> Thoughtful interrogation is the most effective method of ascertaining truthfulness and actualities in the recounted word. As I said, he possessed better than average intelligence. He was a native woodsman and told me that the meeting was in snow. He said he followed the footprints for a couple of miles until it grew dark. Then he grew frightened. *Sotsochtah,* he said, came down from the mountain tops during heavy snows, when prints were often seen.

This is the first major reference I have run across that discusses a crippled sasquatch. In 1969, what are believed by some to be the footprints of a crippled sasquatch were found in Washington. Two scientists who have analyzed the prints and casts made from them are confident the prints were made by a natural foot.

Source: Mike Cramond, 1967. "Province outdoors expert Mike Cramond...looks at the Sasquatch," The *Province,* Vancouver, BC, October 31, 1967.

1967/10/31 – Vancouver. Mining Promoter Considers Sasquatch Expedition: On October 31, 1967, a Vancouver mining promoter said he was considering the financing of a sasquatch hunt, using helicopters, hunting dogs and tranquilizer guns. Richard Angle, who resided in West Vancouver, said he had discussed such plans with Roger Patterson, who along with Bob Gimlin, had recently filmed what they believe was the elusive creature in Northern California. Angle said he and Patterson would get together the next month. He said an expedition of the type he proposed (which would include wildlife experts), taking place for up to one month in the California mountains, could stand a good chance of proving the existence of the creature. "If we got a sasquatch," he said, "I'd like to bring it back to Vancouver and hand it over to a zoo. I'd like to see it on a 5-acre tract somewhere, not behind bars."

As far as I know, nothing became of this initiative.

Source: "Sasquatch Interests Mining Man." The *Sun,* Vancouver, BC, October 31, 1967.

1967/CI/00 – Swindle Island. Typical Ape Tracks: Paul Hopkins of Klemtu reported that he saw "typical ape tracks" in the bush above the beach on Swindle Island in the winter of 1967–68.

Source: John Green, 1973. *The Sasquatch File.* Cheam Publishing Ltd., Agassiz, BC, p. 34, from information provided by Bob Titmus who spoke with the Paul Hopkins.

1968/01/12 – Vancouver. Sasquatch Film Rights Bought by Canadian Researchers: John Green, and René Dahinden reported on January 12, 1968 that they had bought the Canadian rights for what is now known as the Patterson/Gimlin film. They paid $1,500 for the rights.

The pair said they intended to use the film in a one-hour movie they were making in hopes of proving sasquatches do exist on the West Coast.

Source: "Sasquatches Really Exist, It Says Here," *The Associated Press,* January 12, 1968. Also Scott McClean, 2005. *Big News Prints.* Self-published, p. 52, from "Editor Buys Film Clip of Sasquatch." *The Advocate,* Newark, Ohio, USA, January 12, 1968. (Photo: J. Green.)

See the

FASCINATING   CONTROVERSIAL

SASQUATCH MOVIE

Under Auspices of the B.C. Museum

Harrison Memorial Hall
9 PM TUESDAY, THURSDAY
AND SATURDAY
Donations Requested: Adults 1.00 Children 50c

Advertising poster made for showing the film.

1968/WI/00 – Gosnell. Loggers Find Tracks: John Duthie and other workmen reported that they found large bipedal tracks in snow in the Gosnell area in the winter of 1968. The men worked for a logging operation and when they returned to their landing on a Monday, Duthie noted a few tracks near an old bus body used as a tool shack and lunchroom. It was also seen that a full drum of fuel had been placed near the bus doors. The drum appeared to have been tossed aside and the bus doors were open. It was snowing lightly at the time, so the tracks were not detailed. The boss allowed no time to look at the tracks and work soon obliterated them.

Source: John Green, Sasquatch Incidents Data Base, Incident Number 1000105.

1968/02/22 – Alert Bay. Ran As Fast As We Could: Two fishermen, Tom Brown and Harry Whonnock, reported that they saw a sasquatch in the Alert Bay area, Broughton Island, on February 22, 1968. The men were digging clams on the south shore of the island (about 25 miles from Alert Bay, and a short swim to the mainland) when they saw the creature. Brown provided the following information.

> It was late in the afternoon and beginning to get dark. But there was still plenty of light to see that it was not an animal. It was hairy, about 6 feet tall and was looking at us.
> That's about all we saw of it. Harry and I ran as fast as we could for the boat and the sasquatch headed into the bush. I don't know how the sasquatch took it, but all we wanted to do was get away.

René Dahinden said the sighting appeared to be one of the best reported in recent years, and was anxious to head up an expedition to look for the creature or its footprints.

Source: John Green, 1973. *The Sasquatch File*. Cheam Publishing Ltd., Agassiz, BC., p. 41, from "Alert Bay now claims a Sasquatch," The *Province,* Vancouver, BC, February 22, 1968.

1968/03/00 – Golden. Snow Survey—Tracks Found: A British Columbia Hydro employee doing snow surveys reported that he found large unusual tracks east of Golden in March 1968. He telephoned René Dahinden and informed him of the finding.

Source: John Green, 1973. *The Sasquatch File.* Cheam Publishing Ltd., Agassiz, BC, p. 41, from information provided by René Dahinden who talked to the Hydro employee.

1968/05/17 – Sechelt. Ran Away Pretty Fast: Logger Gordon Baum reported seeing a sasquatch in the Sechelt area near Salmon Inlet on May 17, 1968. Baum was working for Fleetwood Logging at Salmon Inlet. He was alone and heading for a waterfall for a drink of water when he saw the creature standing near a logging road. It was about 100 yards from his position. He said it was covered in black hair, and he estimated its height at 5 feet. He thinks he frightened the creature because it put one hand on a log, bounded over it, and ran away "pretty fast." It did not make any noise. Baum remarked that as what he saw was not a bear, it must have been a sasquatch.

He later went to where it had been and found a footprint on the road, but a truck had ran over the back of it. He measured the width of the print and it was 6 inches across from the big toe to the little toe. Judging by the depth of footprint, he estimated the creature's weight at about 200 pounds.

Later a footprint and knee print were seen in earth by Baum and another employee.

The height of the creature in this case would likely rule out an adult. It is noteworthy that the report is provided by a logging company employee, who would certainly know what a bear would look like at any distance.

Map of the Sechelt region and aerial view of Salmon Inlet. Although not far from Vancouver, the area is still considerably remote.

Source: "Hairy Thing No Bear, Says Logger," *The Sun,* Vancouver, BC, June 24, 1968. Also John Green, Sasquatch Incidents Data Base, Incident Number 1000019. (Photos: Map, Wikipedia Commons, please see the Wikipedia site for details, Sechelt area, Image from Google Earth; © 2011 Cnes/Spot Image; Image © 2011 Digital-Globe; Image © 2011 Province of British Columbia.)

1968/06/01 – Mission: No Plan for a Hunt: On June 1, 1968 a newspaper reported that a sasquatch was said to have been sighted near Mission. However, police in Matsqui and Mission said they had no plans to go on a hunt for the creature.

Source: "Beware! Sasquatch Roaming Once More." The *Sun,* Vancouver, BC, June 1, 1968.

1968/06/25 – Vancouver. Hancock Weighs-in On Sasquatch Issue: Noted BC naturalist David Hancock, who has long been involved in the study and preservation of BC's wildlife, waded in on the sasquatch issue on June 25, 1968.

David Hancock

Hancock did not miss the opportunity to see the film of an alleged sasquatch shown at the University of British Columbia, Vancouver in October 1967. Indeed, Hancock is no stranger to sasquatch lore, having previously met the noted researcher Bob Titmus in the wilds of BC. Hancock was after eagles and white spirit bears; Titmus was after sasquatch. They traded notes and Hancock helped Titmus check his network of trip-wired cameras.

Hancock's over-all opinion of the sasquatch in general may be summed up as a "could be." He naturally drew attention to questions on how the creature would be able to find enough food, and also to an adequate population to ensure the on-going survival of the species.

As to the sasquatch film, which to some is the best evidence yet on the existence of the creature, it did not impress Hancock at first, "Oh God! How corny. It's nothing more than a man in a monkey suit," he recalled saying to himself as the furry little image crossed the screen. "But the frame-by-frame analysis wasn't as easily dismissed," he added, and continued, "If it was a hoax, then I join with most of the other observers in saying it was a very clever and sophisticated one."

Hancock then went on to draw a parallel with the yeti:

Still, I can't help thinking about the yeti or abominable snowman of the Himalayas. The story is similar but much longer standing. While no yeti specimen has positively been identified in a scientific institution, the repeated verification of actual sightings of the beast and/or its footprints are well established. So much so that on the basis of this circumstantial evidence, Dr. Bernard Heuvelmans has proposed a scientific name for the creature: *Dinanthropoides nivalis* or "terrible anthropoid of the snows."

As to the general scientific community's "blind eye" with regard to the sasquatch, Hancock said:

It's not fair to shrug off the old sasquatch. Its much better to look for scientific possibilities of his existence. After all, it's only since the turn of the century we have discovered such species as the Mountain Gorilla of Africa. Known for centuries through legend, it was not established as a real creature to science until 1901. The Congo Okapi, short-necked relative of the giraffe, was identified in 1900, the Pygmy Hippo in 1913. The world's largest reptile, the Komodo Dragon, was not discovered until 1912. The loveable Giant Panda was unknown to science and the children of the world until 1936 when the first one was captured. And, in our own time in Canada, we have re-established the range of the Wood Buffalo and Wood Grizzly, when both animals were thought to be extinct.

Drawings of a sasquatch that accompanied David Hancock's article referenced.

In summary Hancock stated:

> I want to make it abundantly clear that I am not saying sasquatches do exist. I think I can say the environment does exist which would make it possible for them to exist.

In discussing this subject with David Hancock in October 2010, he stated:

> The temperate west coast environment of islands, inlets and mountains, with the abundance of rich intertidal areas could, I suspect, readily support sasquatch, as it does the vast array of flora and fauna unique to this region, and has for millennia nourished one of the largest populations of North American Natives and their rich culture.

David Hancock is a publisher (Hancock House Publishers, Surrey, BC, Canada). He has published hundred of books on wildlife, and general subjects, including 23 books on the sasquatch/bigfoot question. Titles by John Green, Thomas Steenburg, Dmitri Bayanov and Chris Murphy are among such works.

Source: David Hancock, 1968. "The Sasquatch Returns." *Weekend* magazine, The *Sun,* Vancouver, BC, No. 2, June 25, 1968, and person communication. (Photos: Portrait of D. Hancock and drawings, Hancock House.

1968/07/21 – Lumby. Sasquatch Fever: On July 21, 1968 René Dahinden headed out from Lumby on a 3-man expedition to find the sasquatch. He as much as admitted that he had "sasquatch fever."

Dahinden was previously the leader of the expedition which conducted a 2-month search for the legendary creature in remote county north of Garibaldi Park.

The other 2 members of the current party were Dennis Primmett of Victoria and Peter Mutrie of Vancouver. Primmett gave up his job as a bank clerk while Mutrie quit an automobile firm job to join the hunt.

Dahinden, estimated the expedition would cost $1,000. He said the group would set up a base camp and make forays into the bush, and added:

We just want to observe a sasquatch. We will be taking a 16-millimeter camera as well as a 35-millimeter so we can make films. We will all carry guns, just in case, but we don't want to shoot the animal.

Dahinden said the group planned to move with caution, however, because of Indian legends that the sasquatch is a man-killer. Dahinden, who said he first started looking for the sasquatch in 1954, estimated his efforts so far have cost him about $10,000. He summed things up as follows:

None of us are sure this thing exists, but we are determined to try and find and film it, hopefully making a better film than Patterson. Sometimes I'd like to stop looking, but I know I never can—it's in my blood.

Source: "Sasquatch Fever—It's in His Blood." The *Sun* or the *Province,* Vancouver, BC, Canada, July 31, 1968

1968/08/00 – Stewart. Ran Up Steep Hill On "Hind Legs": Two hunters reported they saw an unusual creature near Stewart at twilight in mid-August 1968. The men were grouse hunting 4,000 feet up on an old road in their truck. They saw the creature about 25 feet away. It had short dark hair, and they thought it was a bear. However, when it ran up a steep hill on its "hind legs" they then realized it was not a bear and did not shoot at it. One of the men said it appeared to be more than 7 feet tall and very heavily built. It had a short neck, a flat nose, long arms and wide shoulders. A bad odor was sensed in the area.

The next day, 2 other men went to the site to investigate the incident and found some faint footprints near a little stream.

Source: John Green, 1973. *The Sasquatch File.* Cheam Publishing Ltd., Agassiz, BC, p. 34, from information provided by one of the witnesses to John Green. Also John Green, Sasquatch Incidents Data Base, Incident Number 1000012.

1968/09/14 – Chetwynd. Game Guides Sighting: Eddie Barnett, a game guide, and his co-workers saw an unusual animal while working near Chetwynd on September 14, 1968. The men were wrangling horses and saw a large shaggy animal walking on its hind legs. They tried to track the animal, but the terrain became too rugged. There were no clear tracks left by the creature.

Source: John Green, 1973. *The Sasquatch File*. Cheam Publishing Ltd., Agassiz, BC, p. 41, from a letter by Barnett to Mike Amacher of Erie, Pennsylvania. Also, John Green, Sasquatch Incidents Data Base, Incident Number 1000167.

1968/09/00 Courtenay. Silent Watcher: After arriving home in Courtenay with her family at about 2:00 a.m. during September 1968, a young girl noticed a strange, tall man-like figure staring at them from across the road, partially hidden by a telephone pole. The moon was bright and she could make out the shape of the head and the body. The head was to one side of the pole and the shoulders were sticking out from both sides. She could see the shape of its legs down to its knees. The girl alerted her father who quickly had the family go into the house. All of this time, the creature just stood there in silence.

The next day, the girl inspected the area where she had seen the oddity and noted that it had been standing in a ditch. She did not see any footprints.

A few months later when there was about 4 inches of snow on the ground, the girl went into the forest near her home. She found very large footprints that had been made by bare feet.

Source: BFRO website. Witness investigated by Blaine McMillan.

1968/11/16 – Lillooet. In and Out the Bush: A North Delta man reported finding a line of tracks 16 inches long near Lillooet on November 16, 1968. The tracks had a 40- to 40-inch pace in 4 inches of fresh snow. They came out of the bush onto the road, up a steep grade, back into the bush and out again farther up.

Source: John Green, 1973. *The Sasquatch File*. Cheam Publishing Lt., Agassiz, BC, Canada, p. 41, from information provided by René Dahinden.

1968/11/20 – Prince George. Absolutely Speechless: Fort Kells farmer Horst Klein reported that he was ready to believe in the elusive sasquatch after spotting strange tracks near Prince George while on a hunting trip on November 20, 1968. He and a hunting companion spotted the prints in snow on Bowron Forest service road (Sugarbowl Mountain—40 miles east of Prince George). The men did not publicize the finding until they had photographs of the prints developed. Reflecting on the incident, Klein said:

> We were absolutely speechless. We really couldn't believe what we were seeing. There were definite indications of a right and then a left foot, making the tracks alternately. The tracks measured about 16 inches long with a 6-foot stride.
>
> No bear alive could have made those tracks. I know animal tracks. I am not saying they were sasquatch tracks, but I ask you, what were they?

Fish and Wildlife Branch regional biologist Brian Gates said the tracks could have been made by a loping wolverine. This animal can leap that distance and come down on all 4 feet together.

Klein said he doubted Gates' interpretation because the prints appeared to be made by a right and left foot. Also, there was no sign of the snow being disturbed by a loping animal.

The tracks were found in rough country, at the highest point of the forest service road. Fifty prints were counted and they appeared to be a day old. They ended on a fallen log where the trail disappeared into thick bush.

There were, however, signs that a vehicle had stopped at the location. A set of a man's footprints led up to the trackway, then returned to the road. "It looked to me as if somebody else had got out, took one look, and then got scared and took the hell out of there," Klein said.

Source: "Hunters Ready to Believe Sasquatch Alive, Walking." The *Sun,* Vancouver, BC, December 12, 1968.

1968/YR/00 – Kelowna. Walked Upright All the Time: Jack and Dennis Crawford reported seeing a strange animal near Kelowna at Gallagher's Canyon (about 5 miles Northeast of Kelowna) during the summer or fall of 1969. The men said they saw a huge, bulky, hairy creature walking upright about 700 yards away. They thought it might be a bear, but it was walking upright all the time.

Source: John Green, Sasquatch Incidents Data Base, Incident Number 1000166.

1968/YR/00 – Harrison Hot Springs. Sasquatch Provincial Park Named: A park near Harrison Hot Springs called Green Point Park (created in 1959) was expanded and renamed Sasquatch Provincial Park in 1968. The expanded park now includes the adjacent lands containing Hicks Lake, Deer Lake, and Trout Lake. The park was named after the elusive sasquatch said to roam the region.

Source: Wikipedia. (Photo: C. Murphy.)

1969/01/10 – Hope. Turned Upper Body As Well: A man driving with his sister and a friend near Hope reported they saw a sasquatch at about 3:00 a.m. on January 10, 1969. The creature walked across the road from left to right. It came into full view of the car headlights for about 6 seconds. It was between 7 and 8 feet tall, grayish-brown in color, with very long arms, a thick torso, and slanted back ape-like forehead. It walked with long deliberate paces. At one point it turned in the direction of the vehicle, but not as a person would by merely turning the head; it turned the upper body as well. This was for about 3 strides [paces] and then it turned back to the direction it was walking and disappeared from view.

The man's general impression was that it was "Neanderthal looking." He remarked, "There is no way, absolutely no way, that this creature could have been someone in a suit. The temperature was approximately –40 degrees with the wind chill. There were no homes, lights, or structures in the area."

Source: BFRO website.

1969/01/00 Agassiz. First Authoritative Book On Sasquatch: John Green, editor and publisher of *The Agassiz-Harrison Advance* newspaper at Agassiz, BC, and one of the foremost Canadian authorities on the sasquatch, wrote and published a book entitled, *On the Track of the Sasquatch*. This book, the result of 10 years of research on the subject, provides detailed information on sasquatch sightings and related incidents. The book was made available in January 1969.

On the inside front cover of the book there is a letter to Green from Dr. C. Clifford Carl, director of the BC Provincial Museum. It reads as follows:

John Green's first book. The cover continues to the left, and unfolds on the right to show the full size of the 15-inch footprint shown.

October 21, 1968

Do Sasquatches really exist or is the whole business a gigantic hoax?

The numerous sight records and other reports of hairy giants, the photographs of individuals, the innumerable foot prints and other types of evidence all point to the presence of such creatures, but until a specimen is obtained the question must remain unresolved.

In the meantime we owe a great debt to Mr. John Green who has devoted a large proportion of his time and resources to the task of assembling every bit of evidence available concerning this fantastic possibility. The publication of this up-to-date factual account will help greatly to publicize the "Sasquatch Problem" and may speed the day when scientific proof is at last made available.

Source: George Haas, 1969. The Bigfoot Bulletin, No. 1, Oakland, California, USA, January 2, 1969; also, John Green, 1969. *On the Track of the Sasquatch.* Cheam Publishing Ltd., Agassiz, BC. (Photo: J. Green.)

1969/02/00 – Alert Bay. Seen Looking at Them: Tom Brown and Harry Whonnock, reported seeing a strange creature near Alert Bay on Broughton Island in February 1969. The men were digging clams on the south shore of the Island. Just as it began to get dark they saw a hairy creature about 6 feet tall looking at them. They ran for their boat and the creature ran for the bush.

Source: John Green, Sasquatch Incidents Data Base, Incident Number 1000165, from The *Province,* Vancouver, BC, February 22, 1968.

1969/02/00 – Butedale. Ran Off Screaming: Ronnie Nyce of Kitimat reported that he and 2 other men saw and shot at an ape in the Butedale area at Khutze Inlet in February 1969. The men were in a boat traveling the inlet and saw the creature on shore. Nyce shot at it with a shot gun and the creature may have been hit as it ran off into the woods screaming. Nyce said that it was very large.

The Butedale area and Khutze Inlet. The inlet is about 355 miles from Vancouver and about 50 miles from the open Pacific Ocean.

Source: John Green, 1973. *The Sasquatch File.* Cheam Publishing Ltd., Agassiz, BC, p.41, from information provided by Bob Titmus who obtained it from Ronnie Nyce. Also John Green, Sasquatch Incidents Data Base, Incident Number 1000025. (Photo: Butedale are, Image from Google Earth; © 2011 Cnes/Spot Image; Image © 2011 DigitalGlobe; Image © Province of British Columbia; Image; © 2011 TerraMetrics.)

1969/03/00 – Powell River. Cabin Owner Shaken Up: A man who lived in a cabin near Powell River at Lewis Lake (6,000-foot level, above Cheakamus Valley) reported finding frightening tracks in March 1969. He had just returned from an inspection trip via snowmobile, and on top of snow 17 feet deep he found barefoot, 5-toed tracks. He said the stride was 6 feet.The person who talked with the man and provided this report said he appeared quite shaken up.

Source: John Green, 1973. *The Sasquatch File*. Cheam Publishing Ltd., Agassiz, BC, p. 41, from information provided by a Vancouver man who talked with the cabin owner.

1969/04/12 – Vancouver. Ice Man's Canadian Tour Canceled: On April 12, 1969, the Vancouver *Sun* newspaper reported that Frank Hansen, exhibitor of the controversial "Minnesota Ice Man" had cancelled a planned Canadian tour of his exhibit. Hansen said he cancelled rather than fill out customs declarations which might divulge what the "corpse" is and where it was obtained. Commenting on this in a telephone interview he said:"

> I always exhibited it as a mystery. I never called it the Ice Man. I billed it as *Siberskoye Creature*, which I understand means Siberian Creature. Maybe this has some significance. I won't say. But, barring any unforeseen legal problems, we plan to exhibit an interesting specimen as an oddity—as we have done in the past.

The "Ice Man," a frozen something that appeared to be a "corpse" of some kind of hominoid had been on tour in the US for 2 years. It was displayed the previous year at a fair in Chicago and was subsequently examined by zoologist Dr. Bernard Heuvelmans and naturalist Ivan T. Sanderson, who both declared that it was "genuine." As a result, the Smithsonian Institution asked Hansen to provide the corpse for analysis, but Hansen failed to do so. It appears Sanderson thought he could force the issue by getting the Federal Bureau of Investigations (FBI) involved (he thought the creature might have been murdered). Just what the FBI did is not clear.

When Hansen was asked about the Smithsonian's request to see the corpse, he said, "I wrote them and told them the specimen is no longer in my hands, it is in the hands of the owner. I didn't sign anything though." He then said that he had recently been away consulting with the owner.

As to alleged Federal Bureau of Investigation (FBI) involvement in the issue, Hansen stated, "I think the FBI is satisfied no federal laws have been broken."

Commenting on the hasty disappearance of the corpse, he said, "We had to take certain precautions just in case something of this nature developed. We make no claim for it and the burden of proof will be on someone else. Why take a chance on getting it tied up in a legal entanglement?"

The continuing complications in the Ice Man story are certainly a good indication that the corpse was a fabrication. However, the nagging question of the inspection done by Dr. Bernard Heuvelmans and Ivan Sanderson still hangs in the balance. Is it possible that these professionals could have been "hoaxed" with a rubber dummy? There is also the notion that if the corpse were indeed real, then events just might have played out as they had done.

Dr. Bernard Heuvelmans, the "Father of Cryptozoology," is seen here in 1992. It is rather hard to swallow that the technology available in or before 1968 was good enough to fool a man of Heuvelmans caliber.

Source: Ron Rose, 1969. "Ice Man Mystery Deepens." The *Sun*, Vancouver, BC, April 12, 1969, and general knowledge. (Photo: Heuvelmans portrait, Dr. J.P. Debenat.)

1969/04/15 – Vancouver. Canadians "Cold Shoulder" Ice Man: On April 15, 1969 BC's sasquatch researchers John Green and René Dahinden stated they were very skeptical with regard to the "Minnesota Ice Man" (see previous entry). On top of that, Green said it does not appear to be a sasquatch, whatever it is.

Green stated that he was told by a taxidermist friend that it would be fairly easy to mock up a corpse like the Iceman. Green thinks there could be somebody who specializes in making these type of exhibits and putting in gory details. Nevertheless, he added, "But I don't say that it must be a hoax."

As to a sasquatch connection, Green said:

> A sketch made by Sanderson and reports by Heuvelmans give details I would not expect to find in a sasquatch. There is apparently no brow ridge and the nose, if it can be called that, does not project. The thumbs are very long whereas the hand prints we have for the sasquatch show stubby thumbs.

Dahinden was less charitable:

> I don't believe that it will bear looking into. That is why Hansen [Ice Man exhibitor] put it under wraps. He claims he was showing it for educational purposes. If he wants to advance education, why does he not hand the body over to the Smithsonian?

Source: "Sasquatch man doubts U.S. Reports." The *Province,* Vancouver, BC, April 5, 1969. Also Ron Rose, 1969. "Ice Man Mystery Deepens." The *Sun, Vancouver,* BC, April 12, 1969.

1969/06/01 – Merritt. Army Cadet Sighting: David Ludlam, the adult leader of a army cadet group, reported that he and a number of cadets had seen a sasquatch near Merritt at Lilly Lake (12 miles southwest of Merritt) on June 1, 1969. Ludlam dropped into the *Herald* newspaper office on Monday afternoon, June 2, 1969 and excitedly related his story.

The cadet group, consisting of 18 cadets, was camped at the lake over the long weekend. About 6 of the group with Ludlam had sightings of the creature.

As it happened, Sunday night the cadets were having a mock battle in the trees surrounding the camp. One boy, John Szabo, 14, came running back to the camp obviously very frightened. When joined by another boy he told them he heard a noise behind him like someone walking on pine cones. He then heard heavy breathing like somebody was right beside him. He turned around and saw a huge hairy man through the trees. He said it was over 10 feet tall. The creature then ran off through the trees, making a guttural sound while running. Szabo then heard it sloshing around in the creek below were it was sighted. The boy insisted that it was definitely a hairy man, although it had longer arms and legs than a man. He insisted that it was not a bear.

It was now starting to get dark, however, a few of the boys formed a tight little group and went to the location where the creature had been seen. In a sandy spot, they found a 15- or 16-inch footprint, about 5 inches wide, and with a big toe "about twice the size of a chicken's egg." There were no claw marks seen in the print.

That night when the group had gathered close together at their camp, they shone spotlights into the trees. Szabo said that he saw it again in the trees. Two other boys, Jim Ramsey, and Robert Huston also said they saw it. Ramsey looked at it with binoculars. It was leaning against a tree with one arm, watching the camp. He described it as black and hairy with very long arms and with small beady eyes.

One of the other boys (not named) said that he saw it and that it was carrying a tree branch and seemed to be chewing on it. The creature then ran from the area.

David Ludlam, the camp leader, also saw the creature, but did not provide any details. Ludlam believes the creature was a sasquatch and stated, "We positively saw one running and it was no bear…a bear doesn't run in an up-right position.

The group investigated the tree the creature was seen leaning against and found a long hair. They then followed the path it had taken back into the woods and found that it had run across some rotten logs as they were chewed up. They said that the logs had been seen previously and were intact. Then then found a branch on the ground, and it was seen to be "covered in slobber where he had been chewing."

Monday evening Ludlam took a party of *Herald* staff members and family members up to the area and retold the story. What could have been a footprint was observed under the tree mentioned. In the sandy spot were a print was found by the boys, there was just a bare outline of something. Ludlam said that it become messed up when the boys were running around looking for more tracks or to again catch sight of the creature. The party was shown the branch the creature had allegedly chewed on, which appeared as though it had been snapped off, not cut.

At this time Ludlam mentioned that during the camp-out, the group had discovered that canned food and some bread was missing from their groceries.

The *Herald* reporter stated, "As we walked around through the trees we did hear strange noises and sometimes the sound of someone or something walking on pine cones, though it would stop when we would stop to listen. We carefully looked behind every tree as we walked...and we left before dark."

Source: "The Sasquatch Has Come to Merritt." The *Agassiz-Harrison-Rosedale Advance*, Agassiz, BC, from the *Herald*, Merritt, BC, Also John Green, 1973. *The Sasquatch File*. Cheam Publishing Ltd., Agassiz, BC, p. 41. John Green personally investigated this report. He interviewed several of the witnesses. Also John Green, Sasquatch Incidents Data Base, Incident Number 1000013.

1969/06/18 – Klemtu. Fifteen-inch Prints: Sam Brown and 2 other men reported that they found 15-inch footprints on the beach near Klemtu at Kitasu Lake on June 18, 1969. They said the stride [pace] of the prints was 4½ feet.

Source: John Green, 1973. *The Sasquatch File*. Cheam Publishing Ltd., Agassiz, BC, p. 41, from information provided by Bob Titmus who received the information from Sam Brown.

1969/06/00 – Squamish. Three-toed Tracks: Gordon Ferrier reported that he saw a 2-legged, hairy creature across the road from his home near Squamish, Mamquam community, during June 1969. He said it appeared to have a bushy tail. Later, deep 3-toed tracks were found that were as wide as a hand-span and 3 inches longer than a 12-inch boot. Police were contacted and casts were made of tracks, but were then destroyed. We are told that dogs in the area had acted upset on many occasions.

The reference to a "bush tail" and only 3 toes does not indicate that the creature seen was what we believe to be a sasquatch. Nevertheless, there have been other cases of tracks that showed only 3 toes. There is no suitable explanation for this other than the existence of other types of sasquatch-like creatures.

Source: John Green, 1973. *On the Track of the Sasquatch/Bigfoot*. Ballantine Books Inc., New York, New York, p. 163.

1969/07/29 – Agassiz. Green Convinced With US Sighting: A sasquatch sighting on July 29, 1969 at Grays Harbor, Washington convinced John Green that the creature observed was definitely a sasquatch.

The creature was seen by Verlin Herrington, a part-time sheriff's deputy, while he was on duty. Green went to the the area and interviewed Herrington, returning on August 1, 1969. He stated, "There is no doubt about it being real." He summarized the observations provided to him as follows:

- The creature was not a bear.
- It had a snout and its face had a leathery look.
- It was 7 to 7½ feet tall.
- It weighed about 300 to 325 pounds.
- It had hands with fingers, and feet with toes.
- It walked upright.

Reflecting on the incident, Herrington stated:

It's a hard position to be in because people say you're nuts. But when you see it standing right in front of you, with features that don't match with a bear, you know it's something else.

Herrington added that he photographed one track at the edge of the road and it measured about 18.5 inches long.

Source: "Sasquatch believers on the prowl again." The *Sun* (or *Province*), Vancouver, July 31, 1969. Also "Another hairy figure said sighted." The *Columbian*, Vancouver, Washington, USA, July 30, 1969. Also "Sasquatch Expert Sure of Sighting." The *Sun*, Vancouver, BC, August 2, 1969. Also John Green, Sasquatch Incidents Data Base, Incident Number 1000266.

1969/09/20 – Rossland. Observed Movement for 200 Yards: Glen McAuley reported that he saw a huge, unusual creature near Rossland on September 20, 1969. McAuley was hunting south of Rossland when he saw the oddity. He described it as dark brown, heavily built, hairy, upright, and 7 to 8 feet tall. He said it was walking with long steps, and that he observed it as it traveled for about 200 yards across a meadow. The grass in the meadow did not show any tracks.

Source: John Green, Sasquatch Incidents Data Base, Incident Number 1000171.

1969/10/25 – Bella Coola. Sitting by 2 Holes: Game guide Pan Phillips says he saw something very unusual while hunting caribou about 25 miles north of Anahim Lake on October 25, 1969.

At about the 6,000-foot level, he spotted an animal, and then looking through his rifle telescope sight, he saw an unusual brown, hair-covered creature sitting by one of 2

Pan Phillips in 1975.

holes in the snow or ice on the edge of a glacier. The holes had trails (dirty paths) that crossed the snow fringe to bushes on each side. The creature was sitting with its arms folded across its chest. Phillips observed the scene for about 10 minutes. He then started out toward the animal, but upon seeing fresh caribou tracks he decided to pursue that game instead.

Source: John Green, 1973. *The Sasquatch File.* Cheam Publishing Ltd., Agassiz, BC, pp. 41, 42. Also, Tim Padmore, 1975. "Old ranch on trail put on the block." The *Sun,* Vancouver, BC, July 1975. Also John Green, Sasquatch Incidents Data Base, Incident Number 1000064. (Photo: Phillips speaking, Steve Bosch, the *Sun.*)

1969/11/11 – Harrison Hot Springs. Green Goes Forward With Book & Film: John Green returned to his home in Harrison on November 11, 1969 from Los Angeles, after a 2-week trip marketing his book *On the Track of the Sasquatch* in southern California.

While there he gave a talk and showed the Patterson/Gimlin film to zoologists and anthropologists at the University of California, Los Angeles. Furthermore he discussed with representatives of

Walt Disney Studios a documentary on the sasquatch, which they are considering.

Green said he is leaving again later this week to distribute books in Eastern Canada and on the prairies for the Christmas trade. He plans to confer with scientists in New York and at the Smithsonian Institution in Washington, with whom he has been corresponding regarding the sasquatch.

Source: The *Bigfoot Bulletin*, No. 12. Oakland, California, USA, from "Disney Studios See Monster Movie." *The Agassiz-Harrison Advance,* Agassiz, BC, November 13, 1969.

1969/11/20 – Lytton. Wizened Old Man: Ivan Wally reported an unusual experience while driving on the Trans-Canada Highway near Lytton on November 20, 1969. He was about 3 miles east of Lytton, traveling up a hill, when he saw a creature of some sort on the road ahead. It was covered in short, gray-brown hair and stood about 7 feet tall. It had long legs, looked to weigh 300 to 400 pounds, and had the face of a "wizened old man." As Wally approached the creature, "it raised its 2 muscular arms in an attitude of surrender." He said his dog "just about went crazy."

Wally turned his truck around and drove back to Lytton, where he reported the incident to the RCMP. The police were impressed with the report and made a careful search for footprints the next day, but nothing was found.

Source: John Green, 1973. *On the Track of the Sasquatch/Bigfoot.* Ballantine Books Inc., New York, New York, p. 139, from a Royal Canadian Mounted Police report. Also Nat Cole, 1967. "Sasquatch Sighted: Its Face Had a Wizened Look." The *Sun*, Vancouver BC, 967. Also "Sasquatch Seen Near Lytton." *The Agassiz-Harrison Advance,* Agassiz, BC, November 27, 1969.

1969/12/23 – Harrison Hot Springs. Green Returns from Cross-Canada Trip: John Green returned December 23, 1969 to Harrison Hot Springs after a 5-week trip across Canada (as far as Montreal) introducing his book, *On the Track of the Sasquatch.*

During the trip he sold 3,000 copies of the book to wholesalers bringing total sales to about 16,000 copies, and made radio and television appearances in most major cities as well as being the subject of some 20 newspaper articles.

Most of the programs he appeared in were local, but one in the

CBC "Weekend" network series, and a half-hour interview on the Pierre Berton show were also done and are to be broadcast in BC shortly.

Green talked to a reporter in Winnipeg who interviewed 2 men who saw a sasquatch about 300 miles north of that city last year. Also, Green picked up a hitchhiker in northern Ontario who told of seeing a white, upright animal of enormous size, in a field near Toronto 4 years before.

Two Ojibway Indians from different parts of Ontario, both working for the CBC in Toronto, told Green that their people had a word for a man-like monster that lived in the forest of the Canadian shield, although they had forgotten what the name was.

In Ottawa, where he stayed several days with Dr. and Mrs. H.F. Fletcher, formerly of Agassiz, he showed the Patterson/Gimlin film at the National Museum and also at the Russian Embassy.

Green was told by a councillor there that scientific opinion in Russia also tends to frown on reports of hairy, human-like animals, but that government funds are spent on investigating them. A partial translation at the embassy of a scientific paper sent to Green earlier by a Professor Porshnev, head of the Russian program, indicated that the creatures the Russians are interested in are less than 6 feet tall, but that there are many similarities otherwise; for instance that they are good swimmers, see in the dark, keep in contact by high pitched screams, and they appear to hibernate, but do not entirely disappear in the winter.

Green then drove to Toronto for an interview with Pierre Berton. While in this city, he showed the Patterson/Gimlin film to scientists at the University of Toronto. He then drove to Nordegg, Alberta to investigate a sighting at the Bighorn Dam. While there, he got news of a major footprint discovery at Bossburg, Washington and headed for that town.

At Bossburg, he saw a few remaining "cripplefoot" tracks of the more than 1,000 found in the snow on the weekend by René Dahinden and Ivan Marx. These tracks are alleged to have been made by a sasquatch with a deformed foot.

Source: The *Bigfoot Bulletin,* No. 12, Oakland, California, USA, December 31, 1969, from "Sasquatch Book Taken East." *The Agassiz-Harrison-Rosedale Advance,* Agassiz, BC, December 23, 1969.

1969/DE/00 – Turnour Island. Striking on House Posts: Two adults with a group of children on Turnour Island reported that they were awakened during the night in the 1960s by forceful striking on the house posts, which shook the building. They also sensed a foul odor. In the morning, it was discovered that several bags of clams which were on the front porch were missing.

One of the children, Grace Twance, and her friends found unusual footprints on the muddy beach nearby. She stated that although the tracks resembled those of a large human, there was something wrong with them—the big toe stuck out to the side at a 45-degree angle.

Source: John Bindernagel, 1998. *North America's Great Ape: The Sasquatch.* Beachcomber Books, Courtenay, BC. pp. 55, 112, 213, 224, also, same author, 2010, *The Discovery of the Sasquatch.* Beachcomber Books, p.100, from information provided by Grace Twance in the 1990s.

1969/DE/00 – Whistler. Followed Her Husband: A lady reported a frightening experience near Whistler in the late 1960s. The lady was walking with her husband on the logging road to Powder Mountain. At at a fork in the road, her husband went one way and she waited at the fork. The lady said she watched her husband for quite a distance, and was terrified when a creature she thought was a huge bear started to follow him. She then saw that the creature was walking on its hind legs. It went a considerable distance this way and to the lady's relief, turned off the road and went into the woods. Her husband was not aware that he had been followed.

Source: John Green, Sasquatch Incidents Data Base, Incident Number 1000229.

1969/DE/00 – Harrison Mills (Chehalis Reservation). Unusual Prints in Snow: During the 1960s John Green photographed a long line of unusual footprints in snow near Harrison Mills at the Chehalis First Nations Reservation. Although John recalls the incident, he does not have the details. Nevertheless, it can be seen that the prints appear large, have a very long pace, and are in a straight line—a characteristic of the way sasquatch walk.

Large footprints at the Chehalis reservation. The lady seen here is June Green, John's wife.

Source: Christopher L. Murphy, 2010. *Know the Sasquatch: Sequel and Update to Meet the Sasquatch.* Hancock House Publishers Ltd., Surrey, BC, p. 134. (Photo: J. Green.)

1969/DE/00 – Invermere. White & Gray Creature: A sasquatch that was said to be white and gray in color was reported seen west of Invermere in the 1960s.

Source: John Green, 1973. *The Sasquatch File.* Cheam Publishing Ltd., Agassiz, BC, Canada, p. 26, from a report provided to René Dahinden.

1969/DE/00 – Nazko. Along the River: Samial Paul reported that he saw a sasquatch near Nazko along, and close to, the Nazko River in the late 1960s. He was about 20 miles south of Nazko village.

Source: John Green, 1973. *The Sasquatch File.* Cheam Publishing Ltd., Agassiz, BC, p. 41, from information provided to John Green by Samial Paul.

1969/DE/00 – Remote Island. Heard Chest-beating: A homesteader on a remote island off the coast of Vancouver Island stated that he observed sasquatch tracks on the island several times in the late 1960s, and believes he had been followed by a sasquatch. He glimpsed the creature once, and once heard chest-beating. He also believes the creature stole chickens and rabbits from enclosed pens near his cabin. He said that the sliding door of the rabbit hutch had been unfastened, slid open, and the rabbits removed without any damage to the hutch. The twisted-off head of one rabbit was found in a corner.

Source: John Bindernagel, 1998. *North America's Great Ape: The Sasquatch.* Beachcomber Books, Courtenay, BC, pp. 100, 222.

1969/DE/00 – Nelson. Sasquatch Dumps Logs off Truck: A group of Doukhobor loggers reported that they saw a troublesome sasquatch near Nelson on the Kokanee Glacier road in the early 1960s. The men said they watched the creature dump logs off their logging truck.

Source: John Bindernagel, 1998. *North America's Great Ape: The Sasquatch.* Beachcomber Books, Courtenay, BC, pp. 111, 223, from John Green's files, originally from a report provided by the loggers to John Bringsli.

1969/CI/00 – Vancouver Island. Caver Sees Strange Creature: Paul Griffiths reported a strange encounter after exploring a cave on Vancouver Island in the winter during the late 1960s. When he came above ground at about 3:00 a.m. and walked down a roadway, he saw a flat-faced, 7-foot-tall hairy creature with reflecting eyes. When the creature started to walk towards him, Griffiths ran to his camper truck. He noted that that the creature's hands hung lower than its knees.

A short time later when he and others were leaving the area, Griffiths shone his flashlight through the back door of the camper into the spot he had seen the creature and it was still there.

Paul Griffiths in 1969.

Two weeks later he went back to the site and found huge, human-like footprints in snow near the cave.

Griffiths later (1972) remarked that the tracks were the same as he found near a Cariboo cave (see 1971/06/00 – Wells. Tracks Near Cave).

Source: John Green, 1973. *The Sasquatch File.* Cheam Publishing Ltd., Agassiz, BC, Canada, p. 41, from personal communications with Paul Griffiths. Also David Stockand, 1972. "Fresh Track: Sasquatch cave sealed by water." The *Sun,* Vancouver, September 6, 1972. (Photo: Griffiths portrait, Ray Allan, the *Sun.*)

1969/CI/00 – Gordon River (VI). Deer Heads Appeared Twisted Off: A hunter reported that he found 2 buck deer heads about 100 feet apart near the Gordon River on Vancouver Island in October or November 1969 (but could have been 1970). The head appeared as though they had been twisted off the rest of the bodies. They were found at a gate house on a logging road. Unusual tracks were found in the area, which were referred to as sasquatch tracks.

Source: John Bindernagel,1998. *North America's Great Ape: The Sasquatch.* Beachcomber Books, Courtenay, BC, pp. 73, 216, from information provided by a BC forest industry worker.

1969/CI/00 – Keremeos. Green-blue Eyes Reflecting: Wendy Edelson reported that she saw a strange creature near Keremeos in 1969 or 1970. Edelson was driving East on Highway 3 between Keremeos and Penticton. She came around a curve by a lake and saw greeny-blue eyes reflecting on the roadside. When she drove past the eyes she saw a 7-foot creature covered with longish dark brown hair standing very still as if hoping not to be seen. She said that it had proportions like a human. The creature was on the lake side of the road. A high bank, rocks and dirt were on the other side.

Source: John Green, Sasquatch Incidents Data Base, Incident Number 1000964.

1970/01/07 – Squamish. It Was Carrying a Fish: Bill Taylor, a BC Department of Highways worker, reported that he saw a large, hairy, bipedal creature on Highway 99 near Squamish at Cheakamus Canyon on January 7, 1970.

As Taylor rounded a sharp turn near the top of Cheakamus Canyon he saw, about 200 feet ahead on the left shoulder of the road, what he first took to be a bear. The creature appeared to step towards the highway, then it stumbled and fell on all fours with its back to Taylor, who had stopped his truck to watch. By this time he was only 80 feet away from the oddity.

The creature stood up and glared at Taylor. It then walked upright with swinging arms across the 40-foot-wide highway in 4 or 5 quick paces. On the other side, it climbed a 10-foot bank, using its left hand once to gain a hold, and then disappeared over the bank.

During this time, Taylor noticed that it was carrying a fish in its right hand. He could see the head and tail protruding on each side of the creature's closed hand.

As to the creature's size and other aspects, Taylor estimated (after re-visiting the scene) that it was at least 8 feet tall and weighed between 700 and 800 pounds (minimum). Its body was covered in reddish brown hair about 3 to 4 inches long. Its forehead sloped to the rear and the head came to a point. The hair on its face was either much shorter or much lighter than that on the rest of its body because, "I could see very clearly an expression—of either anger or fear." The eye brows were prominent, the nose was flat, and the eyes were like a human's. It was about 3 feet across the shoulders and had a very prominent stomach. Taylor summed up by saying, "Generally, it was monkey-like with human eyes; it was simply a hairy, large creature that scared the daylights out of me."

Source: Don Hunter with René Dahinden, 1993. *Sasquatch: The Search for North America's Incredible Creature.* McClelland & Stewart Inc., Toronto, Ontario, pp. 7–9. Also John Green Sasquatch Incidents Data Base, Incident Number 1000010.

1970/02/00 – Klemtu. Parted the Salal With Its Hands: Andrew Robinson report seeing a strange creature near Klemtu in early February, 1970. Robinson said that he was out in his rowboat about a mile south of Klemtu and was rowing to shore when he saw the animal on the beach. He said it was about 6 feet tall, light-brown, had heavy legs, and long heavy arms. He watched it as it walked away and parted the salal bushes with its hands as it entered the forest. Salal bush grows quite high and is difficult to penetrate.

Source: John Green, 1973. *The Sasquatch File.* Cheam Publishing Ltd., Agassiz, BC, p. 51, from a report provided by Andrew Robinson to Bob Titmus. Also John Green, Sasquatch Incidents Data Base, Incident Number 1000050. (Green notes that Robinson had poor eyesight.)

1970/02/00 Skidgate (QCI). Small Beady Eyes: Tina Brown and Herman Collison reported to police that they saw what they believe was a sasquatch near Skidgate (Queen Charlotte Islands) in March 1970. The pair was in a car driving through a church camp about 7 miles north of Skidgate Mission. It was about midnight and, as Brown stated, "It was standing there in the bush and our headlights went right on it, then it walked away into the bush." They estimated that they were only about 20 feet away from the oddity.

Tina Brown

As to its description, Miss Brown said the creature was about 7 feet tall, had hair all over its body including its face, and small beady eyes. She said it looked more like a gorilla than a man.

The police went to the area but did not find any trace of the creature.

Source: Lorraine Shore, 1970. "Queen Charlottes now boast of their own Sasquatch," The *Sun,* Vancouver, BC, June 24, 1970. Also, John Green, Sasquatch Incidents Data Base, Incident Number 1000232. (Photo: The *Sun,* Vancouver.)

1970/03/00 – Juskatla. Lifted Up Its Arms: A boy reported seeing a sasquatch at the logging camp in Juskatla during March 1970. Upon being sighted, it lifted up its arms and ran. Footprints were found in the sighting area.

Source: Lorraine Shore, 1970. "Queen Charlottes now boast of their own Sasquatch," The *Sun,* Vancouver, BC, Canada, June 24, 1970. Also John Green, Sasquatch Incidents Data Base, Incident Number 1000236.

1970/04/02 – Harrison Hot Springs. Man or Gorilla: Keith Shepard reported that he had to slam on his brakes to avoid hitting what he thought was either a man or a gorilla at Harrison Hot Springs on April 2, 1970. The creature was in the middle of the road at what was for many years a well-known animal crossing within the village boundary. Shepard said it was about 8 feet tall, heavy, and covered in dark hair except for its dark and wrinkled face. Despite the near collision, the creature ambled on across the road.

Source: John Green, 1973. *The Sasquatch File*. Cheam Publishing Ltd., Agassiz, BC, p. 51, from personal communication with the witness. Also "Hotel Guest Sights Sasquatch Crossing Hot Springs Road," The *Sun* (?), Vancouver, BC, April 2, 1970. Also John Green, Sasquatch Incidents Data Base, Incident Number 1000058.

1970/04/23 – Klemtu. Ape-like Face: The crew of the Vancouver seiner *Bruce I* reported seeing a strange creature near Klemtu on April 23, 1970. They saw the oddity on the mossy shore about a mile north of Klemtu at a distance of 50 to 100 yards, and observed it for about 2 minutes. They described it as being about 6 feet tall, very heavy, erect, covered in beige-colored hair, and with an ape-like face. It stood still for awhile and then walked off upright. Tracks were later found where the creature had been. Because of the moss, they were poor. They were estimated to be about 14 inches long.

Source: John Green, 1973. *The Sasquatch File*. Cheam Publishing Ltd., Agassiz, BC, p. 51, from information provided to people in Klemtu by the crew of the ship. Also John Green, Sasquatch Incidents Data Base, Incident Number 1000231.

1970/05/00 – Nazko. Watched It Go By: Samial Paul reported that he observed a strange creature while hunting near Nazko in May 1970. Paul saw the creature coming and got out of sight as best he could so that he could watch it go by without noticing him. He said he was about 50 feet from the oddity as it walked by and was able to observe it for about one minute.

He said that it walked erect with sloped shoulders. It was about 8 feet tall, hair-covered, and very heavy. Of its facial features, he noticed that it had heavy brow ridges, a short nose, projecting jaw, and long canine teeth. There was no hair on its face. It had a high forehead that sloped back

Later, with others, whom he named as Patrick and Leonard Paul, Stanley Boyd, Norman Jack and Harvey Jack, he followed the creature's tracks for 10 miles, finding one place where it sat down. The trail was then lost on hard ground.

The observance of "long canine teeth" in this report is unusual. Such are not mentioned in any other BC report in this work.

Source: John Green, Sasquatch Incidents Data Base, Incident Number 1000032.

1970/06/00 – Trail. Track in Dry Mud: Dennis Merlo, a city employee, reported that he and another employee found a 5-toed track, 12 to 14 inches long and 6 inches wide, in a patch of dry mud near Trail in June 1970. The track was found on a hillside in the city's watershed area.

Source: John Green, 1973. *The Sasquatch File*. Cheam Publishing Ltd., Agassiz, BC, Canada, p. 51, from information provided to John Green by Dennis Merlo.

1970/08/05 – Harrison Hot Springs. Green & Dahinden At Odds: On August 5, 1970 it was announced that John Green and René Dahinden had a "parting of ways" over Green's plan to use a computer to analyze sasquatch-sighting information and to provide information to others. Green is working with American National Enterprises (film producer) in connection with a film documentary on the sasquatch and is assembling data for his own information, and other possible purposes

Dahinden considers the computer a "gimmick." Also, he said that he does not like passing hard-earned information, "gathered at our own expense," to American National Enterprises which "just wants to make a fortune by making a film and selling it."

Green said the computer survey, using a questionnaire sent to hundreds of people who claim to have seen a sasquatch, is nearly completed. As to Dahinden, Green said:

> He left the computer program and decided he wants to get the sasquatch himself. He doesn't want to share his information as I do. I'd be happy to see anyone prove the existence of the sasquatch. René wants everything for himself.

At this time, Dahinden announced plans for a expedition into the Pitt Lake area, which was detailed in a newspaper report on August 7, 1970 (see next entry).

Source: Maurice Chenier, 1970. "Man, computer hunt Sasquatch." The *Province*, Vancouver, BC, August 5, 1970.

1970/08/05 – Richmond. Dahinden Announces Pitt Lake Expedition: René Dahinden announced plans in Richmond on August 5, 1970 for an expedition into the Pitt Lake area to search for the sasquatch. He will be joined by friends from Winnipeg who accompanied him on a 1968 Washington State expedition near Colville. Later, others will swell their ranks, including Roy Fardell of Kimberley, Roger St. Halair, a New York zoologist, and Ivan Marx, Colville guide/photographer.

If the team does find evidence of the sasquatch, Dahinden said a group of Seattle businessmen, 12 Canadians, and U.S. scientists will form a corporation called *Probe* to raise $50,000 for scientific research into the sasquatch issue.

Source: "Richmond man begins new Sasquatch search. The *Sun*, Vancouver, BC, August 7, 1970.

1970/AU/00 – Clearwater. Tracks in Remote Area: A British Columbia Hydro employee stated that he found large, barefoot tracks in the fall of 1970 near Clearwater at the north of Wells Gray Park. He had been flown into the area on a project. It is believed the tracks were in snow.

The fact that the employee had been flown to the spot indicates it was remote and probably not serviced by a road. That a person would be in the region walking barefoot is hardly probable.

Source: John Green, 1973. *The Sasquatch File.* Cheam Publishing Ltd., Agassiz, BC, Canada, p. 51, from information provided to John Green by George Harris who spoke with the Hydro employee.

1970/11/00 – Princeton. Small Round Head: Stuart Syme, outdoors columnist for the *Courier,* Kelowna, reported seeing a strange animal in the Princeton area at Allison Lake during November 1970. Syme was grouse hunting in the hills west of Allison Lake. He saw at the bottom of a gully, about 75 feet away, a creature 5 feet tall, covered with long, straight, cinnamon-brown hair. It had a small round head, sloping shoulders, and stood with its arms dangling loosely. He put a cartridge in his shotgun and aimed it at the creature, but upon gazing into its reddish-brown eyes he had a feeling of tranquility and lowered the gun. Syme's dog whimpered, hair standing on end from nose to tail, and pressed against his master's leg. Syme bent down to attend his dog and in doing so lost sight of creature.

Syme went down into the gully but did not see the creature again. He did not look for tracks. He mentioned that his dog went to

Allison Lake. Syme was hunting in the hills seen on the left.

where the creature had been without hesitation, and there tried to climb a tree. He didn't try to find out the reason.

This report is noteworthy for 2 reasons. Firstly, it comes from an outdoors newspaper columnist who would certainly be familiar with the various animals in BC. Secondly, we have yet another case where a hunter could not bring himself to shoot the creature.

Source: John Green Sasquatch Incidents Data Base, Incident Number 1000172, from the *Courier*, Kelowna, BC. (Photo: Image from Google Earth; Image © 2011 DigitalGlobe; Image © 2011 Province of British Columbia.)

1971/01/00 – Masset (QCI). Hairs More Than 7 Inches Long: Four hunters reported that they saw a sasquatch near Masset at the north tip of Kumdis Island in January 1971. Bob Titmus and Dr. Jim Proctor went to the sighting location to investigate the report. It had snowed since the sighting; however, when Titmus and Proctor arrived, they still found some evidence. In the bare ground at the base of a windfall they found a single large footprint and 2 partial prints. Titmus found reddish-brown hairs more than 7 inches long.

Source: John Green, 1973. *The Sasquatch File*. Cheam Publishing Ltd., Agassiz, BC, p. 51, from information provided by Bob Titmus.

1971/WI/00 – Hope. Miles of Tracks: Mrs. Eileen Yerxa reported that miles of large tracks were found about 14 miles east of Hope in the winter of 1971. The tracks, about 14 inches long, with a pace of 30 to 36 inches, were in deep snow. She said that men trying to follow the tracks, at one point sank in so far that they had to be helped out, but the tracks went right on through.

It is evident that the size of the track-maker was large enough to allow it to travel in deep snow without any problems.

Source: John Green, 1973. *The Sasquatch File.* Cheam Publishing Ltd., Agassiz, BC, p. 51, from a letter written to John Green from Mrs. Eileen Yerxa.

1971/05/00 – Squamish. Tracks On a Sandbar: G. Conway of Delta reported that he found large tracks on a sand bar in the Squamish Valley in May 1971. They were found more than 30 miles up a logging road. He said they were 14 inches long and looked like the photographs of sasquatch tracks, but were fairly old.

Source: John Green, 1973. *The Sasquatch File.* Cheam Publishing Ltd., Agassiz, BC, p. 51, from a letter written to John Green from G. Conway.

1971/06/00 – Wells. Tracks Near Cave: Paul Griffiths reported finding sasquatch tracks near Wells at Bowron Lakes Park in June 1971. The tracks were at the entrance of a cave which he said was previously undiscovered.

He went back to the cave the following year and again found a track at the cave mouth. The print was 13 to 14 inches long.

Griffiths previously had a sasquatch sighting on Vancouver Island (see 1969/CI/00 – Vancouver Island. Caver Sees Strange Creature.)

Source: John Green, 1973. *The Sasquatch File.* Cheam Publishing Ltd., Agassiz, BC, p. 51. Also David Stockand, 1972. "Fresh Track: Sasquatch cave sealed by water." The *Sun,* Vancouver, BC, September 6, 1972.

1971/07/00 – Lake Louise, Alberta. Shoulder-length Very Blond Hair: A lady riding in a camper reported seeing a strange creature in the Lake Louis area in July 1971. The camper was traveling east on the Trans Canada Highway, west of the Highway 93 junction, and east of the spiral tunnels. Although this puts the sighting in Alberta, it is only by a few miles.

The lady saw something sliding in the loose gravel (presumably above the road). "It was walking standing, but grabbed with a hand to save a fall," she said. It walked about 20 feet down to what looked like a cave entrance, sat down, rested its chin on its knee and wrapped its arms around its legs. The lady estimated that the creature was 8 or 9 feet tall. Its body was covered with short, reddish hair like a red deer (fine and shiny or glossy). On its head, there was shoulder-length, very blond hair. She noticed 2 very woman-like breasts, indicating the creature was female. The lady said she yelled "sasquatch" through the camper intercom, but the driver didn't know what she meant. After she told the driver of the sighting, they went back to the spot where the creature was seen, but this took about 10 minutes, and during this time the creature had left.

Source: John Green, Sasquatch Incidents Data Base, Incident Number 1000175.

1971/07/00 – Houston. Long White Hair: Mrs. Byron Sheldon reported seeing a strange creature on 2 occasions near her recreational trailer, Houston area, in the vicinity of Grizzly Lake (18 miles south of the town) in July 1971. After breakfast one day, she saw a creature estimated to be 8 feet tall come out of the bushes on the other side of a creek (Buck Creek), about 40 feet away. It was covered with long, white hair and had long arms with hands hanging down to its knees. It simply walked by Mrs. Sheldon without looking her way.

The next morning, she saw a similar creature, brown in color, come out of the bushes and go down to the creek.

Source: John Green Sasquatch Incidents Data Base, Incident Number 1000173.

1971/SU/00 – New Denver. Tracks Impressed 2 Inches Deep: Robin Flewin and Rick Dankoski said they found large tracks on a dirt road in the New Denver area, along Silverton Creek, in the summer of 1971. The tracks were 17 inches long, 6 inches wide and were impressed 2 inches deep.

Source: John Green, 1973. *The Sasquatch File.* Cheam Publishing Ltd., Agassiz, BC, p. 51, from a letter to John Green from Robin Flewin and Rick Dankoski.

1971/09/00 – Princeton. Walked Slightly Stooped: Harvey Kirby reported that he saw a sasquatch near Princeton in Manning Park at Allison Pass, during early September 1971. Kirby was a passenger in truck that had been parked to allow the driver to get some sleep. At about 4:00 a.m. he noticed movement in the rear-view mirror and upon turning around watched the creature walk across the paved and lighted highways department yard at a distance of fewer than 100 feet. Kirby said the creature was about 8 feet tall, covered with dark hair, had a small head, long large body (longer than its legs), long arms (like an ape's), and walked slightly stooped. The light was such that Kirby could not see its face clearly. It went down a slope, slowed down and looked back and forth. It then disappeared into the bush.

Source: John Green, 1973. *The Sasquatch File.* Cheam Publishing Ltd., Agassiz, BC, p. 51, from a report provided by Harvey Kirby. Also John Green, Sasquatch Incidents Data Base, Incident Number 1000176.

1971/AU/00 – Sechelt. Color of a Collie Dog: John Stewart, a prospector, reported seeing 2 large bipedal animals near Sechelt (north of) in the fall of 1971. He said that at about the 3,500-foot level he came over a hill and saw the oddities, which were of human shape and covered with hair the color of a collie dog. They were near some swamp holes. Apparently surprised at being seen, the creatures immediately left. Stewart inspected the area where they had been and found footprints in the mud, but they were not clear. The tracks appeared to indicate that the creatures had been pulling up lily roots.

Source: John Green, 1981. *Sasquatch, The Apes Among Us.* Hancock House Publishers, Surrey, BC, p. 420, from a report provided by the witness, also John Green, Sasquatch Incidents Data Base, Incident Number 1000234.

1971/11/12 – Richmond. Dahinden Goes to Europe: Disgruntled with the indifference and inaction of scientists in North America regarding his sasquatch-related evidence, René Dahinden of Richmond took the matter to Europe, where he hoped to arouse more scientific interest. Before leaving on November 12, 1971, Dahinden stated:

Scientists in Canada and the United States totally reject the mere idea of the sasquatch being real. They are smug, self-satisfied and totally indifferent. It's okay to say there's something in the Himalayas, but not in our own back yard.

Dahinden hoped that European scientists would be more open-minded, and even planed to go to Scotland Yard. "I might find an expert on footprints there," he stated, "a man who could give a technical opinion as to how artificial feet and live feet behave under certain conditions on certain materials."

Source: Scott McClean, 2005. *Big News Prints,* (self-published), p. 53, from "Legendary Creatures: The sasquatch rivals them all." The *Dominion News,* Morgantown, West Virginia, USA, December 4, 1971. Also "Hunter of 'Creature' Seeks Help in Europe," The *Times,* Los Angeles, California, USA, December 12, 1971.

1972/01/15 – Yakima, Washington, USA. Roger Patterson Passes On: Roger C. Patterson, who along with Bob Gimlin, filmed an alleged sasquatch at Bluff Creek, California, died on January 15, 1972. Patterson was laid to rest in the West Hills Memorial Park, Yakima, Washington.

The film taken was first shown to scientists at the University of British Columbia on October 26, 1967 (see 1967/10/26 – Vancouver. UBC Scientists First to See Film of Alleged Sasquatch).

Source: Christopher L. Murphy, 2008. *Bigfoot Film Journal.* Hancock House Publishers, Surrey, BC, p. 54.

1972/01/00 – Moscow, USSR. Dahinden Visits *Moscow News:* While in Europe (early November 1971 to early March 1972), René Dahinden went to Moscow, Russia, and visited *Moscow News* (probably in January, 1971) and explained the purpose of his visit to that country. The newspaper people asked him to submit an article. He did so, explaining in considerable detail the situation with the sasquatch in North America. He closed his article with the following statement:

> If by way of international cooperation we manage to determine the quality of the material we possess as well as to evaluate the possibility and probability of the evidence of the above mentioned creatures [sasquatch/hominoids] in certain parts of the world, then we'll make a big step forward

in solving the problem which I consider one of the greatest scientific problems of all time.

Source: "Need for International cooperation in search for relic hominoids." *Moscow News,* November 5, 1972.

1972/03/25 – Richmond. Dahinden Declares Trip to Europe Great Success: René Dahinden announced in Richmond on March 25, 1972 that his trip to Europe to provide sasquatch-related evidence to European scientists was a great success. He stated, "I accomplished everything I wanted. Now we have some international co-operation in analysis of the evidence."

He said that the European scientists were amazed that such a body of evidence should be rejected by scientists in North America.

Dahinden gave a copy of the Patterson/Gimlin film to the Russian hominologists Dmitri Bayanov and Igor Bourtsev. They performed the first detailed study of the film and reported that it definitely showed a natural creature. Bayanov provided a full report on

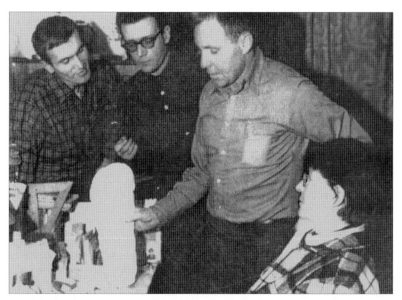

René Dahinden is seen here showing a plaster cast of one of the footprints left by the creature in the Patterson/Gimlin film. Igor Bourtsev (far left), Dmitri Bayanov, and Dr. Marie-Jeanne Koffmann (far right) are seen looking on.

their findings for a conference at the University of British Columbia held in 1978 (see 1978/05/09 – Vancouver. Anthropology of the Unknown Conference). He went on to write a book on the subject entitled, *America's Bigfoot: Fact Not Fiction; U.S. Evidence Verified in Russia* (1997).

Source: "New Sasquatch tracks wanted." The *Province*, Vancouver, March 25, 1972. also, D. Bayanov, 1997. *America's Bigfoot: Fact Not Fiction; U.S. Evidence Verified in Russia.* Crypto Logos Publishers, Moscow, Russia. (Photo: I. Bourtsev.)

1972/07/00 – Hope. Came Out Of a Cave: A man and his companion hunting in the Hope area reported seeing a strange creature during July 1972. The men were looking for mountain goats on a cliff face just east of the Hope Slide. They saw a very large, reddish-brown figure come out of a cave about a thousand feet above the highway. It stood in front of the cave, then walked on its hind legs around to the side of the cave. It then went across the open rock face using all 4 limbs until it entered some trees. The cave was later investigated and found to be only a few feet deep, floored with broken rock. It did not shows any sign of use by any creature.

Source: John Green, Sasquatch Incidents Data Base, Incident Number 1000178.

1972/SU/00 – Houston. Something Seen & Filmed: A resident on her recreational property near Houston at Grizzly Lake, saw and filmed an odd creature in the summer of 1972. The resident took an 8-mm movie of something unusual through a window of her home. The movie is said to show what appears to be a large, white, upright creature walking up out of a neighborhood creek (known as Buck Creek) at a distance of about 50 feet.

John Green saw the film. He said that a white creature of some sort is seen, but no details can be determined.

Source: John Green, Sasquatch Incidents Data Base, Incident Number 1000173, from information provided by Mrs. Byron Sheldon who lived at the same location (a neighbor) and who had 2 sightings in July 1971 (see 1971/07/00 – Houston. Long White Hair).

1972/12/24 – Castlegar. 150 Yards of Tracks: Mrs. Gail Davidson reported that she and her husband found 150 yards of unusual tracks near Castlegar on December 24, 1972. She said the tracks were 16 inches long, and more than 6 feet apart. They were found on a trail on the hill behind the couple's home.

Source: John Green, 1973. *The Sasquatch File.* Cheam Publishing Ltd., Agassiz, BC, p. 51, from information provided to John Green by Mrs. Gail Davidson.

1972/CI/00 – Stewart. Ten Feet Tall: A truck driver reported seeing a huge, strange creature near Stewart in about 1972. The man was an ore truck driver for Granduc Mine. He saw the oddity, which he estimated was 10 feet tall, on the road from the mine to Stewart during a snow storm. He was so shaken he had to radio for someone to come and drive his truck back.

Source: John Green, Sasquatch Incidents Data Base, Incident Number 1000177.

1973/03/21 – Big Bay. Watched Taking Steps from Large Rock to Large Rock: Three Vancouver herring fishermen, Peter Spika, Luko Burmas and Nick Piscas, returned to the city considerably excited about seeing a sasquatch near Big Bay at Bute Inlet on March 21, 1973. They were aboard the seiner *Tracy Lee* running up Bute Inlet, near the head on the east side, when through the galley window Nick Piscas saw the creature walking erect on the shore line rocks. The three men went on deck and watched it until their boat went past a point. It was man-like, very tall, walking with its arms swinging as if for balance, taking steps from large rock to large rock—far too great a distance for a human. The men waved and it turned towards them, but just watched them. The skipper rejected the crew's urging to go ashore and pursue the creature.

Source: "Not All Quiet on the Western Front," published in *Pursuit* magazine, Gulf Breeze, Florida, USA, Volume 7, No. 4, 1973. Also, John Green, 1973. The Sasquatch File. Cheam Publishers Ltd., p. 73. Also John Green Sasquatch Incidents Data Base, Incident Number 1000034.

1973/06/00 – Sechelt. Appeared to Have a Goatee: Don Solinski reported seeing a strange creature in the Sechelt area at Chapman Creek in June 1973. Solinski, a cat skinner-powderman, working on a logging road extension, said he saw the oddity at Mile Nine along the creek at 7:30 a.m.

He had just driven a drilling rig to the end of the road and climbed on top of a huge log to look over the grade ahead. Across the clearing he saw a shaggy-haired (hair 3 to 4 inches long) creature about 150 to 200 feet away. It was standing on 2 legs and appeared to have a goatee. It looked to be larger than a man and was jumping up and down with its arms bowed out at its sides. Its legs were not totally visible and the worker thought that it was standing on a log.

Solinski looked around for somewhere to run and as he did so, he thought he saw the creature do a somersault off the log it was on to the lower side of the road cut. Solinski then jumped into the drill rig and started to drive it away in reverse, but got stuck in a mud hole. He jumped out, ran down the road and cut across to where he had left a tractor parked on a lower switchback. There he had a clear view all around him, so he waited for about an hour until other workers arrived. He then told his boss that he would not work alone on the road anymore. Men investigated where the creature had been seen and found some footprints in mud. They said the prints were more human-like than bear-like.

Source: John Green, 1981. *Sasquatch, The Apes Among Us.* Hancock House Publishers, Surrey, BC, p. 420, from The *Sechelt-Peninsula Times,* June 1973 and from an investigation by Joel Hurd. Also John Green Sasquatch Incidents Data Base, Incident Number 1000235.

1973/07/10 – Ocean Falls. Rooting for Vegetation: Olle Ek, a BC Fisheries patrolman, reported that he saw a strange creature near Ocean Falls at Roscoe Inlet on July 10, 1973. Ek was on his way to start opening a trail up the Roscoe River. While dragging his boat onto a mud flat he noticed an animal hunched over, dirty white to gray in color, apparently foraging in the salt water grass, which was about knee high. Ek estimated the creature was about 150 to 200 yards away. After trying to figure out what it was, he clapped his hands to get a reaction and it quickly left the scene walking upright. Ek did not see its legs, and he felt that it left "floating." He said it looked like a big man in a buffalo coat. When he eventually went over where it had been, the tide had covered the area.

Source: John Green, Sasquatch Incidents Data Base, Incident Number 1000039. Also "Not All Quiet on the Western Front," *Pursuit* magazine Volume. 7, No. 4, p. 98., 1973.

1973/07/00 – Lumby. "Plinkers" Shoot at Sasquatch: Three young boys reported that they encountered and shot at a strange creature near Lumby in July 1973. The boys had gone "plinking" with a .22-caliber rife in the Hunters Range area (local name) of the nearby woods. They intended to hunt small animals.

The leader of the group, Pete Nab, aged 13, was carrying the rifle, while the other 2 boys, Brent Shoutley and Jeff Brockton, kept an eye out for targets. As they pressed deeper into the forest, they encountered "game" that will haunt their memories for many years to come. There, standing just outside the tree line beside a pile of dead-fall logs, was a nightmarish, gigantic, ape-like creature.

The boys froze and stared—they had stumbled upon the elusive sasquatch. The spotters yelled at Nab to shoot it. After a few moment's hesitation, he fired 4 rounds into the middle of its chest. Its only reactions were to raise its right arm to grab hold of the log pile, and to make a swatting motion with its left hand, as though brushing off a bug.

Nab was the only one to see this reaction, for his courageous companions were almost out of sight, homeward-bound. The creature seemed to be pondering what the little stings it was feeling were all about, rubbing with its hand at the location where the bullets had hit it.

Nab then ran home as well. He looked back only once, in time to see the creature turn and walk into the trees, apparently no worse for wear.

In 1980, Nab, then 20 years of age and newly wed, was living in Calgary, Alberta. He had not given his brief sasquatch sighting much thought until he spotted an ad in the Calgary press placed by Thomas Steenburg asking for information on sasquatch encounters.

Nab talked the matter over with his wife, telling her of his experience and asking if she thought he should call Steenburg. She encouraged him to do so, and a few nights later he made an appointment to be interviewed.

Nab described the creature as between 8 and 9 feet tall, standing on 2 legs, and covered from head to toe with black hair. He described its eyes as deep-set, and the nose as flat with big nostrils. Skin color was dark, but lighter than the hair color. No sound of any kind was heard coming from the creature and no odor was detected. He estimated that the entire incident happened over a period of maybe 2 minutes. He has never since encountered any other animal like this animal.

An illustration (artwork by RobRoy Menzies) on this incident is provided in the color section.

Source: Thomas Steenburg, 2000. *In Search of Giants: Bigfoot/Sasquatch Encounters.* Hancock House Publishers Ltd., Surrey, BC, pp. 98–102. Report provided by T. Steenburg. Also John Green, Sasquatch Incidents Data Base, Incident Number 1000179.

1973/08/00 – Agassiz. I Saw a Sasquatch That Night: Ralph Bobb was astounded by what he saw on the roadside in the Agassiz area, near the Ruby Creek Bridge, in August 1973. Here is the story in his own words:

> I have been hunting all my life, and I have seen a lot of strange things. But nothing could compare with the sasquatch I saw that night on the side of the road. I had heard a lot of stories about the sasquatch living where I do, but I never thought I'd ever see one. I always thought if I ever saw one of those things, it would be in the bush on a hunting trip or something like that. Not of the side of Highway 7 watching my car go by.
>
> We [wife, Jennifer, and children] were driving home from the Chilliwack drive-in late at night. I'd say between 1:00 a.m. and 1:30 a.m., along Highway 7. I had an uneasy feeling like I get when I am out hunting sometimes. Hunter's

Ralph and Jennifer Bobb.

214

feeling. I can't explain it, the hair on the back of your neck stands up, you know.

Anyway, we were approaching Ruby Creek Bridge. The kids were asleep in the back of the station wagon. Jennifer was covering them with a sleeping bag. I was just driving along, when I saw this huge creature standing on the left side of the road looking at us. I don't remember what I said, but I slammed on the brakes and brought the car to a halt. Lucky for me it was late at night and there were no other cars on the road or I probably would have been rear-ended for sure. Jennifer didn't say anything, she just sat there and wondered what the hell I was doing. "Did you see it?" I said. She replied, "No, I was covering the kids."

I did a U-turn and went back to where the creature was standing, but it was gone. I searched the area with the car lights. I did not have a flashlight with me, but I found nothing. The creature probably moved back into the trees the moment I hit the brakes. I don't believe it crossed the highway after I drove by, but then again it might have. It was so big it probably could have crossed the highway in 2 or 3 strides [paces].

The next day Jennifer and I went back to the area to look for tracks, but we didn't find any. We looked everywhere, but the area was too rocky for footprints to have been left behind. Nobody believed me when I told them about what I saw. I really don't care what people say, I know I saw a sasquatch that night.

Source: Thomas Steenburg, 2000. *In Search of Giants: Bigfoot/Sasquatch Encounters.* Hancock House Publishers Ltd., Surrey, BC, pp. 34, 36. Also John Green, Sasquatch Incidents Data Base, Incident Number 1001516. (Photo: Ralph and Jennifer Bobb, T. Steenburg.)

1973/11/26 – Vancouver. Stood About 10 Feet Tall: Fishermen Nick Pisac and 2 friends, all of Vancouver, trying their luck off the Vancouver coast say they saw a sasquatch on November 26, 1973. Pisac said they spotted the thing walking slowly along the beach. "It stood about 10 feet tall, and was sort of lightish gray," he said.

Source: Scott McClean, 2005. *Big News Prints* (self-published), p. 54, from "Creature Sighted," Edwardsville Intelligencer, Illinois, USA, November 26, 1973 (AP).

1973/YR/00 – Vancouver. Hunter with Dahinden Book Released: In 1973 a book written by *Province* (newspaper) deskman Don Hunter, with René Dahinden, entitled *Sasquatch* was released.

The book is as much about sasquatch hunting and the hunters as it is about the sasquatch itself, and as such is very entertaining. This observation was expressed by Tom Paterson of the *Province,* who wrote a review on the book. To quote from Paterson:

> We could, we suppose have reviewed this book in a more serious vein, but we have done that before with Green's books. And as one who has followed the Sasquatch puzzle for years, we can truthfully say that Hunter's (Dahinden's) version is without equal as far as entertainment goes."

Nevertheless, the book does contain a lot valuable information.

Source: Tom Paterson, 1973. "Sasquatch Puzzle." The *Province*, Vancouver, BC. Also, Hunter/Dahinden, 1973. *Sasquatch*. McClelland & Stewart, Toronto, Ontario. (Photo: McClelland & Stewart.)

1974/01/00 – Terrace. Seen in Garbage: A landfill worker reported seeing an odd creature crouched in garbage at a Terrace landfill during January 1974. The worker had returned to the site at about 2:00 a.m. to retrieve keys that had been left in a piece of equipment. As he walked past the area of the last dump he saw a figure that he thought was a person, about 100 feet away in the garbage. He called out, "Hello there, what are you doing in the dump at this time of night?" The figure then stood up and towered some 10 feet tall. It looked at the worker, and then calmly walked away to the bush at the rear of the landfill. It looked back several times, apparently to see if it was being followed.

After it had travelled a safe distance away, the worker went to where it had been in the garbage. In the snow he found footprints that were 5 feet apart. He remarked that the creature had walked in 2 feet of snow as easily as a human could in snow ankle deep.

Source: BFRO website, from "The Sasquatch in Northern BC, Canada," HBCC UFO Research, April 7, 2001.

1974/01/00 – Jasper, Alberta. One Was a Step Ahead of the Other: Lucien Lacerte reported that he saw 2 strange creatures near Jasper, Alberta on the Yellowhead Highway (in BC) during January 1974. Lacerte was driving west from Jasper. There was snow on the ground and it was fairly deep off the road. He rounded a curve, and to the south there was a clearing. He saw the creatures about 200 feet ahead on the road by the clearing. One creature was larger than the other. Lacerte described them as human-like, black and hairy. They were striding very fast down the road. The larger one was a step ahead of the other, and would look back. The oddities crossed the road from left to right, in a hurried fashion and went into the bush. Lacerte remarked that they definitely were not bears.

Source: John Green, Sasquatch Incidents Data Base, Incident Number 1000180.

1974/04/00 – Terrace. Eating Pine Needle Tips: Children reported that they saw 2 sasquatch, a large and smaller one, from their home in the Terrace area during April 1974. There was still about 4 feet of snow on the ground that had frozen hard enough to walk on. The snow had melted at the base of small jack pine trees, leaving them in snow holes. The 2 sasquatch were seen at about 100 yards behind the house.going from tree to tree picking and eating the needle tips on the tree branches.

Source: BFRO website, from "The Sasquatch in Northern British Columbia, Canada," HBCC UFO Research, April 07, 2001.

1974/05/00 – Shawnigan Lake (VI). Five Young Girls See Sasquatch: Five young girls who went out for a walk in the Shawnigan Lake area stated they saw what they believe was a sasquatch in May 1974. The girls heard something crashing through the bush in the distance across a meadow. Being afraid that it might be a bear, they hid behind a fallen tree.

They watched in the direction of the sound and near the edge of the trees they saw a strange creature. It was very large, dark, blackish-brown, shaggy haired, and was walking on 2 legs. It had long arms that swayed, and was stooped at the shoulders. Its head was round, and its face dark. It did not have a snout; they did not notice ears. Its height appeared to be about 6.5 feet. The girls said they then noticed a foul odor.

When one of the girls sneezed, the creature appeared to smell the air in the direction of the group. It then walked back into the woods and the girls ran home.

Source: BFRO website. One of the witnesses was interviewed by Cindy Dosen.

1974/07/00 – Harrison Hot Springs. Sasquatch Seen at Youth Camp: Wayne Jones, a well-known Vancouver businessman and director of a youth camp outing, reported that he saw a sasquatch near Harrison Hot Springs at Stokke Creek (37 miles up the east side of Harrison Lake) on July 18, 1974. He said he is now a firm believer in the creature.

Jones was sitting alone by the campfire when he saw the oddity walk around the corner of a camp building and stop about 20 to 25 feet from his position. Jones stated:

> I had a good 4 or 5 minutes look at it and it was definitely more human than animal, I remained still, but it soon saw me and would look at me every few seconds. It had a rounded head and long arms and legs. It was covered with dark hair—possibly black—and it had more hair on its arms and legs than on its chest. It was definitely hair and not fur. Half of it was wet as though it had just walked through a creek and it was muddy down one side. It was licking mud off its fingers.

The creature stayed until several other campers arrived with flashlights, at which time, "It took off, using its hands to pull its way through the bushes. It didn't run, but it sure got out of there fast, and it obviously knew where it was going," Jones said.

It was later determined that the top of the creature's head reached the eaves on the building. This made it about 9 feet tall.

The camp building at Stokke Creek as it presently appears. Wayne Jones said that the top of creature's head reached the eaves of this building.

Furthermore, a few days later while Wayne Jones was away, 15-year-old Claire Nicol says she saw the creature at about 11:30 p.m. She said it cross the road about 50 feet in front of her, "It was dark and all I could see was a silhouette, but the thing was 8 to 9 feet tall and he definitely wasn't a bear," she stated. That same night, about 5 minutes later several other campers said they also saw the oddity.

Source: Don Hunter with René Dahinden, 1993. *Sasquatch. The New American Library, Inc.,* New York, New York, p. 196, from "Director Sees Creature: Sasquatch visits youth camp," the *Sun,* Vancouver, BC, July 23, 1974. Also Moira Farrow, July 24, 1974. "Sasquatch hunters unwelcome," the *Sun,* Vancouver, BC. Also Fred Curtain, 1974. "Face to Face with Sasquatch," the *Province,* Vancouver, BC, July 23, 1974. Also, John Green, Sasquatch Incidents Data Base, Incident Number 1000182. (Photos: Camp building, T. Steenburg; Claire Nicol, the *Sun,* Vancouver.)

Claire Nicol

1974/07/00 – Powell River. Pushing the Trees & Branches: Patrick Kennedy, a teenager, reported seeing a strange animal in the Powell River area in July 1974. Kennedy was babysitting at an isolated house in a wooded area at the back of the Yukon gravel pit. At about 6:00 p.m. he heard something moving in the bush behind the house. Fifteen minutes later there was a loud rustling noise down the road about 100 yards away. He went to the area of the noise with the homeowner's dog and saw that the noise was coming from around the corner of a stand of small alders where there was a wild blackberry bush.

Kennedy threw a rock in the direction of the noise and it stopped. He then yelled, and an 8- to 9-foot-tall creature stood up and walked out of the bush, looking straight at him. The dog walked forward about 10 feet and stopped dead. The creature turned and walked into some small alders about 40 feet away, pushing the trees and branches out of its way.

Kennedy ran to the house and watched from an upstairs window as it moved about 200 yards into the bush, again pushing branches out of the way as it went. He said the creature was really heavy, had long arms, and was covered with long, dark-brown hair with lighter brown ends. It had a flat face and no neck.

Source: John Green, Sasquatch Incidents Data Base, Incident Number 100018.

1974/07/00 – Okanagan Lake. Standing On the Patio: An 8-year-old boy living in a house that overlooked Okanagan Lake reported that he saw a strange creature in July 1974. The boy awoke at about 5:00 a.m., and saw the oddity standing on the house patio looking towards the patio doors. It just stood there motionless with its arms hanging by its side. From the profile, the body appeared thick. It was tall enough that it was stooped under the patio eave. He estimated its height at 8 feet.

The boy described the creature as being hair-covered and dark in color. The hair was long on its body and shorter on its head. The head was cone shaped or sloped towards the back. After about one minute, it turned around 180 degrees and walked away from the house towards the lake.

Source: BFRO website. The witness was interviewed by Blaine McMillan.

1974/SU/00 – Hope. Inside the Town Limits: Two young women believe they saw a sasquatch in the town of Hope—actually within the town limits—during the summer of 1974.

When the sun had gone down, the women, who were visiting the area with their boyfriends, strolled down to the banks of the Fraser River. One of the women saw what she thought was a man sitting on a large log watching them. She asked her friend, "Who's that?" The other girl looked and exclaimed "My God, that's a sasquatch!"

They called to their boyfriends, who were back at their car, and the "sasquatch" stood up. The women then raced back to the car. One of them stated she couldn't tell how tall the creature was, but that when it stood up, it appeared taller than anyone she knew. She said that it was covered in hair that looked brown in color. She got the impression that it didn't want to have anything to do with them.

Source: Thomas Steenburg, 1990. *Sasquatch/Bigfoot: The Continuing Mystery.* Hancock House Publishers Ltd., Surrey, BC, pp. 70, 71.

1974/08/25 – Chilliwack. Seen On a Rocky Cliff: Edward Vaughan and Donald Nikkel reported seeing a strange creature near Chilliwack from their private plane on August 25, 1974. The men were about 16 miles east of Chilliwack and saw the oddity on a rocky cliff, 50 feet or so in height. It was standing about 30 feet from the edge of the cliff, and they flew past it at about the same level at a distance of 200 to 250 feet. Vaughan said that there are no roads or trails in the area.

They described it as like a man but much larger (about 7 feet tall), covered in chocolate brown hair, and having long arms. Nikkel said he recalled that it had a white patch on its

Edward Vaughan (Top) and Donald Nikkel.

right side. Vaughan said it looked very much like the creature in the Patterson/Gimlin film.

The creature appeared to have looked at the plane and then apparently walked into the trees. They did not see it walking, but were positive it was standing upright when they saw it. They circled back to have another look but by this time it had disappeared. Both men were reasonably sure they saw a sasquatch.

Source: Thomas Steenburg Sasquatch Incidents file, Number BC 10140. (Photos: T. Steenburg.)

1974/09/00 – Harrison Hot Springs. Went Into Shock: A woman visiting from Sedro Wooley, Washington with her husband was shocked at the sight of a strange creature near Harrison Hot Springs in September 1974. She had stayed in the car while her husband got out to look at what he called "beaver ponds" with jumping trout seen from the road (near Hicks Lake). The woman saw a large erect creature in the brush near the car and went into shock. She estimated that it was about 6 feet tall. The actions of the couple's dog indicated some disturbing animal was nearby.

The incident was investigated by researchers; however, the description of the location given by the couple did not match well with either the beaver pond near Hicks Lake nor with Trout Lake which is in the same area.

Source: John Green, Sasquatch Incidents Data Base, Incident Number 1000183.

1974/09/00 – Cougar, Washington, USA. Canadians Meet With Morgan: Noted Canadian sasquatch researchers John Green and René Dahinden met up with their esteemed US counterpart, Robert Morgan, in Cougar, Washington, in the summer of 1974. Morgan and his partner, Ted Ernst, formed The American Anthropologi-

(Left to right) Robert Morgan, René Dahinden and John Green, 1974.

cal Research Foundation in 1972 which launched 4 expeditions to find the sasquatch. Two of these were co-sponsored by the National Wildlife Federation.

Source: Communication with Robert Morgan. Photo: J. Green.

1974/10/00 – Mackenzie. Took a Shot Over Its Head: Barry Jones, a hunter, believes he saw a sasquatch in an area southwest of Mackenzie at about 2:00 p.m. during the last week of October 1974. He originally thought it was a stump or a bear, but as he got closer, he realized such was not the case. He fired a shot over its head and it turned and ran down a steep ridge. He recounted the incident as follows:

> We [Jones and his uncle] had gone up to the firetower, which is quite high, and I remember it was a windy day. We drove back down to the first plateau and stopped. We were bear hunting. That's why I thought it might be a bear. I was walking along the ridge and my uncle, he went further down the hill. I was walking north, northeast, kind of looking around, and at first I noticed something. Well, it looked like a big, black stump standing against another tree, and as I was walking along, it moved and light came between it and the tree. Again, I thought maybe it was just the wind moving the tree back and forth. As I got closer to it, I thought I saw it shift again. Well, I brought the rifle off my shoulder thinking maybe it's a bear, and that's when it moved back in behind the tree. That's when I realized that this was no bear. I started walking towards it, and it moved again, farther behind the tree. And that's when I brought my rifle up and took a shot over its head. The thing turned and ran and it was moving very fast.

The witness was thoroughly interviewed by Thomas Steenburg. When the creature was first noticed, the hunter was between 500 and 600 feet away. As he moved closer he saw that the creature was standing upright and looking at him. "It looked like it was shifting from one foot to the other." The hunter was about 200 feet away when he fired his rifle. He described the creature as being covered in black hair, and very tall (well over his own height of 6 feet 4 inches). He could see its eyes, which he said were "mostly white. It had very big arms, "His right arm came around, and I bet it was bigger than my upper leg. His bicep was bigger than my leg...and I have big legs," Jones stated.

After the creature had gone, Jones went to where it had ran down the ridge. He stated the following:

> I ran up to where I saw him and looked over the other side of the ridge down the hill. I couldn't see it anywhere. One thing that surprised me was that there was no slide marks, so that means it just ran down that hill. There were no places where he had skidded or fell, like a man would running as fast as he could down a steep hill.

Jones told his uncle of the sighting and they went to where the creature had been and looked for footprints, but none were found.

They reported the incident to the RCMP. The officer that took the report chuckled to himself and said, "Well, we don't believe in them here."

Source: Thomas Steenburg, 2000. *In Search of Giants: Bigfoot/Sasquatch Encounters*. Hancock House Publishers Ltd., Surrey, BC, pp. 81–84. Also John Green, Sasquatch Incidents Data Base, Incident Number 1000185.

1974/10/00 – Princeton. King Kong or Something: Two brothers reported that they saw an ape-like creature in Princeton in late October 1974. The boys were jogging around the Princeton airport. It was about 8:00 a.m., and after jogging most of the 3-mile distance, they decided to walk the last 200 feet or so. At this point they were approaching the bridge that spans the Similkameen River. At about 150 feet in front of them they saw some sort of black or brownish creature on 2 legs crossing the road. They stopped and watched as the creature went up a bank and onto a dirt road. The boys decided to follow the oddity and noted that its head brushed against tree branches that overlapped the road. They did a rough measurement at a spot and estimated that the creature had to be 7½ to 8 feet tall. They said it appeared to be very heavy and guessed at 400 to 500 pounds. They noted that its arms were very long, "past the knee area."

The bank that the creature ascended and the dirt road. Its height was estimated based on trees that overlapped this road.

One of the boys yelled "Hey!" at the creature and it turned around, "...and it was just like you were seeing, I don't know, King Kong or something. It was definitely an ape-like face."

As the creature was heading into a bushy area, the pair decided not to pursue it any further. They went back to where it had climbed up the bank and found footprints. They measured one print and determined it was 13 inches long. They said that the foot that made the print was definitely a bare foot—no shoe marks.

Source: Thomas Steenburg, 2000. *In Search of Giants: Bigfoot/Sasquatch Encounters.* Hancock House Publishers Ltd., Surrey, BC, pp. 85–91, also John Green, Sasquatch Incidents Data Base, Incident Number 1000184. (Photo: T. Steenburg.)

1974/YR/00 – 100 Mile House. Crouched & Looked Back Down: A man reported that he saw a sasquatch at 100 Mile House during 1974. The creature walked across the road in front of the man near what he said was the Gang Ranch. It went up the bank on the other side of the road. It then stood semi-crouched and looked back down at the man. The man also noted that it was on 2 legs as it went up the bank, so it appears it was also walking erect the whole time it was in view. Where it went up the bank it left tracks 20 inches long.

Source: John Green, Sasquatch Incidents Data Base, Incident Number 1000237.

1974/YR/00 – Castlegar. Not At All Bear-like: A young couple driving the Castlegar-Silverton Highway reported that they saw a huge figure standing at the edge of the road as they came around a corner in 1974. They said it was covered in dark brown or black hair and had its arms clearly hanging by its side. Its head was well-rounded and not at all bear-like. The creature just stood perfectly still as they passed by. The couple did not stop or go back, thinking that it was best left alone.

Source: John Green, Sasquatch Incidents Data Base, Incident Number 262 from information provided by the witnesses to the Bigfoot Information Center in The Dalles, Oregon.

1974/YR/00 – Terrace. Odor Like a Camel: A prospector reported that he saw a large bipedal animal near Terrace at Lorne Creek in 1974. He spotted what appeared to be a moving patch of brown a short distance away. The prospector moved closer and saw that the creature had its back to him and was standing on 2 legs. It was eating saskatoon berries, using its front paws like hands. Moving closer, to about 50 feet, the prospector saw that it was some kind of hair-covered human, a little over 6 feet tall. He could now see that the paws were hands. He now sensed a strong odor that reminded him of the smell of camels that he experienced when he once worked in a circus. He watched the creature for about 10 minutes, after which it apparently sensed that it was being observed. At that point it turned sideways and the prospector saw that it had breasts so was obviously a female.

Source: BFRO website, from "The Sasquatch in Northern British Columbia, Canada," HBCC UFO Research, April 07, 2001.

1974/CI/00 – Sicamous. Snow Angels: A man and his girlfriend out cross-country skiing near Sicamous at Shuswap lake reported odd and amusing findings in December 1974 (or 1975). They were first surprised to see 2 sets of human-like footprints in the snow. One set was larger than the other and they appeared to

A child making a "snow angel."

have been made earlier in the day. The prints in both sets were considerably larger than the man's boot prints. However, the space between the prints was not much larger than his prints.

The couple followed the prints for about one-half mile and came to a spot where it appeared that whatever had made the prints had laid down in the snow and made "snow angels" like children do, i.e., lay in the snow on their backs, wave their arms up and down, and move their legs from side to side.

The man laid down in both impressions to see how he would "measure up." He was 5 feet, 10 inches tall and weighed 190 pounds. He could not reach across the smaller snow angel, and the "arms" of the larger were at least 12 inches beyond his arms. The "body" in both imprints was much wider than his body, and he could not reach the farthest part of either imprint with his feet. The man estimated that the smaller print was at least 6.5 feet long, and the larger 8 feet. The pair looked for hair in the imprints but did not find any.

The couple then continued following the footprints, which went a considerable distance down to the beach where there was a dock. They found a large fish that had several bites taken out of it and was then left in the snow. The fish was not yet frozen, so had not been there very long. As it was getting dusky, they left the area and went back to their resort.

It is likely that the sasquatch got the idea of making "snow angels" from watching children playing. We might also reason that the small pace for the prints was the result of the larger creature simply taking shorter paces in order for the smaller to stay by its side.

Source: BFRO website. (Photo: Snow angel, Wikipedia commons; credit: http://www.flickr.com/photos/kerys/111308573/ Kerys, see the Wikipedia website for further details.)

1975/01/00 – Kamloops. Three Teens See Sasquatch: Teenagers reported seeing a sasquatch on a hillside north of Kamloops in late January 1975. Two of the boys, Shawn Olszewski (13) and Ken Spivak (12) said they saw the creature while tobogganing on Saturday, January 25. The creature simply ran away, but then later reappeared about 10 yards from the boys. "I saw the creature make a hole in the brush with its hands," said Olszewski. I ran to the bottom of the gully because I was scared of it.

On Sunday the boys went back to the area with another boy, Rodney Hopkins (14), and all of them spotted the creature.

The boys found footprints about 16 inches long and 7 inches wide. They were about 9 feet apart in an area where the creature walked downhill, and about 4 feet apart on level ground. Olszewski said the creature was about 9.5 feet tall, and looked to weigh 400 pounds.

An experienced animal tracker was to visit the area on Monday, January 27.

Source. "Sasquatch spotted by teenagers." The *Sentinel*, Kamloops, BC, January 28, 1975.

1975/06/20 – Cherryville. Crossed 10 Feet in Front of Witness: Wilfred Morris had a close encounter with a strange creature in Cherryville on June 20, 1975. Morris was walking home when a huge, hair-covered creature crossed the road 10 feet in front of him. It covered the distance in 4 steps, and went up a high bank. Morris described the creature's height as, "tall as a man with someone riding on his shoulders."

Source: John Green, Sasquatch Incidents Data Base, Incident Number 1000187.

1975/07/00 – Cranbrook. Strange Footprints: A young boy reported that he found a series of strange footprints near Cranbrook along the Joseph River in July 1975. He was on a fishing trip and had wandered a short way down the river and then traveled up into the tree line due to the high water level. As he walked along the wet ground, he looked down and saw large tracks that looked like bare human feet had made them, but they were about 15 inches long. The prints sank into the ground about one inch.

He followed the prints and they led down to the river bank. At the water's edge, whatever made the prints appeared to have stopped and squatted as there there were 2 prints side-by-side with toes pointing to the river.

Source: Thomas Steenburg, 1990. *Sasquatch/Bigfoot: The Continuing Mystery*. Hancock House Publishers Ltd., Surrey, BC, pp. 69, 70.

1975/SU/00 – Lake Cowichan (VI). Threw a Rock to Get Its Attention: Julius Szego and his cousin, Edward, both around 15 years old, reported seeing a sasquatch near Lake Cowichan at Skutz Falls (Cowichan River Provincial Park) in the summer of 1975.

They first saw the creature crouched at the edge of the water, about 40 feet away, pulling up roots. It was on the same side of the falls as they were. They said it was brownish black in color and had "mossy" hair on its body. There was no hair on its face, which was human-like with a flat forehead. The boys watched it for about 30 seconds, and then threw a rock towards it to get its attention. The creature stood up, grunted, and simply walked off. The boys estimated that it was about 7 feet tall.

They left the area and reported the incident to the RCMP and park rangers. Officials went to the site but did not find anything.

Source: Thomas Steenburg, 2000. *In Search of Giants: Bigfoot/Sasquatch Encounters.* Hancock House Publishers Ltd., Surrey, BC, p. 76. Also John Green, Sasquatch Incidents Data Base, Incident Number 1000185

1975/08/00 – Strathcona Provincial Park (VI): Black Human-like Form: Robert Alley of Ketchikan, Alaska reported an encounter with a sasquatch in Strathcona Provincial Park in August 1975.

Alley was camped with 3 friends. On the first night at around 1:00 a.m. he was awakened by what sounded like a piece of gravel dropping off the side of his tent. He peered out but did not see anything. The same thing occurred again and 3 more times over a 20-minute period with no other unusual sounds.

Robert Alley

Alley grabbed his penlight and ventured outside, half thinking some teenagers had wandered upon the campsite and were playing a prank. He shone his penlight toward the dark mass of hemlocks and saw a tall (over 6 feet), black, human-like form (like a burnt stump) standing at the edge of the trees. The creature immediately charged off into the second growth and in a few seconds was no longer visible. The commotion woke his friends, who quickly assembled and asked what was happening. Alley calmly told them he had disturbed an elk while attending needs, rather than upset them with what he had actually seen.

Source: Christopher L. Murphy, 2010. *Know the Sasquatch: Sequel and Update to Meet the Sasquatch.* Hancock House Publishers Ltd., Surrey, BC, p. 262. (Photo: R. Alley.)

1975/09/00 – Colwood (VI). Reflecting Yellow Eyes: Youths (ages about 16 to 20) reported to the RCMP that they saw a strange creature near Colwood on the evening of September 10, 1975. Some of the group saw the creature before dark on the semi-open hillside above the E & N Railroad just east of Humpback Road (near Goldstream Park). Others saw it there shortly after dark. The descriptions given for the oddity were that it was a dark-haired creature, 8 to 12 feet tall (later 9 to 10 feet tall), with reflecting yellow eyes. It moved very fast on its hind legs, hid when it saw the teenagers, came out of hiding and then hid again. More youths came out the next night when the news evidently got around. Some youths climbed a hydro tower, some sat in a car. The creature was said to have chased a pickup truck with youths riding in the back. None of the witness could provide a detailed description of the creature.

Source: "Unidentified hairy object haunts Humpback area." The *Colonist,* Victoria, BC, September 14, 1975. Also, John Green, Sasquatch Incidents Data Base, Incident Number 1000190 and 1000191.

1975/09/00 – Marble Canyon. Pronounced Long Neck: A man driving between Marble Canyon and Eisenhower Junction, (Alberta) reported that he saw a sasquatch during September 1975. He provided the following account:

> The light was good and there was not yet any snow. There in a clearing about 15 feet from the trees stood a creature about 7 feet tall. It was rusty tan in color and was covered with hair. It stood erect and had long arms which hung down at the sides. It had an ape-like face with no prominent ear features, but the whole body was very slim, and it had quite a pronounced long neck. I know that this is in contrast with the general concept of sasquatches which are usually observed a being dark in color with heavy bodies, and a head setting into the shoulders, with virtually no neck.

The creature slowly backed into the timber with a kind of half-a-step, full-step motion, and then disappeared in the trees.

Source: Ron Ede, 1977. "Man Claims He Saw "Bigfoot." The *Valley Echo*, Lake Windermere, BC, April 7, 1977. Also John Green, Sasquatch Incidents Data Base, Incident Number 1000189

1975/09/00 – Kimberley. Did Not Make Any Noise: Rene Phillips, reported that he saw a sasquatch in the Kimberley area in late September 1975. He said the creature was from 7 to 10 feet tall. It was ebony in color, and did not run like a human, but shuffled its feet slowly—although it moved quickly. It did not make any noise. It appears other people were with Phillips at the time, but no details were provided.

Michael McLelland, another Kimberley resident, said he also saw a 2-legged animal that was 7 feet tall. It was black with a tan colored chest. He said it ran like a human, but much faster. Again, it did not make an noise.

The time frame for the second incident is not certain, but appears to have been directly following the Phillips sighting.

Source. Jack Holmes, 1975. "More Sasquatch Sightings?" The *Daily Bulletin*, Kimberley, BC, October 6, 1975.

1975/10/00 – Princeton. Came Within 12 Feet: Frederick Kirkwood and his wife Millie came very close to a strange creature near Princeton in October 1975. The couple were traveling east on the Hope-Princeton Highway at 2:30 a.m., 2 miles west of the Allison Pass works yard. When they rounded a curve they saw a dark creature come up onto the road about 100 yards ahead. Mr. Kirkwood slowed their car almost to a stop while the creature crossed the road angling towards the car. It came within 12 feet of the car, looked straight ahead and kept going.

The Kirkwoods said that its eyes looked black and did not reflect. Its face was light brown and it looked human. It was covered in hair about 1.5 inches long, light in color "like a dark golden retriever." It did not have prominent ears. It had hair on the backs of its hands, but not on its feet, which were shiny black and looked wet. They estimated that it was 5 feet 8 inches tall. It was very heavy, thicker from back to front than a man. Its arms were of normal length. Its head hair was the same length as that on its body. Hair came down onto its forehead but not on its face. Its nose was human-like, and its neck was short, but not exceedingly so.

Source: John Green, Sasquatch Incidents Data Base, Incident Number 1000192

1975/10/17 – Agassiz. Ran On All Fours: C.L. Butler reported seeing a strange animal in the Agassiz area at Seabird Island at about 6:45 p.m. on October 17, 1975. He said that while he was driving east in a heavy rainstorm, he saw the creature run across the Lougheed Highway, three or four miles east of the Agassiz overpass. It was covered with long, reddish-brown hair, and ran on all fours far faster than a human could run. In Butler's own words it was "just flying." It seemed to run on its hind legs after crossing the road.

Butler estimated its standing height would have been 6 to 7 feet. Its "back legs" were 3.5 to 4 feet long. Butler insisted that it definitely was not a bear, saying that its legs and feet were big, and hairy right to ground. He noted that the soles of its feet were pink.

Source: John Green, Sasquatch Incidents Data Base, Incident Number 1000193. Butler telephoned in his report to a local radio show (Pat Burns show).

1975/11/11 – Sooke (VI). Long Line of Tracks: John Barr and a friend reported finding unusual tracks in snow near Sooke at Tugwell Lake on November 11, 1975. The story, which follows (edited for this work), is both interesting and amusing as it expresses the surprise and feelings people experience when they find large human-like tracks in the wilderness.

It was a typical winter day in Victoria, dark and rainy. I was 20 years old and had a job as an equipment operator running backhoes and excavators in the construction industry. We put in all the underground services for new housing subdivisions complete with roads. The rain proved too much for us to work, so we were sent home in the morning. One of my co-workers was my hunting partner and we decided we would go home, get our rifles, and hunt for the afternoon.

I had hunted from the age of 16 and enjoyed being in the outdoors as much as possible. We hunted for deer and had a favorite spot that we had hunted for years. I really didn't care if I saw a deer because I just loved being out in the wilderness.

At the time I owned a 1969 "beetle" that we used for going hunting. It proved to be an excellent car and could almost go anywhere we wanted, and was good in the snow.

We headed out of Victoria to Sooke, which was on the West coast of Vancouver Island, about a 40-minute drive. Sooke was a small logging community at the time, and the surrounding area was a favorite spot

John Barr with the sasquatch sculpture located on the Crowsnest Highway, just east of Osoyoos.

for many hunters. The main logging road in the area was called Butlers Main. Once on this road, it was exactly 11.5 miles to where we hunted.

On the left side of Butlers Main, there was a small spur road that climbed up a hill to a very small lake at the top— this was our spot, and it was perfect. We had been successful there many times before.

It had snowed the night before, so when we arrived the ground was covered with about 2 to 3 inches. This was perfect for hunting because if there were deer around, and they had crossed the road, they left tracks.

We parked at the bottom of the spur road, grabbed our rifles and started to climb to the top of the hill to hunt. It was cold enough that you could see your breath. The snow was crisp and easy to walk in.

As we walked up the incline, I continually scanned the

timber line and the road in front of me. There was no evidence of tracks of any sort. When we arrived at the top of the hill, we split up to go around the lake. My partner went to the right and I went to the left, just as we had done many times before. We had agreed to meet back at where we started at a certain time. This gave us at least 3 hours to hunt and get back to the car before sunset.

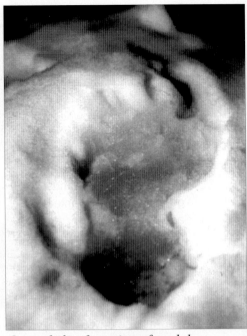

I was the first one to get back to the meeting place, so sat and had a cigarette while waiting for my partner to return.

One of the footprints found by John Barr and his partner.

As I waited, I sensed an odor that I thought was a bear. I fully expected to see one come over the crest of the hill. A short time later, my partner arrived and after some conversation, we headed back to the car.

As we walked down the hill, I noticed not only our tracks in the snow, which we made walking up, but other tracks as well. We stopped to take a closer look. The tracks were about 17 inches long and were made by someone walking in bare feet! The hair on the back of my neck stood up, and the fear of God went through me. We both wanted to get out of there as fast as possible.

Whatever had made those prints was much larger than a man, and besides, what kind of an idiot would be walking around in the middle of nowhere bare-footed in snow? There was no doubt in our minds that whatever had made those prints was not human.

Whatever it was, it had walked up the road after we had

almost reached the crest of the hill. It then turned around and headed back down the road. Perhaps it had heard us talking, or smelled my cigarette, I really don't know. However, it turned around and went back down the hill, possibly to stay clear of us.

We both reloaded our rifles and continued the decent to the car, carefully watching the timber and the road ahead of us. We were ready if anything came out at us.

We were both relieved to see the car at the bottom of the hill and couldn't get into it fast enough. I remember thinking that I sure didn't want to come face to face with whatever had made those prints. I even thought that if 3 or 4 men could roll over a '69 "beetle," there is no telling what that thing could do to my car and us.

My poor little car never worked so hard getting us out of there. Both of us have never been so relieved as when we saw the main highway intersection ahead of us.

I dropped my partner off at his house, and told him I would see him at work the following morning. I then headed home to my wife and a good home-cooked meal. During dinner, of course, I was asked about what had happened on our hunting trip.

I was a little apprehensive about saying anything. Who would believe what I said? It would be like telling someone you saw a flying saucer. However, there was a difference, the prints we had seen were physical evidence. I had, of course, heard of Bigfoot, but all it was to me at the time was an Indian legend—a story.

Nevertheless, I relented and told my wife of the events. I then decided to call and tell a friend of ours who was getting his teacher's degree at the University of Victoria. He was an intelligent and reasonable person—someone we respected, and an all-around nice guy.

I think he was more excited than I was about the footprints. He told me that he and I should go back out there as soon as possible (before it rained) and make a cast of one of the prints. He said he had a box of poly filla and would bring warm water. We needed flashlights, a camera with a flash, and also a Coleman lantern.

I was a little nervous about going back out there, but if I wanted proof, it had to be done. I telephoned my partner and told him what we were going to do. He agreed to come with us. I told him to bring his rifle, and said I would bring my shotgun. I certainly did not want to go back out there without some protection.

We all met at an agreed location and set out in my friend's 4x4. The bush looked very different at night, and I could taste the fear that was brewing in my stomach. It took a lot of courage to be out there at night.

We arrived at the spur road and walked to where the footprints were. I took a photograph of a left footprint which was pointing uphill. We then made the plaster and poured it into one of the prints. Once the plaster had set, we pulled out the cast, wrapped it in an old blanket and then headed for home. I was again relieved to get out of the area.

Footprint cast made by John Barr and friends.

Some weeks later, I made 2 copies of the cast, one for each of my friends. I then decided to do some research on Bigfoot. I purchased books by John Green and found that my partner and I were certainly not the only people to find unusual footprints in wilderness areas.

I contacted John Green and he came to see me. I made another copy of the cast and gave it to him. I did not tell anyone else of this experience until 2007. I figured that after 30 years, it was time to share with all the "believers" out there.

Source: From a written report provided by John Barr. His cousin, Alex Solunac, told me about the finding and arranged to get me the full account, which was provided in June 2011.

1975/YR/00 – Natal. Seen Near Bridge: Joe Lastwick reported seeing a sasquatch in the Natal area (Indian Prairie) in 1975. He saw it near the bridge crossing the Fording River and the CNI road which leads to Grave Lake.

Source: "No Optical Illusion. Sasquatch sighted near Harmer Ridge." The *Free Press,* Fernie, BC, April 12, 1976.

1975/CI/00 – Telegraph Cove (VI). Stones Scare Off Boys: Two young boys reported a strange occurrence on the coast near Telegraph Cove in about 1975. While the boys were climbing a leaning totem pole in an abandoned First Nations village, they were showered with small stones from the adjacent forest. The stones were small, less than an inch in diameter, but were enough to make the boys leave the area. They did not see the stone thrower, but reports of sasquatch in the region might indicate that the creature was involved.

Source: John Bindernagel, 1998. *North America's Great Ape: The Sasquatch.* Beachcomber Books, Courtenay, BC, pp. 116, 225, from information provided by Tom Sewid of Telegraph Cove.

1975/CI/00 – Kingcome Inlet. Unusual Nest: A prospector stated that he found an unusual nest structure at Kingcome Inlet (Wakeman Sound) in about the mid 1970s. The structure was located in the center of a clump of small mountain hemlock trees. Branches had been woven in and out around the outside rim of the nest.

Source: John Bindernagel, 1998. *North America's Great Ape: The Sasquatch.* Beachcomber Books, Courtenay, BC, p. 69, 214.

1976/01/22 – Victoria (VI). Shoulders 3 to 4 Feet Across: Harvey Maser and Jack Phelps reported seeing an odd creature near Victoria on January 22, 1976. The men were traveling on Metchosin Road near Duke Road at about 6:25 a.m. The car headlights illuminated a creature about 100 feet ahead walking beside the road. The men estimated that it was 8 to 9 feet tall and 3 to 4 feet across the shoulders. It seemed to walk stooped over. It was covered in hair 4 to 5 inches long. Upon being seen, it quickly turned right and disappeared in the bushes. A search was later undertaken for footprints, but none were found.

Source: John Green, Sasquatch Incidents Data Base, Incident Number 1000194.

1976/04/02 – Natal. Big Monkey: A partsman for a private contractor reported that he saw a strange creature near Natal close to the Indian Prairie area on April 2, 1976. He spotted it on the road to the Fording Coal operation at about 8:30 a.m. He referred to the creature as a "big monkey." "I saw it," he said, "I saw it and it wasn't an optical illusion. I saw it. I know I saw it and I swear I saw it." He went on to state:

> If only I had a camera with me. From now on I'm going to have one in the pick-up, for I'm sure the big monkey lives in this area somewhere, and I know darn well that he or it will be back.

He described the creature as having "ape-like" proportions, about 7 feet tall, and brown in color. He said that it "walked with a gait." He continued:

> It didn't seem startled when it spotted me. It just looked at me, grunted, and headed for the boondocks with a fairly slow and measured pace.

Source: "No Optical Illusion: Sasquatch sighted near Harmer Ridge. The *Free Press,* Fernie, BC, April 12, 1976. Also John Green, Sasquatch Incidents Data Base, Incident Number 1000195.

1976/04/03 – Harmer Ridge. Webbed Feet: Jim Van Maarion, a truck driver, reported over his radio that he saw a strange "monster" walking over one of the dumps on Harmer Ridge on April 3, 1976. He said another man on the hill also saw it. "It's hard to explain what it looked like," Van Maarion said, "because it was crouched over. Later, I looked for footprints and found some which appeared as though the creature had webbed feet."

Source: "No Optical Illusion: Sasquatch sighted near Harmer Ridge. The *Free Press,* Fernie, BC, April 12, 1976. Also John Green, Sasquatch Incidents Data Base, Incident Number 1000195.

1976/05/00 – Wycliffe. Taller Than the Horses: Gale Shay of Nelson reported seeing a huge sasquatch in the Wycliffe area during May 1975. Shay said she saw the creature running across a field about 100 yards from her home. She said the horses in the field panicked and galloped off. The sasquatch simply loped past them, arms swinging, without apparent effort. Shay said that the creature was taller than the horses, and that it was in sight for 2 minutes. She continued watching the creature as it ran off up a butte.

John Green, Sasquatch Incidents Data Base, Incident Number 1000196.

1976/06/11 – Grand Forks: Kids See Sasquatch: Several boys in a group of about 70 children out camping reported that they saw a sasquatch near Grand Forks on June 11, 1976. The campers, aged 9 to 11, had been stirred to frantic excitement during the afternoon of the sighting, causing some concern. The boys took adults to the spot where the creature was seen and pointed out footprints as evidence of their claim.

Source: "Fact or Fiction." Grand Forks Gazette, June 16, 1976.

1976/06/00 – McBride. Over a Dozen Witnesses: Over a dozen people are said to have witnessed a large black creature run on 2 legs across the road in front of their vehicle near McBride in June 1976. The oddity simply disappeared in the woods on the other side of the road.

Source: BFRO website.

1976/07/17 – Terrace. Major Track Finding: Young boys playing near Terrace close to the Skeena River reported that they found giant, human-like footprints on July 17, 1976. Sasquatch researcher Bob Titmus was contacted and he went to the location. He photographed and made plaster casts of some of the prints.

Titmus reported that the tracks crossed over a pile of stumps and root systems near a slough just off the Skeena River. The tracks were 3 or 4 days old and had been exposed to heavy rain for a couple of days. The tracks measured 15.5 inches long, 6 .5

Bob Titmus, 1990s.

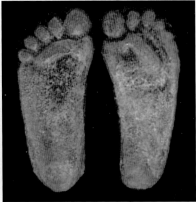

(Left) One of the Skeena River prints in the soil. (Right) copies of casts (right and left foot) made from the Skeena River prints. This set of casts is the best ever obtained.

inches wide at the ball and 4 inches wide at the heel. The walking stride from toe to heel was 78 inches. The heel depth was approximately 1.63 inches, and the toe depth approximately. 1.13 inches. Five casts were made in all of the 12 or 15 tracks found.

The following year in the same area, Titmus found many branches in the adjacent bush up to the size of his wrist that were broken off or twisted off about 6 feet above the ground.

Bob Titmus (left) and other researchers on location at the site where footprints were found on July 17, 1976.

240

Source: John Green, 1981. *Sasquatch: The Apes Among Us.* Hancock House Publishers Ltd., Surrey, BC, pp. 364, 365. Also John Bindernagel, 1998. *America's Great Ape: The Sasquatch.* Beachcomber Books, Courtenay BC., p. 76. (Photos: R. Titmus portrait, J. Green; Print in soil, R.Titmus; Casts, C. Murphy; Group at site, R. Titmus.)

1976/07/00 – Bella Coola. Sasquatch Follows Witness: A man reported seeing a sasquatch in the Bella Coola area during the first week of July 1976. He said the creature followed him down a river. He went back to where it had been and found footprints 16.5 inches long. The sighting was to be investigated by members of The North American Wildlife Research Association (NAWRA) which was doing research on sasquatch in that area.

Source: Scott McClean, 2005. Big News Prints, (self-published), p. 54, from "'Bigfoot' Reported in B.C." The *Union-Bulletin*, Walla Walla, Washington, July 7, 1976.

1976/09/03 – Kimberley. Sasquatch Chased, Followed by Car: Mickey McLelland, a member of the Kimberley Fire Department, reported that he, a friend, and several other witnesses saw a sasquatch in a field about 9 miles from Kimberley on September 3, 1976 (near Kimberley Trap and Ski Club). McLelland described the creature as about 7 feet tall, black with a tan-colored chest, and long arms.

McLelland and his friend got out of their car and chased the creature. At this time it was about 125 yards away. Unable to catch up, the men returned to their car and followed it by car driving on the highway. They caught up with it, and clocked it about 50 miles an hour. The creature then veered off into the forest.

Two other cars were parked at the roadside with 6 or 8 witnesses in all. They watched for perhaps 3 or 4 minutes before the creature started to run very fast.

Source: ""Sasquatch Active in East Kootenay," the *Daily Freepress,* Nanaimo, BC, September 4(?), 1976; "Sasquatch legend fueled by multisightings in BC," the *Herald*, Bellingham, Washington, USA, September 9, 1976; "Sasquatch sightings reported," *The Globe and Mail (?),* Toronto, September 10, 1976; "Sasquatch sighting has Kimberley wondering if it's been ad-hoaxed," the *Sun,* Vancouver, BC, September 10, 1976; "Tall, tan, hairy," the *Province,* Vancouver, BC, September 10, 1976; "Many Sasquatch sighted in BC", *The Associated Press* (newspaper not known), September, 1976, "Sasquatch haunting area?", the *Daily Bulletin,* Kimberley, September 10, 1976. Also John Green, Sasquatch Incidents Data Base, Incident Number 1000197.

1976/09/03 – Wycliffe. Polaroid Photo Taken: Barbara Pretula, a merchant, reported that she saw a sasquatch behind her store in Wycliffe on September 3, 1976. She said it was black with a light stomach and stood 6 to 7 feet tall. She managed to get her camera and take a polaroid photograph of the oddity. The camera flash frightened the creature away—it ran into the bush. "I was about 15 feet away from it when I took the picture," Pretula said, "It didn't see me come up, I snuck around beside the building. It was just looking around while down on its hind legs."

Although the photograph came out quite blurry, she said it still gives a good indication of the animal's size and shape.

This sighting took place about 15 minutes or half an hour after the same or a similar creature had been sighted and chased by Mickey McLelland, a fireman, and his friend (see previous entry).

An RCMP officer, Constable Hugh Steward, who saw the photograph taken by Pratula said, "It's a good photo, though a little blurred." The photograph was loaned to Roy Fardell of Cranbrook.

As far as I know, the photograph has never been made public. I don't know of any sasquatch researcher, other than Fardell, who has seen the photograph. I do not personally know Fardell, but he was a good friend of René Dahinden. I am sure he would have seen it. However, René never mentioned anything to me about the photograph.

Source: ""Sasquatch Active in East Kootenay," the *Daily Freepress*, Nanaimo, BC, September 4(?), 1976; "Sasquatch legend fueled by multisightings in BC," the *Herald*, Bellingham, Washington, USA, September 9, 1976; "Sasquatch sighting has Kimberley wondering if it's been ad-hoaxed," the *Sun*, Vancouver, BC, September 10, 1976; "Tall, tan, hairy," the *Province*, Vancouver, BC, September 10, 1976. "Many Sasquatch sighted in BC, *Associated Press* (newspaper not known), September, 1976. Also John Green, Sasquatch Incidents Data Base, Incident Number 1000198.

1976/09/07 – Kimberley. It Simply Moved: Three young people reported seeing a strange creature near the Kimberley-Radium Highway (across from the Department of Transport airport) on September 7, 1976.

Trudy Phillips, 18, her brother Rene, 12, both of Kimberley, and their sister, Sebrina Lucas, 19, of Cranbrook were driving up the hydro right-of-way to find mushrooms. Trudy Phillips stated:

Up the road was this big black thing. We thought it was a

burnt stump, then it started to move. It didn't seem to walk, it simply moved. It didn't go fast, just at a man's pace.

She said that the creature's head did not seem to project from the body, and that where shoulders would be the animal was rounded or hunched over.

The creature was first seen at about 200 yards and it moved towards the group's car. They never got out of their vehicle and left the scene immediately for their safety.

Source: "Sasquatch seen in Kimberley." The *Daily News*, Nelson, BC, September 9, 1976. Also John Green, Sasquatch Incidents Data Base, Incident Number 1000199.

1976/09/09 – Kimberley. RCMP Reports 8 Sightings In a Week: The RCMP here stated on September 9, 1976 that during this last week 8 people in Kimberley have come forward and reported sasquatch sightings. The reports were consistent as to the creature's shape, size and color. It appears the sightings were all of the same individual.

Source: Scott McClean, 2005. *Big News Prints*, (self-published), p. 55, from "B.C. Residents Report Sightings of Bigfoot." The *Union-Bulletin*, Walla Walla, Washington, September 9, 1976.

1976/09/29 – Deroche. Put Its Face In the Creek: Chris Bekkering and Ken McConnell reported seeing an unusual creature near Deroche at Norrish Creek on September 29, 1976. The youths were fishing in the creek when they saw a black figure, bigger than a man, standing beside the water about 100 yards away. It appeared to be looking at them. It then went down on all fours and put its face in the creek. It seemed to have something long and dark in its hand, about the size of a rifle.

Source: John Green, Sasquatch Incidents Data Base, Incident Number 1001351.

1976/11/29 – Bella Coola. Huge Man-type Tracks: Silas King reported that he found sasquatch tracks near Bella Coola on the road between Red Hill and Young Creek on November 29, 1976. It was about 3:00 a.m. and snowing heavily when he noticed the tracks. He said they were huge, man-type tracks taking short steps coming down the hill. He stopped and got out of his vehicle and examined the tracks with a flashlight. He reasoned that the sound of his motor had spooked the creature and it ran off into the woods.

Source: "Sasquatch reported in valley." The *News*, Bella Coola, BC, January 27, 1977.

1976/11/00 – New Masset (QCI). Long Dark Hair: Reggie Clayton reported that he saw a sasquatch near New Masset in November 1976. The creature was described as being close to 8 feet tall with long, dark hair completely covering its body and face.

Another sighting in December 1976 was reported to have occurred at the same spot as this sighting (see next entry).

Source: Gerry Williams, 1976. "Sasquatch or Gogeet?" The *Observer, Queen Charlotte Islands,* December 19, 1976. Also John Green, Sasquatch Incidents Data Base, Incident Number 1000200.

1976/12/18 – New Masset (QCI). Same Individual?: Pat Widen, a taxi driver, and his passenger Paul Williams reported that they saw a sasquatch at 6:30 a.m. near New Masset on or about December 18, 1976. The location given was about the same spot as a previous sighting by Reggie Clayton in November 1976 (see previous entry). Also, the description given of the creature was the same—being close to 8 feet tall with long, dark hair completely covering its body and face.

Source: Gerry Williams, 1976. "Sasquatch or Gogeet?" The *Observer, Queen Charlotte Islands,* December 19, 1976. Also John Green, Sasquatch Incidents Data Base, Incident Number 1000201.

1976/12/21 – Bella Coola. Light-colored Chest: Mrs. Doris Anderson said she saw a sasquatch beside the road near Bella Coola, at Anahim Lake at about 11:00 p.m. on December 21, 1976. She saw the creature as it ran across the road and up the side hill. She said the oddity was generally covered in long, dark hair, but had a light-colored chest. Mrs. Anderson was with her husband, George, at the time, but he did not see the creature.

Source: "Sasquatch reported in Valley." The *News*, Bella Coola, January 27, 1977, from information provided to Roy Lack of Oregon who was in the Bella Coola area. He was provided the information by Clayton Mack. Also John Green, Sasquatch Incidents Data Base, Incident Number 1000202, from the *Tribune*, Williams Lake, BC, January 27, 1977.

1976/12/00 – Bella Coola. It Stinks: Philip Binder, a baker, reported that his son, about 9-years-old, said he saw a strange animal near the family home in Bella Coola in December 1976. Binder's children had come in the house from playing at about 9:00 p.m., and the boy said he had seen a big animal and that "it stinks." The family live on the road that goes to the wharf in Bella Coola.

Source: John Green, Sasquatch Incidents Data Base, Incident Number 1000202, from The *Tribune*, Williams Lake, BC, January 27, 1977.

1976/12/00 – Bella Coola. Beside Garbage Dump: A Bella Coola resident reported that he (or she) saw a sasquatch near Bella Coola in December 1976. The witness said the creature was seen near the pond beside the garbage dump, across the road from the BC Hydro office.

Source: John Green, Sasquatch Incidents Data Base, Incident Number 1000202, from The *Tribune*, Williams Lake, BC, January. 27, 1977.

1976/YR/00 – Chilliwack. Smiling Sasquatch: Robin Kelly reported seeing a sasquatch near Chilliwack, along a tributary of the Chilliwack River, in 1976 (Soowahlie First Nations Reservation). He said the creature was taller than he was, even though it was standing in a creek bed beside the road about 3 feet below him. Kelly said that the sasquatch was smiling at him.

Source: John Bindernagel,1998. *North America's Great Ape: The Sasquatch.* Beachcomber Books, Courtenay, BC, pp. 108, 223.

1976/YR/00 – Harrison Hot Springs. Sasquatch Carving History: The first thing that once caught your eye as you drove into Harrison Hot Springs was the sasquatch carving in front of The Springs RV Resort. Photographs of the carving have gone around the world.

The carving was commissioned by Bob Clark in 1976. Clark bought the cedar log and then had a local artist create the sculpture for a total cost of about $700. At that time, Clark owned a rock shop in Harrison and put the carving in front of his shop. The large rock the creature holds over his head was painted jade green, to represent BC jade.

In 1981 Clark bought the campground where the carving is

Author with the famous sasquatch carving at Harrison Hot Springs, 1993.

presently located and placed the carving at the entrance. A friend, Waldo Fischer, suggested that Clark call his new facility "The Bigfoot Campground," which he did, and it operated under than name until about 1991. At that time Clark sold his campground to John Fogliato who operated it under the same name until 2009 when it again changed hands. The new owner changed the name to The Springs, but left the carving in place. However, in September 2011 it was hit by a car and destroyed.

Source: From information proved to author by Bob Clark and Thomas Steenburg. (Photo: C. Murphy.).

246

1976/CI/00 – Lillooet. Legs & Feet Seen In Trees: A lady reported a strange incident that occurred near Lillooet around 1976. She said that she and others were on a mountain bench above Lillooet. She saw among the pine trees what at first appeared to be 2 tree trunks moving slowly. She then realized they were huge legs and feet. Only about 4 feet of the lower half of the creatures could be seen below the foliage. She estimated the distance at about 200 yards. She pointed out the strange sight to some relatives who were with her.

Source: John Green, Sasquatch Incidents Data Base, Incident Number 1000121.

1977/01/00 – West Kootenays. Splashing the Water: Two young boys who were out in the West Kootenay woods reported seeing a strange, black, man-like creature in January 1977. The boys were having fun trying to break through ice in a frozen lake. They were out a little way on the ice. The creature was bending over in an ice-free section, splashing the water. They said it may have been washing its hands. The boys yelled several times to get its attention, but it did not respond. Then it suddenly stood straight up. The boys said it appeared to be 6 or 7 feet tall. They immediately ran for shore, and when they looked back, the creature was still watching them. They kept running.

Source: BFRO website.

1977/03/28 – Spuzzum. Witness White As a Ghost: Sixteen-year-old Richard Mitchell of Hatzic, who was working at the Miss Spuzzum Cafe during the school spring break, reported that he saw a sasquatch near Spuzzum on the afternoon of March 28, 1977. Mitchell saw the creature ahead of him on the old Trans-Canada Highway, just to the north of the new Alexandra Bridge, while he was out riding his motorcycle some time after 4:00 p.m.

He turned his motorcycle around as fast as he could and fled, but while he was breaking and turning (now closer to the oddity), he observed the animal for several seconds.

The creature was walking erect down the road away from Mitchell at a distance (when first seen) of about 100 yards, but was angling from one side of the road to the other. It turned enough so that he saw its profile although he did not notice that it ever looked back at him.

It could not have been a bear, he said, because no ears were visible and it did not have a snout or a tail. It was covered, except possibly the hands and face, with shiny, greasy-looking black hair. The hands and face were also black.

The old road is blocked with a high pile of boulders just beyond where the creature was walking, and using those as a standard, the youth estimated that the creature was over 7 feet tall. It was very heavily built, with wide shoulders, prominent buttocks and very thick legs. The torso was long in relation to the legs, and the feet were large and covered with hair. No neck was evident. It swung its arms and shoulders from side to side in an exaggerated fashion as it walked.

The old road is paved and most of the surrounding area is rocky. Where the creature was seen, the old road is just below the CPR track, and there is nothing but woods above the track as the new highway had already crossed to the other side of the river.

Mitchell fled at high speed to the nearest house, which is occupied by the owners of the cafe, Gordon and Linda McCarten. Mrs. McCarten said Mitchell was white as a ghost, shaking and almost incoherent at first.

She said there was no doubt that something had frightened him; she had never before seen anyone so frightened.

After they heard his story, she and Larry White, who was staying with the McCartens, went to the place where the creature was seen, but nothing was found. Other people looked around later but still nothing was noticed, and a pet German shepherd "just sniffed around."

However, Bob Titmus made a thorough search for tracks on April 17, 1977 and found some good partial tracks up the hill above the railroad that goes through the area.

The Alexandra Bridge area facing south.

Source: "Sasquatch Sighted Close to Spuzzum," probably the *Province,* Vancouver, BC, March 30, 1977. Also "Something hairy sighted on the road to Spuzzum," the *Sun,* Vancouver, BC, April 2, 1977. Also John Green, Sasquatch Incidents Data Base, Incident Number 1000238. (Photo: Image from Google Earth; © 2011 Cnes/Spot Image; Image © IMTCAN; Image © Province of British Columbia; Image U.S. Geological Survey.)

1977/05/15 – Mission. Misguided Misfits Pull Off Silly Hoax: A creature bearing the resemblance of a sasquatch was seen on a highway near Mission on May 15, 1977 by the driver and passengers on a Pacific Stage Lines bus. The incident was investigated by the RCMP and sasquatch researchers. Footprints were found, and along with testimony, the police and researchers were led to believe that an actual sasquatch sighting took place. However, on May 27 a group of pranksters came forward and provided conclusive evidence that they staged the incident.

I am not going to give those who were responsible for this silly stunt the satisfaction of seeing their names in print. (Hopefully they have grown up a little by this time.)

It might be noted that the incident took place nearer to Norris Creek or Deroche than Mission. As it has commonly been called the "Mission Hoax," I have used that location.

Source: "RCMP convinced Sasquatch no hoax." The *Columbian,* New Westminster, BC, May 16, 1977. Also, "Mission sasquatch sighting just 'man in gorilla suit.'" The *Sun,* Vancouver, BC, May 27, 1977.

1977/05/19 – Victoria (VI). Sasquatch & Politics: Premier Bill Bennett said in Victoria on Thursday, May 19, 1977, that the sasquatch would be extended the same protection by the province as is given Ogopogo. He made the statement in reply to a reporter's question.

However, Ogopogo, the snake-like creature said to inhabit Okanagan Lake, is not protected under any provincial laws.

Source: "Now sasquatch gets protection." The *Sun,* Vancouver, BC, May 20, 1977.

1977/06/17 – Vancouver. Titmus Says He Will Never Give Up: After 19 years trying to catch up with the sasquatch, researcher and former taxidermist Bob Titmus was as firm as ever in his conviction that that the creature exists. "I shall always hunt the sasquatch, any time I am able, physically and financially," he stated on June 17, 1977.

Titmus has personally sighted the sasquatch twice in his life. The first time, he said he was close enough to "see the wrinkles on it lips." But that was before he was involved in hunting the creature and says he flatly refused to believe what he was seeing. Many years later, while seeking the creature, he says he saw 3 of them climbing a sheer granite cliff face (see 1963/07/00 – Kemano. Titmus Sees Unusual Figures).

His interest in sasquatch began in 1958 when a friend, Jerry Crew, asked him how to make footprint casts and then brought him a plaster cast of a sasquatch footprint. From his experience, Titmus knew of no animal that could make a footprint of that nature, whereupon he embarked on a passionate search for the creature.

It needs to be mentioned that Jerry Crew took his footprint cast to a newspaper and the resulting story, which used the word "Bigfoot," established this name as the common term for the creature in the US.

Titmus's search has taken him from north of Fairbanks, Alaska, to Mazatlan, Mexico, and he has since amassed the largest collection of original sasquatch footprint casts in the world. He has also found possible sasquatch hair and fecal matter which he sent for analysis; professionals were unable to identify the source.

Titmus said that scientists do not generally believe in the existence of sasquatch and stated:

> This is a giant primate, perhaps the first cousin of man, and people won't believe in it. Some scientists dismiss it out of hand, without even looking at the evidence. Personally, it's not a question of thinking they exist. I know positively. I'm not trying to convince anyone.

He believes the creatures are being driven further into the wilderness as more people move into areas they frequent.

Source: Scott McClean, 2005. Big News Prints (self-published), p. 55, from "Sasquatch hunter on trail 19 Years." The *Herald*, Lethbridge, Alberta, June 17, 1977.

1977/08/08 – Chetwynd. Four-toed Prints: Five vacationers from Fort St. John on a fishing trip found unusual 4-toed footprints about 40–50 miles south of Chetwynd, beside Bull Moose Creek, on August 8, 1977. The prints measured about 17 inches long, and 10 inches wide. They were about 6 feet apart. Marilyn Wiles, one of the group, said, "They were really big…so big that no one else could do them. They weren't fresh, so we didn't get scared."

John Ross, another of the group, said the prints were about a mile away from the nearest road, and went from a creek across sandy soil, and ended in a rocky, bushy area. He added, "They'd been rained on once; they weren't all that fresh. If anybody wanted to try a hoax, they would have done them near the bridge or near the road."

Plaster casts were made and the prints were reported to the area fish and wildlife office. Fred Harper, a biologist with the wildlife branch, and another biologist, Brian Churchill, inspected the prints on August 28, 1977. Although the tracks were still visible, they were now at least a month old, and 3 of them were obscured by plaster used to make casts. Harper said the prints were about 9 inches wide and were impressed about 1.5 inches into the soil. He added that Churchill, who weighs 200 pounds, did not make any footprint impressions in the same soil, so it was concluded that something very heavy made the prints. Harper stated:

> As biologists we're skeptical about the Sasquatch because all one ever sees or hears [about] are footprints or blurry movie film. These creatures have to reproduce, eat, take shelter, and leave feces but there are never any signs of these things. However, there is no question that these tracks were found. But we don't know what they are and there is no way to find out.

Here we have another case of only 4 toes being seen in the impression. As previously stated, from my own research I have concluded that this happens sometimes because the little toe fails to register as deeply as the other toes (i.e., it is not subjected to the same weight). When the prints get old, as in this case, the little toe impression virtually disappears. Harper's comment about other evidence indicates he was not well-informed on the subject.

Source: "Latest Sasquatch find reports only four toes." The *Province* or the *Sun*, Vancouver, BC, August 1977, from an A.P. release. Also "4-toed 'Bigfoot' tracks found in B.C.," probably one of the same papers, August 28, 1977. Also "Prints 'Sasquatch-like': But Biologist Skeptical," the *Sun*, Vancouver, BC, September 2, 1977.

1977/08/12 – Victoria (VI). Government MLA Introduces Sasquatch Protection Bill:
Sasquatch in the Province of BC found a new friend in government during August 1977—George Mussallem, the Social Credit Member of the Legislative Assembly (MLA) for the Dewdney area.

Mussallem introduced a private member's bill on August 11, 1977 aimed at protecting the mysterious creature. The MLA said there is cause for "special concern over the potential harassment and possible extinction of the creature."

His bill called on educational institutions to increase efforts "to document the existence and habits of Bigfoot," and urges citizens to use caution when they encounter the creatures to ensure their preservation.

Although Mussallem admitted he has never seen a sasquatch, he recalled a visit some years ago to a logging camp near Harrison Bay and recounted the following incident:

As dusk thickened the loggers—50 or 60 strong—huddled in the stifling heat of their bunkhouse. When I asked the reason for this inexplicable behavior, my query met with silence. Only with the greatest reluctance did the men admit their fear of the great sasquatch.

Alone outside, I was eerily aware of something or someone watching me from behind the dense cover of bush...no one else ventured out into the open that night.

Mussallem said provincial agencies "with employees most likely to encounter Bigfoot" should develop special procedures to be used in such instances.

Source: "Sasquatches find a friend in Mussallem." The *Sun,* Vancouver, BC, August 12, 1977

1977/AU/00 Fall – Prince George. Slightly Stooped Posture: A man traveling on a motorcycle about 50 miles east of Prince George reported that he saw a large bipedal animal in a ditch on the left side of the road during the autumn of 1977. He originally thought it was a bear, but as he approached, it walked out of the ditch on 2 legs and crossed the road. He described it as being covered in dark brown or black hair, muscular, having long arms (longer than a human), and standing well over 6 feet tall. He noted that it walked with a slightly stooped posture.

Source: BFRO website. Witness was interviewed by Blaine McMillan.

1977/AU/00 – Harrison Mills. Possible Sasquatch Corpse: A man found what he believes was a sasquatch corpse in the Harrison Mills area (north end of the Chehalis River) in the autumn of 1977. He and his companion, a hunter, were following deer tracks at the time. In his own words:

> I thought at first that it was the remains of a grizzly bear, but the size did not fit with that of a bear, of which I have seen at close quarters in the same area. The remains were decomposed and badly mutilated, probably by cougar, raccoons, lynx, etc., of which there are plenty in the area.
> By the appearance of the surrounding area of the body,

The north end of the Chehalis River. It can be clearly seen in this image how roads built by logging operations and the government forest service operations provides wilderness access for hunters and others.

it had been there a long while. I recall we were both very cold and drove out of the area to the Inn [probably the Sasquatch Inn] and called Fish & Wildlife, and the Forest Service; both were closed, and I called the next day and reported what I had seen. I was told that it was probably the remains of a bear. When I refuted that, they, the F&W, said they would investigate. Subsequent calls to them led me to believe they took no action.

The 2 men stayed in the area for about 20 minutes, so definitely had time to have a good look at the remains.

The creature was laying on its back. The man estimated the height of the corpse at 7 feet. He described the general shape of the body as "ape-like." The hair was matted and dirty. Very little of the skull remained, or any of the face. The arm skin appeared to be extra long.

Source: Bigfoot Phenomenon Anecdotal Reports "A" Classification, Volume 1. North American Science Institute, Data Publication, September 15, 1977, Questionnaire #138, Sighting.

1977/YR/00 – Vancouver. UBC to Look at Sasquatch Evidence: Dr. Marjorie Halpin of the University of British Columbia, Vancouver reported to the press in 1977, "It's time the academic community came out of the closet and seriously addressed the question of the sasquatch." Halpin, curator of the university's Museum of Anthropology, said that the museum would sponsor the world's first academic meeting on monsters, to be held in May 1978.

Canada Council financing would provide transportation for Canadian scholars, Halpin said, but more money was needed to bring in experts from the US and the Soviet Union.

Source: "Academia To Discuss Sasquatch."*Associated Press* article probably featured in the *Sun* or *Province*, Vancouver, BC, 1977.

1977/CI/00 – Black Creek (VI). Sasquatch "Bower" Fred Bunnell, a forestry professor doing field research on grizzly bears, found an unusual structure near Black Creek at Knight Inlet-Ahnuhati River, in the summer of 1977 or 1978. He described the structure as a sasquatch bower. It was so designated by him because of bent, broken, and over-arching branches placed up against a rock face. When the professor was asked why he attributed the stricture to sasquatch, a wildlife species not yet treated seriously in academic circles, he replied, "No bear makes a day bed like that."

Source: John Bindernagel, 1998. *North America's Great Ape: The Sasquatch*. Beachcomber Books, Courtenay, BC, p.69.

1977/CI/00 – Hope. Like the Back of a Moose: A group of campers experienced several unusual possible sasquatch-related incidents near Hope in about 1977. Ron Glover and Aaron Lengyel, then teenagers, were clearing an area on their grandparents' mining claim just west of Silver Lake. They were camped near Silverhope Creek along with friends Don Price and Tom Harrison. The second night as they were sitting around the fire a log about a foot thick and 10 feet long was thrown near the fire. No one saw how.

Lengyel went to bed and his lean-to collapsed on his tent. He scrambled out to find Price, who had been sitting on a log by the fire, acting hysterical, saying there was a bear in the camp area. Lengyel threw some gasoline on the fire and as it flared he saw something going away, "like the back of a moose." (There are no moose in this area). Glover said that as Price was sitting by the fire he rose and fell forward as if lifted from behind. Glover said he saw a creature on the other side of the log behind Price, about 15 feet away, with red eyes and a well-defined shape. Its arms were long and seemed covered with matted, dirty, brown fur. Its legs and shoulders were huge. Glover said it was gone in 3 steps.

Next day the group found some apparent footprints, one quite clear, on a path. It was like a human print with 5 toes, but estimated to be 16.5 inches long, 9 inches maximum width, and 8 inches at the heel.

All or some of the group went back to the spot 2 weeks later. They heard a lot of noise in the bush and found more footprints in sand by the creek, which they think were made by the same individual that was at their campsite.

I would have expected that the log-throwing incident would have caused the group to immediately leave the area, not to mention the "lifting" of Price and then the sighting of the creature. If the story is true, then it appears we are missing some information.

Source: John Green, Sasquatch Incidents Data Base, Incident Number 1000239.

1978/02/16 – Berkley, California, USA. George Haas Passes On: On February 16, 1978, George Haas, the editor of the *Bigfoot Bulletin* passed away. He was 72 years old. Haas founded a formal newsletter for sasquatch-related incidents, which he distributed free to serious researchers. The first newsletter was sent out on January 2, 1969, and the last in June 1971. Haas's initiative was of great service to researchers in both the US and Canada.

George Haas

It needs to be mentioned that Roger Patterson started a newsletter to members of his research association in 1968. It appears this was the first newsletter on the bigfoot subject. However, it was not a regular publication. He simply prepared it when he had time.

Source: Author's files.

1978/05/05 – Vancouver. Sasquatch Poll: On May 5, 1978 the *Sun* published a preliminary report on the results of a poll taken on belief in various cryptozoology creatures and other phenomena. It was stated that some 43% of British Columbians were either sure sasquatch exist or they thought the creature might possibly exist. Remarkably, prairie people came out higher at 47%.

On belief in extra-sensory perception (ESP), British Columbians came out at 74%, far greater than any other province.

Source: "B.C. tops poll with believers in fantastic: Sasquatch, ESP." The *Sun*, Vancouver, BC, May 5, 1978.

1978/05/09 – Vancouver. "Anthropology of the Unknown" Conference: From May 9–13, 1978 the UBC Museum of Anthropology, Vancouver, hosted a conference, "Anthropology of the Unknown—Humanoid Monsters: Sasquatch and Similar Phenomena." Most, if not all, of the major sasquatch researchers in the Pacific Northwest were in attendance and some of them presented papers.

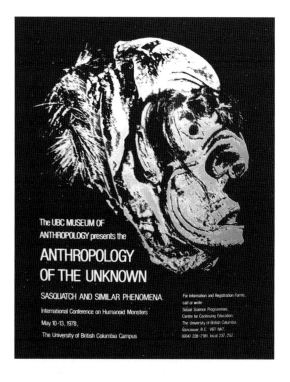

The UBC MUSEUM OF ANTHROPOLOGY presents the

ANTHROPOLOGY OF THE UNKNOWN

SASQUATCH AND SIMILAR PHENOMENA.

International Conference on Humanoid Monsters

May 10 -13, 1978,

The University of British Columbia Campus

For Information and Registration Forms, call or write:
Social Science Programmes.
Centre for Continuing Education.
The University of British Columbia.
Vancouver, B.C. V6T 1W7.
(604) 228-2181, local 237, 252.

I recently asked John Green, who was at the conference and presented a paper, to give me his opinion. He said the following:

It seemed that it would be a real breakthrough at the time, and in some ways it was, for which Halpin in particular and Ames [event organizers] deserve our gratitude, but it turned out to be two conferences in one. The sasquatch people and the mythology people basically ignored each other.

It is apparent that the "sasquatch people" were not happy with mythological explanations, and "myth" people were not open to any other explanation. (You can lead a horse to water, but you can't make it drink.)

Some of the papers (only *some*) presented at the conference on the subject were compiled into a book edited by Marjorie Halpin and Michael Ames (both anthropology professors) published by the University of British Columbia Press in 1980. The book, *Manlike Monsters on Trial: Early Records and Modern Evidence* is discussed under 1980/YR/00 – Vancouver. UBC Publishes Book on Humanoid Monsters. I explain in this entry the university's great injustice towards, and unprofessional treatment of, many of the scientists and others who participated in or contributed to the conference.

Source: Author's file and communication with John Green. (Photo: Poster, UBC.)

1978/05/00 – Vancouver. Dr. Koffmann Reports on Russian Snowman: Dr. Marie-Jeanne Koffmann, the noted Russian authority on the almas (also almasty) or Russian snowman, provided a report to the UBC conference on sasquatch and similar phenomena held in May 1978 (see previous entry). Dr. Koffmann was not in attendance; the report was delivered to the conference. I am providing what she said as it is possible some BC sasquatch sightings are actually of almasty.

Dr. Koffmann said that the almas is now rarely seen; however, a generation or two ago it was a part of the landscape.

People in remote areas would offer the creature food and even clothing. There was a special sympathy towards their females with babies, which were said to be pink, like human babies. They did not become hairy until they were about a year old.

The adults of the species are somewhat sasquatch-like, although generally smaller (about human size).

They eat fruit, berries, a variety of wild and cultivated plants, small animals, birds' eggs and food they take from humans—dairy products, meat, honey and porridge.

Marie-Jeanne Koffmann is seen on the right of the classic photograph showing the founders of Russian hominology. From left to right: Boris Porshnev, Alexander Mashkovtsev, Pyotr Smolin, and Dmitri Bayanov.

In winter, the almas rest in chance refuges—a grotto, a haystack, or an abandoned hunter's cabin. It is believed they may semi-hibernate at this time. In summer, they sleep in trees, or build nests on the ground. On this point Koffmann stated:

> On the ground he makes a lair with a bedding of rags and soft grass. He ties up tops of tall weeds—making knots is one of his favorite pastimes—and covers the frame with a canopy of burdock leaves.

Koffmann also stated that although the almas seems to have no natural enemies (other than man), its numbers are declining as man's activities increase.

Source: Tim Padmore, 1978(?). Newspaper article (no heading). The *Sun,* Vancouver, BC, (actual date not known). (Photo: I.Bourtsev.)

1978/SU/00 – Quesnel. Face Like an Ape: Randall Colelaugh believes he saw a sasquatch on Highway 97 just outside the town of Quesnel during the summer of 1978.

Colelaugh was delivering newspapers in the early morning hours between the towns of Cache Creek and Prince George. His helper, a lady, was with him. He described his experience as follows:

> There was a large curve in the road. And as I came around this curve my headlight filled the ditch and I startled this sasquatch creature. It was obviously scared and ran up the bank, 30 feet, at an incredible rate of speed. It was gone before we knew it. Incredible speed. I slowed down and the lady I was with said, "Do you want to stop?" I said, "Are you crazy?" I just kept on driving. I didn't stop on the way back either because I was scared about stopping in that area. I did stop there on another trip, about 3 weeks after, to check for tracks, but I did not find any.

Colelaugh estimated that the creature was about 200 feet away when it was spotted and that it appeared to be about 8 or 9 feet tall. He said that the face was like that of an ape. He went on to say that the creature had "strange ears." They looked like "large lynx-type ears."

There have been no other reports I have seen that mention ears of this nature. Indeed, ears are seldom noticed. It is probable that what Colelaugh took as ears was something in the background that sort of blended in with the creature's figure. At 200 feet this type of illusion could definitely occur.

Source: Thomas Steenburg, 2000. *In Search of Giants: Bigfoot/Sasquatch Encounters.* Hancock House Publishers Ltd., Surrey, BC, pp. 37, 38.

1978/08/00 – Rocky Mountains. Thump on Roof of Car: Three young women traveling the highway to Alberta reported a man-like creature encounter in the Rocky Mountains in August 1978. They pulled off the road to get some sleep, covering themselves with their sleeping bags to stop car lights from bothering them. Two of the group had not fallen asleep yet when they heard a thump on the roof of the car. They looked out and saw a hairy (brownish), man-like creature running across the road. It was illuminated by approaching car headlights. They said it looked taller than a man, but not over 7 feet. Also, it was not "bulky."

Source: BFRO website. One of the witnesses was interviewed by Gary Cronin.

1978/09/00 – Rock Bay (VI). Eye-watering Stench: Jesse Gordon stated that he encountered a sasquatch on a forest trail near Rock Bay in September 1978. He said that the stench of the creature was "eye-watering."

Source: John Bindernagel, 1998. *America's Great Ape: The Sasquatch.* Beachcomber Books, Courtenay, BC, pp. 77, 217, from information provided by Jesse Gordon.

1978/12/00 – Kaslo. Large Footprints Very Convincing: Rex Alexander, 19, reported that he found a series of 17-inch, human-like footprints near Kaslo in December 1978. He found the prints in a heavily wooded area while hunting deer. Roy Green, a long-term Kaslo mayor and alderman, went to the location and made a plaster cast of one of the prints. Both he and Alexander were convinced the prints were not hoaxed. Alexander stated, "They were way off in the bush. It was just a fluke that I came across them."

Source: John Robert Colombo, 1989. *Mysterious Canada.* Doubleday Canada, Toronto, Ontario, p. 352 from the *News*, Nelson, BC, (Canapress), no details. Also, "17-inch footprints raise Sasquatch scare." The *Tribune*, Lewiston, Idaho, USA, (*Associated Press* release) December 8, 1978.

1978/YR/00 – Midway. Big Black Figure: Robert Ziesman, 14, reported that he had several sasquatch-related experiences in the Midway area in 1978. In the first case, he was riding his motorcycle near his home at Kettle Valley. He stopped on top of a small hill beside a pond and saw at a distance a big, black figure standing on its "hind legs."

In the second case, while he and another youth (believed to be Mark Farrier, also 14) were climbing a cliff at twilight, they saw another big, blackish figure walking on 2 legs along the cliff top. It had very long arms, took big steps, and walked sort of hunched over. It stopped almost at the edge of the cliff and gazed towards the youths. Ziesman said its cheeks appeared fat.

In the third case the youths found tracks (10 or so) about one-quarter-inch deep, crossing a dirt road. The tracks indicated 5 toes, definitely no claw marks. The length was guessed at 18 to 24 inches. Some unusual droppings were found that were greenish in color. Ziesman remarked that he found tracks whenever he went "up there."

In the fourth case, Ziesman found large, human-like tracks in snow beside a friend's house. He looked around and saw a big black figure up on hill. He later went up to the spot, but did not find any tracks.

Source: John Green, Sasquatch Incidents Data Base, Incident Number 100241 and 1000249.

1979/04/28 – Little Fort. High-pitched Squeal: Tim Meissner, 16, and 3 friends reported that they saw a sasquatch in the Little Fort area at Dunn Lake on April 28, 1979. After hearing a high pitched squeal they saw the creature on the other side of the lake, about 300 yards away, with its arms raised. The youths went over to where the creature was seen to investigate. They ventured into the bush about 100 yards and found a dead dear concealed under branches and moss. The animal's neck had been broken and its back legs were missing.

Meissner had another sighting 2 days later which is detailed in the next entry.

Source: Janet and Colin Bord, 1982. *The Bigfoot Casebook.* Granada Publishing Ltd., London, England, p. 139. Also John Green Sasquatch Incidents Data Base, Incident Number 1000242. This incident was featured in many newspapers in conjunction with a second sighting (and shooting of the creature) by Meissner provided in the next entry. The sources are detailed under that entry.

1979/04/30 – Little Fort. Sasquatch Shot & Drops: Two days after seeing an upright animal near Little Fort at Dunn Lake, and finding the concealed body of a deer with a broken neck (see previous entry), Tim Meissner and friends returned to the site (April 30, 1979) to do a further search.

They separated and Meissner, now alone, heard branches breaking then found himself face to face with a sasquatch 56 feet away

Dunn Lake. It is about 193 acres in size and is a popular fishing spot.

(measured by investigators). He shot at the creature and later provided the following account:

> He was about 9 feet tall, black, and hairy. He had a human-like face with great big glaring bright eyes and shoulders 4 feet wide. He stood there glaring at me for at least 3 seconds. He was 50 feet away—so close I could smell him. I don't even know why I shot. I was just scared, really scared. I was aiming for right between his eyes and he went down on one knee and one hand. At first I thought he was dead, but I guess I only grazed him, because he got up and ran away at about 30 miles an hour. It couldn't have been any other animal, and it wasn't a human because no human can run that fast, especially straight up an embankment.

One of his friends had a camera, but was below an embankment and could not see the creature.

Subsequent reports on the incident provided more details, although somewhat conflicting. One account states that Meissner fired at the creature in panic and it dropped on one knee (no hand mentioned), then ran into the bush. However, Meissner's father, Ken Meissner, stated that the animal "fell" (implying that it fell completely to the ground). Another source quotes Tim as saying that the creature stood silently watching him (not glaring) from the trees until he fired at its head. Whatever the case, all the stories are essentially the same.

After Meissner shot at the creature he went to get his friends, but when they returned to where the animal was shot, it was gone. Ken Meissner (father) stated that he had investigated at the scene and found clear tracks, human in shape, 16 inches long and 9 inches wide, sinking in soft soil more than an inch. He made casts, and measured the stride [pace] at 5 feet.

Tim Meissner added that the creature appeared to weigh about 400 pounds. It was covered with long, curly, dark brown hair, except for a bare chest, had a squarish face, looked more human than ape, glared at him with huge, shining eyes, and stank like a rotten garbage dump.

René Dahinden investigated the incident and stated that Tim Meissner had said he could see neither female breasts nor a penis on the creature.

Another source quotes Ken Meissner (father) as saying that Tim estimated the animal to weigh 900 pounds, and that its shoulders were 4 feet wide.

Jack Wood, an agriculture teacher, went to the site area and found a 5-toed footprint 16 inches long, 9 inches at the ball of the foot, and 6 inches across the heel There was no indication of claw marks. The print was found on the side of a mountain trail. Wood mentioned that one toe was much smaller than the others. He made a plaster cast of the print.

Source: Janet and Colin Bord, 1982. *The Bigfoot Casebook.* Granada Publishing Ltd., London, England, p. 139 from, articles featured in the *Columbian,* Vancouver, Washington, USA, May 6, 7, and 9, 1979, and the *News Review,* Roseburg, Oregon, USA, May 7, 1979. A further article, "'Hurt Bigfoot Hunted; Youth says he shot Sasquatch'" was in the *Journal,* Portland, Oregon, USA, May 7, 1979. Also "Canada town hunts for injured 'Bigfoot.'" The *Daily Herald*, Arlington Heights, Illinois, USA. May 7, 1979. Also the *Sentinel*, Kamloops, BC, May 4, 1979; the *Herald*, Calgary, Alberta, May 11, 1979; and the *Advocate*, Red Deer, Alberta (no details). Also John Green Sasquatch Incidents Data Base, Incident Number 1000243. (Photo: Image from Google Earth; Image © 2011 Province of British Columbia; © 2011 Cnes/Spot Image; Image © 2011 DigitalGlobe.)

1979/05/20 – Haney. Tracks Old & New: Lance Barnett submitted a report to John Green on track findings in the Haney area on May 20, 1979. He and 4 companions had climbed up Gold Creek valley to a 2,000-foot level pass leading to Pitt Lake. He said they observed an alder sapling with the top partially broken off and twisted around, as if with hands, and later found a hollow of crushed ferns.

While going down to a creek, one of the group in the lead saw on the far slope, among the trees, a large furry animal—not a bear. He said he first thought it was a burnt stump, but on second look it was gone. He did not mention it until after fresh 13-inch tracks were found in forest litter and dirt at the top of a bank a short time later. Another of the group then found a print 18 inches long in moss. The tracks were photographed. As the group traveled on through the pass they found many tracks, old and new, in moss, dirt and snow. Returning to the area 3 days later they found more tracks.

Source: John Green, Sasquatch Incidents Data Base, Incident Number 1000244.

1979/06/18 – Salmon Arm. Standing In the Shadows: Jon Clark, 15, reported that he saw a strange creature in the Salmon Arm area near Pillar Lake on June 18, 1979. Clark was walking towards his school bus stop when he saw the creature standing in the shadow of trees by the road. It was standing on 2 feet, hunched over. He noted that it was covered in brown and black hair and had long arms. He could hear it breathing heavily—like groaning. He watched it for about 30 seconds at a distance of about 10 feet. When the creature noticed him, it walked away upright.

Source: John Green, Sasquatch Incidents Data Base, Incident Number 1000205, from the *News*, Kamloops, BC, June 20, 1979.

1979/07/00 – Hope. Ten Feet Tall: Darwin John of the Northern Native Development Corporation reported a frightening encounter near Hope in July 1979. John told people that he had fled from "a large, hairy object 10 feet tall" standing erect, while he was clearing trees for a pipeline right of way parallel to the Coquihalla highway. John's experience (heard third hand) was related by Peter Stewart, a Westcoast Transmission Company employee.

Source: John Green, Sasquatch Incidents Data Base, Incident Number 1000206, from information provided by Peter Stewart of West Vancouver.

1979/07/00 – Port Alberni (VI). Eyes Gleaming Orange: Claude and Dorothy Martin of Nanaimo reported seeing an odd creature between Port Alberni and Parksville in July 1979. They were traveling in a truck at night and had stopped for water at a little creek. They thereupon saw a big, hair-covered creature standing upright like a human, eyes gleaming orange, in the headlights of their truck. They said it was apparently black. The possibility of a black bear standing upright was suggested.

Source: John Green, Sasquatch Incidents Data Base, Incident Number 1000207, from the *Free Press*, Nanaimo, BC, July 27, 1979.

1979/YR/00 – Nelson. Whole Race of Them: Brent Hastings reported that he saw a sasquatch while hiking near Nelson in Kokanee Provincial Park in 1979. He said that he believes there is a "whole race" of the creatures in the Kootenays. "I don't doubt them at all," he said, "whatever contact they've made with humans hasn't been good." He said that the shrinking wilderness will likely provide more sightings.

Source: Suzy Hamilton, 1993. "Ape-man alert: Kootenays awash in Sasquatch sightings": The *Province,* Vancouver, BC, January 12, 1993.

1979/YR/00 – Etolin Island, Alaska, USA. Fifteen Feet Away: Dana Jacallen reported that he came within 15 feet of a sasquatch while hunting with a companion on Etolin Island in 1979. The island is off the coast of BC, and is part of the state of Alaska. He found large tracks in the snow at his campsite when he awoke before dawn the first morning on the island. He followed the tracks up a hillside and provided the following account:

I stopped and looked up. Fifteen feet away, one hand on his hip, the other gripping a branch of hemlock that was just

Etolin Island is 30 miles long and 10–22 miles wide, with a land area of about 339 square miles, making it the 24th largest island in the United States. As of the 2000 census, Etolin had a population of 15 people.

above my head, was the biggest, hairiest guy on two legs I ever want to see. Without thinking I started to thumb the safety off. He turned his head to stare at me straight in the eyes. His teeth were showing but it was not a snarl or a grin. I want to say it was a knowing glint in his eye. He had intelligence. I knew I could place a killing shot, but also knew it would be wrong, even criminal.

[I experienced]…an unmistakable sense of warning as clearly as if someone had spoken. It wasn't logic, it wasn't fear—it was communication. He was saying to me with all the power of the spirit, "Leave me alone."

The creature then turned and fled. When relating his story much later (late 1980s), Jacallen said that the creature looked just like the bigfoot in the movie *Harry and the Hendersons,* and then continued:

This one was equally as large, but more lean. He had long, silky hair on his arms, it was a good 6 inches long. He walked erect and he was quiet. He made no sound.

Jacallen scrambled down the hillside and met his partner on the beach. After daylight they retraced Jacallen's steps. The creature's footprints measured 17 inches long and 5 or 6 inches wide. The impressions found in mossy ground convinced them that it had weighed 600 to 800 pounds.

Source: David Hulen (editor), 1993. "Hairyman to Hunter: 'Leave us Alone." *Daily News,* Anchorage Alaska, USA, March 1, 1993, from Nell Waage, the *Daily Mirror,* Kodiak, Alaska, USA (date not known). Also Wikipedia. (Photo: Image from Google Earth; Image © 2011 DigitalGlobe; Image © 2011 TerraMetrics.)

1979/DE/00 – Sooke (VI). Long Gray Hair On Its Chest: A man reported a very close encounter with a sasquatch near Sooke in the 1970s. He lived in a cabin near the Sooke River. He said his cabin was slapped firmly by an unknown animal at night. Looking out his window, he was able to see only the long gray hair on the creature's chest. When the man's dogs howled, the creature responded with similar howls.

Source: John Bindernagel, 1998. *North America's Great Ape: The Sasquatch.* Beachcomber Books, Courtenay, BC, p. 123, 227, from a personal interview with the witness.

1980/03/00 – Victoria (VI). Sasquatch Protection Put Aside: In March 1980 one of BC's government "sasquatch protection" advocates, George Mussallum, put aside further efforts in this regard. He had attempted several times without success to accord official protection for the creature in BC. Mussallum stated:

> I do believe there is something to the sasquatch theory. If they exist, they definitely must be protected like any endangered species. There is no doubt that there is something there beyond our understanding at this time.

He said he planned to pursue a campaign on the federal level to have the sasquatch declared off limits to firearms.

Source: "Sasquatch protection not a government priority." *Alaska Highway Daily News,* Fort St. John, BC., March 16, 1980.

1980/03/00 – Dallas, Texas, USA. Scientists a Virtual No-show: BC's sasquatch seeker, René Dahinden, sent out invitations in March 1980 to 238 scientists in the Dallas, Texas area to attend a screening of the Patterson/Gimlin film and discuss the sasquatch. The Texas-based Bigfoot Research Society sponsored Dahinden's appearance. Unfortunately, none of the invited scientists showed up. Dahinden ended up chatting with a half-dozen people and wondering about the lack of interest in the scientific community. "Don't be so damn closed-minded," Dahinden relayed to scientists in general. "I want people who are qualified to study our evidence. If it isn't anything, tell me. But let's have cooperation."

A second invitation was sent for a session to be held 5 days later at East Texas State University in Commerce. This time 2 scientists showed up. One was a sociologist and the other an anthropologist, who expressed interest.

Source: Doug Domeier, 1980. "Hunter seeks comment, but Bigfoot crowd small." The *Morning News,* Dallas, Texas, March 16, 1980.

1980/05/00 – Vavenby. Odd Figure at Water's Edge: Ken Karateew, a pilot, reported that he saw a strange creature while flying near Vavenby in May 1980. The oddity was sighted along the East shore of Adams Lake (northern half). Karateew was looking for a possible farm site in an area where there were no roads. From about 150 feet up he saw a figure stooped over at the water's edge. There was nothing in sight to give a size comparison. The figure stood up, looked at the plane, turned 180 degrees and disappeared into the woods. Karateew circled the area several times to try and see it again.

He thought at first that it was a person, but it was brownish black all over and no sign of clothes, pack or a boat. He also thought of a bear, but it ran upright with long strides [paces], swinging its arms and shoulders. No one else in the plane saw it, but Mrs. Karateew who was with him said that her husband immediately asked whether anyone else had seen the strange animal. Karen Karateew, the pilot's sister, also in the plane, said in later years that her brother never got over seeing the oddity. He died in 1996.

Source: John Green, Sasquatch Incidents Data Base, Incident Number 1000208. Tom Steenburg Sasquatch Incidents File, Number BC 10124,

1980/07/21 – Cumberland (VI). Something Big Shaking the Alders: A summer park ranger reported seeing an odd creature near Cumberland at McBride Lake on July 21, 1980. The ranger, on his day off, had hiked up a creek from the head of Great Central Lake to McBride Lake to fish. As he sat on a rock at the northeast corner of the lake assembling his fishing rod, he heard a motion in the alder brushes near the shore across a small tongue of the lake. Something big was shaking the alders and throwing chunks of slide rock, which he could hear rolling downhill. He shouted, but there was no answer. He also heard some alders snap. The alders averaged about 10 feet tall. After about 10 minutes he saw a head and shoulders of some sort of creature sticking up above some lower alders, about 50 yards from him. Over a 20-minute period this happened several times as the creature stood up.

The creature was moving farther away, and at a greater distance he saw it from the waist up with one arm raised. It was black, wide at the shoulders, no neck, and head shaped like the small end of an egg. He didn't see any facial features. He said it was at least 7 feet tall. The ranger abandoned his plan to go to the head of the lake and returned down the creek. The following summer, the ranger took John Green to the site.

Source: John Green, Sasquatch Incidents Data Base, Incident Number 1000209.

Drawing provided by the witnesses.

1980/12/13 – Cumberland (VI). Like a Man In Torn Rain Gear: Two hunters, Terry Kerton and Ken Berkeley, reported that they saw what they believe was a sasquatch near Cumberland at Comox Lake on December 13, 1980. The men were dragging a deer out of the bush when they saw the creature. It was leaning against a tree with one arm raised above its head and resting on the tree. The other hand was on its hip. The men observed the oddity for about 5 minutes. They estimated that it was about 50 yards away. There was not enough light to see which way it was facing or make out any details. However, what appeared to be long hair could be seen hanging off its arms. This led the hunters to initially believe that they were looking at a man in "torn rain gear."

Source: John Green, Sasquatch Incidents Data Base, Incident Number 1000210. (Photo/Artwork: Terry Kerton and Ken Berkeley.)

1980/12/23 – Rossland. Poor Old Trapper: A boy (12) out snowmobiling with his uncle (15) near Rossland, Nancy Green Lake area, reported a possible sasquatch sighting on December 23 or 24, 1980. The pair sensed an odor like rotten meat, and then the boy (alone) saw what he thought was a trapper run across the trail ahead and go into the bush, knocking the snow off bushes as he proceeded. The "trapper" was about 6 feet tall and looked like he was wearing a very long fur coat covered with snow. The boy went to where the trapper had been and found human-like naked footprints in the snow. He later met up with his father and told him that he saw a poor old trapper running around with no boots on. The father then explained to the boy that what he saw could have been a sasquatch.

Source: BFRO website

1980/YR/00 – Harrison Mills (Chehalis Reservation). Chehalis (Sts'ailes) Band Adopts Sasquatch Logo: Long known for their connection and association with the sasquatch, the Chehalis First Nations Band in the Harrison Mills area, BC, adopted a sasquatch image for their logo in 1980. Created in the traditional First Nations art style by Ron Austin, the logo shows a sasquatch looking straight at the viewer, reminiscent of the creature seen in the Patterson/Gimlin film.

Source: Information and logo photo provided by the Chehalis Band.

Chehalis Band logo.

1980/YR/00 – Vancouver. UBC Publishes Book on Humanoid Monsters: Following a conference on "humanoids" in May 1978, the University of British Columbia published a book in 1980 entitled, *Manlike Monsters on Trial: Early Records and Modern Evidence,* edited by Marjorie Halpin and Michael Ames, both professors of anthropology. The book is divided into three sections as follows:

I   Monsters in the Forests of the Mind

II   Manlike Monsters of the Native New World

III Contemporary Sasquatch Investigations

UBC book resulting from the conference on "humanoids" held in May 1978.

Although the sasquatch was given coverage at both the conference and in the book (publication of *some* conference papers), only about 20% of the book pages directly discuss the sasquatch as a reasonably possible real species. The rest of the book (80%) deals with "mind and myth."

The foregoing was the result of the highly disappointing and somewhat underhanded decision to *not* include all of the sasquatch-related papers presented. The presented papers omitted were by Dr. Grover Krantz, Professor Vladimir Markotic, Dr. James Butler, Dr. Carlton Coon, Grant Keddie, Jay Miller, and papers sent to the conference by the Russian hominologists Dr. Marie-Jeanne Koffmann, Dmitri Bayanov and Igor Bourtsev. The Russian delegates were invited to the conference, but were not allowed to attend by the Russian government. It might be noted that in view of these omissions, the book cover hardly reflects its contents.

Just why Halpin and Ames excluded the work of these these highly authoritative academics is difficult, if not impossible, to reconcile. I can only conclude that they did not want to totally bury the "myth" people. The official statement from Marjorie Halpin as sent to Dmitri Bayanov and Igor Bourtsev, and apparently the others, was as follows:

> The problem is essentially that the form of argument, and the evidence presented, is not adequate to resolve the problem as formulated. The canons of scientific argument are simply more rigorous than that.
>
> In order to give the problem the greatest possible chance to be taken seriously by the academic community, we cannot risk weakening the book by presenting an argument that can be dismissed by the scientist.

Every possible effort was made to get to the bottom of this professional "atrocity," and find out who was really responsible, but Halpin, who had so strongly supported all of the participants, was apparently cornered and could not say anything or do anything. The person who made the decision still remains unknown.

The epilogue of the book was written by Michael Ames. The following is the first paragraph, which I think more or less sums up the "forced" feelings of the editors:

> If monsters did not exist we would invent them, because we need them. And there lies the problem regarding the Sasquatch and other humanoid monsters. It is easy to make a case for monsters being inventions of culture: it is more difficult to demonstrate that they might also exist as creatures of nature, roaming real forests as well as the forests of the human mind. However, the urge to "prove" the natural existence of these anonymous creatures is a persisting one. Though neither the U.B.C. Conference on Humanoid Monsters nor this publication that results from it will likely satisfy that urge, the papers and discussions do suggest directions for further research and, the editors hope, will also help to establish parameters within which that work can be carried out.

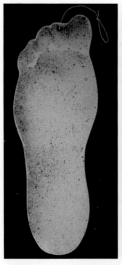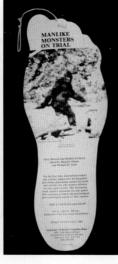

Promotional material used by UBC to sell their book. The footprint cut-out is 15.5 inches long.

Although the sasquatch as a "real creature" did not seem to impress Halpin, Ames and unknown others, there were apparently no misgivings in their using a sasquatch image on the front cover of the book; also, use of a frame from the Patterson/Gimlin film and a footprint cast image (actual size, but not from the film) to advertise the book (as shown on the right). This is highly ironic because they disallowed publishing the papers that explained and analyzed the film. Indeed, the paper by Bayanov and Bourtsev was the first to provide firm confirmation that the film was authentic. Their findings have never been refuted.

The omission of the papers mentioned resulted in Dr. Markotic and Dr. Krantz publishing a book entitled, *Sasquatch and Other Unknown Hominoids,* which included the papers omitted in the UBC book plus a great deal of other important information. *This book is far superior to the UBC book.*

That the Krantz/Markotic book resulted from the UBC conference needs to be acknowledged. In other words, had Halpin and Ames not held the conference, we would not have had the second book. All in all, we ended up with a university book (albeit poor to mediocre) and a proper book.

The detailed story of this event is provided by Dmitri Bayanov in his book *America's Bigfoot: Fact Not Fiction (1997).* There can be little doubt that the "world of science" can be very devious. This brings to mind a discussion I once had with René Dahinden. I asked him what he would do if he found a dead sasquatch. He replied, "I might just bury it. To hell with the scientists." I can still see him sort of holding his chin and looking up. He then said that having satisfied himself that the creature was real, he could stop looking for it and go on to something else. In some ways, I get the feeling he was not alone here. Indeed, such might already be the case.

Source: Dmitri Bayanov, 1997. *America's Bigfoot: Fact Not Fiction.* Crypto Logos, Moscow, Russia. pp. 81–83, 168, 170–172, 178, 179, plus discussions with John Green, Thomas Steenburg and Dmitri Bayanov. Also Marjorie Halpin and Michael M. Ames, editors, 1980. *Manlike Monsters on Trial: Early Records and Modern Evidence.* University of British Columbia Press, Vancouver, BC. Also Vladimir Markotic and Grover Krantz (editors), 1984. *The Sasquatch and other Unknown Hominoids.* Western Publishers, Calgary, Alberta, (Photos: Book cover and cut-out, UBC Press; the image of the sasquatch is public domain.)

1980/CI/00 – Kitimat. Appeared To Be Waving: Two men driving near Kitimat at William Creek Flats reported that they encountered a strange creature in about 1980. A ground fog covered the swampy area, and as they came around a slight bend they saw what they thought was a man waving his arms on the roadside. Thinking someone had driven off the road and needed help, they pulled off onto the road shoulder and stopped along side of the waving figure which was at a distance of several car lengths. They then saw that it was not a person. It was a very tall, hair-covered, human-like creature standing in the water at the foot of a grade that faced the highway.

The creature's eyes and mouth opened wide in fright. It threw its head back, crossed its arms over the front of its head, as if to be warding off a blow from a club, and then turned and dashed into the swamp. The men watched it go through waist-deep holes in the swamp until it disappeared in the fog.

The men said the creature's face was covered with hair, much the same as a full beard would on a man. They said its facial features were human-like, not like that of a chimp or ape.

An investigator reasoned that the creature was feeding at a bush, pulling the branches towards itself, thus giving the impression that it was waving its arms. The way the road traveled made it appear to be on the roadside.

Source: BFRO website, from "The Sasquatch in Northern British Columbia, Canada," HBCC UFO Research, April 07, 2001.

1981/06/14 – Vancouver. Dr. Halpin Provides Her Thoughts: Dr. Marjorie Halpin of the University of British Columbia provided her personal feeling on the sasquatch in a news article published on June 14, 1981. It will be recalled that Halpin, along with Michael Ames, organized the conference on "man-like monsters" in 1978 (see 1978/05/09 – Vancouver. Anthropology of the Unknown Conference) and the subsequent book, *Manlike Monsters on Trial* (see previous entry).

Dr. Marjorie Halpin with the footprint "cut-out" promoting the UBC book (see previous entry).

Halpin stated that she caught the sasquatch "bug" when she went to work for the Smithsonian Institution in 1968. She provided the following statement:

> The scientists there arbitrarily [Smithsonian] decided it didn't exist and that attitude got me more curious. Then when I moved here—right to Sasquatch county—I really got involved.
>
> If the Sasquatch is not there, how can we explain the fact that people report on it? All the witnesses are emphatic and sincere. They are convinced themselves so why should I doubt them? The question is, what is reality and what is perception.
>
> History shows that people have always wanted to believe in anomalous beings. There is legitimate anthropology of monsters. They've always been with us.
>
> People want to believe in mystery. But why they want to is a mystery in itself...

The world—let's face it—gets humdrum. We want something to play with and wonder and marvel about. We want some fantasy, imagination, mystery.

I'm especially fond of the Sasquatch because it has a way of confounding science.

The scientist's job is to explain the mysteries of the universe and the Sasquatch confounds that. By believing in it, we're protecting the mystery.

Source: "She's a believer." The *Province,* Vancouver, BC, June 14, 1981. (Photo: the *Province,* Vancouver.)

1981/11/00 – Squamish. Crouched By Water: A motorist reported seeing a strange creature after stopping at a rest area at night near Squamish in November 1981. He said his headlights illuminated the area across a small lake where the creature was seen crouched by the water. The motorist thought it was a bear, but it stood upright and walked into the bush on 2 legs with arms swaying and taking rather long strides [paces].

There is a discrepancy as to the location of this sighting. It has been reasonably taken to be Squamish because of the location description, however, there is a reference to the Sunshine Coast, a totally different area, across the strait from Squamish. It is likely the Squamish area was thought to be the Sunshine Coast.

Source: John Green, Sasquatch Incidents Data Base, Incident Number 1000215.

1981/YR/00 – Kaslo. Crossed Close to Car: Jim Wallington reported that he saw a sasquatch about 2 miles from Kaslo at about 11:15 p.m. during 1981. Wallington was driving into town when the creature crossed the road, close to his car. It was definitely walking on 2 legs.

Source: John Green, Sasquatch Incidents Data Base, Incident Number 1000213.

1982/01/30 – Saanichton (VI). Saw It & Ran: Vance Webster, 13, Tony Gruber, 9, and Larry Lambert, 12, reported that they saw an unusual creature near Saanichton on January 30, 1982. The boys were playing in a patch of woods when they encountered the oddity, which they described as 10 feet tall with hairy arms and legs. They ran from the area and reported what they had seen to the police who did not investigate the incident.

John Green was contacted and he interviewed the boys and went to the scene of the incident. The boys related that they were were investigating a growling they had heard. Webster first saw clearly an arm and a leg of the creature as it stepped out from behind trees onto the trail. When it was in full view, all of the boys saw it and ran. Larry Lambert said the creature came towards them before they ran away. The heights the boys provided were closer to 8 feet than 10. Webster said the creature was greyish black, Lambert that it was dark brown. Both boys, independently, pointed out the same exact location where the creature was seen.

Source: John Green, Sasquatch Incidents Data Base, Incident Number 1000216.

1982/06/10 – Fernie. I Don't Think a Man Could Leap Like That: Robert Harrison and Frank Mier reported seeing a probable sasquatch near Fernie at about 6:00 p.m. on June 10, 1982. The men were traveling in their truck along a dirt forestry road about 10 miles north of Fernie. Both men were fishermen and hunters. They came around a bend and saw a large, hair-covered creature standing on 2 legs in the middle of the road, facing the left side of the road. As the truck approached, the creature turned its head and looked at the oncoming vehicle for about 2 seconds then turned and started running down the road ahead. The men chased the creature for a brief time and when they got closer, it leapt off the road to the left and disappeared in the trees.

The men stopped at the place where the creature left the road and got out of their truck to see if they could again spot the oddity, but it was nowhere in sight. They found skid marks 9 feet down the embankment, which they believer were caused when the creature's feet touched down. On this point Harrison remarked, "It was an incredible leap. I don't think a man could leap like that..." The men searched around for actual footprints but did not find any.

Harrison described the creature as being just over 6 feet tall, and covered with short, reddish-brown hair. The face and hands were not hair-covered; the skin was a dark color, either dark gray or black.

They contemplated informing the RCMP of the sighting, but decided against doing so.

Source: Thomas Steenburg, 2000. *In Search of Giants: Bigfoot/Sasquatch Encounters.* Hancock House Publishers Ltd., Surrey, BC., pp. 27, 28.

1982/06/16 – Golden. Seemed To Be Looking At the Back Of the Train: Francis Rand reported that she saw a sasquatch from a train as it traveled through Glacier National Park, near Golden (between Revelstoke and Golden) on June 16, 1982. She wrote the following letter to Dr. Vladimir Markotic:

Sketch provided by Francis Rand of the creature she saw from a train.

I saw a sasquatch on June 16, 1982 at about 1:30 p.m., from a moving train on my way home to Calgary from Vancouver. It [the incident] was in Glacier National Park British Columbia, between Revelstoke and Golden. The sasquatch was about 75 feet away at the edge of the dense forest, sitting on a fallen log. I could see his left side, he seemed to be looking at the back of the train which, by the way, was very slow in that particular spot.

The sasquatch was very tall and very lean, I would say 7 feet or more. I remember thinking that he was much taller than my son-in-law who is 6 feet, 3 inches, and also very lean. He was covered with short, dark-brown fur, that resembled "Borg lining," used in winter wear. I noted the shape of his head because I marveled at the roundness of it, that fact eliminates a lot of wild game.

I did not see any ears. He looked like a very tall human all covered with fur. His left hand was up to his face as if he was eating something. His left leg bent at a 90-degree angle, his right one extended. I saw him for 4 to 5 seconds only. But what I saw was a sasquatch, it could not have been anything else, of that I am firmly convinced. I refer to the sasquatch as a he as I did not see any breasts.

We do not know if anyone else on the train saw the creature. Francis Rand provided a sketch of what she saw as shown here.

Source: Thomas Steenburg, 2000. *In Search of Giants: Bigfoot/Sasquatch Encounters*. Hancock House Publishers Ltd., Surrey, BC, pp. 79, 80, from a letter to Vladimir Markotic from Francis Rand, date not known. Also John Green, Sasquatch Incidents Data Base, Incident Number 1000217. (Photo/Artwork: Francis Rand.)

1982/10/00 – Vancouver. Scientific Proof Presented:

Dr. Grover Krantz, an anthropologist with Washington State University, held a special press conference at the University of British Columbia (UBC) in October 1982 to present new evidence on the existence of sasquatch. The press conference was organized by the International Society of Cryptozoology, which had its semi-annual meeting at UBC during which Krantz revealed his findings.

Dr. Grover Krantz

Krantz presented evidence of dermal ridges detected in alleged sasquatch footprint casts taken from prints found in the Blue Mountains, Washington. He said the footprint casts and photographs presented were made by the staff of the Walla Walla, Washington, office of the U.S. Forest Service and an Oregon search and rescue volunteer. Twenty-one prints were found and no human footprints were on, or anywhere near, the scene. He further stated that some of the prints are believed to have been made by an ape-like creature seen June 10, 1982 by forest service employee Paul Freeman near Walla Walla in the Blue Mountains, close to the Washington-Oregon border. Freeman said that the creature he saw was about 8 feet tall and weighted between 600 and 800 pounds. The footprints appeared to confirm these estimates. Other prints were obtained in the same general area on 2 subsequent occasions a week or so later. (Note: Although not stated in the source information for this entry, I believe all of the footprints were originally found by Paul Freeman.)

Referring to the dermal ridges, Krantz stated, "These are the same kind of ridges we have in our fingerprints, it is beyond the ability of anyone to fake the ridges."

Krantz said he called in a police fingerprint expert who concluded that the prints were not human. Also, he said that the toes were not those of an ape, nor those of a human. He said that the digits themselves are consistent in size, unlike the human foot with its significantly larger first toe.

Further analysis of the footprints was performed by police and other anthropologists. Krantz added that a police expert told him that whoever made the tracks had walked barefoot for a long time because some of the ridges were worn.

To my knowledge, no footprints or casts taken in BC have shown dermal ridges. The main evidence for such comes from footprints found by Paul Freeman and subsequent casts taken by him in the Blue Mountains, Washington (1982 and 1984). Nevertheless, what could be dermal ridges were observed in a print found on Blue Creek Mountain, California in 1967 and casts from Hyampom, California (1963), and Blanchard, Idaho (1977).

Unfortunately, Paul Freeman was considered by some (many?) researchers as a hoaxer. There is one incident where he submitted alleged sasquatch hair for analysis and it turned out be synthetic fibers. As I understand, Freeman admitted to the hoax. He claimed that he had submitted other samples but never got a reply, so decided to send in some hair from a toy (doll) to see what would happen.

Paul Freeman

Nevertheless, Dr. Jeff Meldrum and his brother, Michael, visited Paul Freeman in 1996 unannounced. Freeman took them to an area and showed them some tracks. They all left the area, and then Jeff and Michael went back and found additional prints in the same area but beyond those shown to them by Freeman. Analysis of the actual prints together with analysis of photographs and casts taken did not support that the prints were fabricated.

It also needs to be noted that Freeman took casts of an alleged sasquatch knuckle print and hand print. Both casts were deemed probably authentic by Dr. Krantz. The hand print alone was implied to be probably authentic by Dr. Henner Fahrenbach.

Thomas Steenburg, who knew Freeman well, is convinced that he was a hoaxer and dismisses any evidence associated with him. John Green is also of the same opinion. René Dahinden went on record in 1983 declaring footprints found by Freeman as being fabrications (see 1983/08/04 – Richmond. Dahinden Disputes Blue Mountain Footprints).

Paul Freeman died in 2003. There are those in the sasquatch/bigfoot field who claim he was a great researcher, and others, a blatant hoaxer.

Source: Moira Farrow, 1982. "Foot skin patterns 'prove existence of Sasquatch.'" The *Sun,* Vancouver, BC, October 23, 1982. Also Bob Chamberlain, 1982. "Toe Print specimens proof' of Sasquatch." The *Province,* Vancouver, BC, October 24, 1982. Christopher L. Murphy, 2010. *Know the Sasquatch.* Hancock House Publishers Ltd, Surrey, BC, pp 153, 154, 162. Also "Physical and Morphological Analysis of Samples of Fiber Purported to be Sasquatch Hair." *Cryptozoology, 10,* 1991, Allen Press, Inc., Lawrence, Kansas, USA pp. 55-65. Photos: Grover Krantz.

1983/07/00 – Hazelton. It Was Just Curious: Four children playing in the yard of a farmhouse near Hazelton (Kitwanga First Nations Reservation) reported they saw a sasquatch staring at them from nearby bushes in July 1983. It appeared to be about 7 feet tall, and very broad. It didn't make a sound or any movements, it just stood there watching the children.

The creature had caused some commotion with farm geese, and caused the family's dogs to run under the house. One of the children ran into the house and told his father and grandmother what was happening and they came out onto the porch. The father was shocked to see the creature and ordered all the children into the house. When one of the children asked his grandmother what it was, she simply said, "Oh, it's probably a bigfoot" and went into the house without a second look.

In later years one of the children remarked that he did not feel frightened, and thought that the creature was just curious about all of the children laughing and playing, and then went on his way.

Source: BFRO website. One of the witnesses was interviewed by a BFRO researcher.

1983/SU/00 – Gibsons. Tall Figure & Roars: A Young Men's Christian Association (YMVCA) camp leader, assistant, and 9 boys reported a strange experience near Gibsons, Tetrahydren Ridge area, during the summer of 1983. As they hiked, they saw a tall, black figure in the distance that appeared and disappeared in the bush. From what could be seen, its head was proportional to its body, but it did not have a neck. As the group proceeded, they found large footprints where the figure had been and tree branches broken off at the 8 or 9 foot level.

The camp leader became worried and wanted to turn back, but it was too late to make their home base before nightfall. The group therefore made camp almost at the summit of Tetrahydren Ridge. That night they heard terrifying roars. The camp leader said the following:

> We banked the campfire and began to enter our tent when the first of many "roars" (5 to 10) echoed off the surrounding mountain sides. It was unlike anything I have heard before or since and the power behind them was disturbing. The sound seemed to be generated effortlessly. We had to reassure the boys many times, and slept by the campfire, siting back to back, debating the risk of hiking down in the dark. Finally the sounds stopped, and although we thought we heard something in the bush, about 200 meters below us, the rest of the longest night I have ever experienced came to an end.

They broke camp after breakfast and headed back to home base. Nothing was seen or heard during the hike back.

TETRAHYDREN RIDGE

S

The Tetrahydren Ridge area, facing south.

Source: BFRO website. Report supplied by Kyle Mizokami. (Photo: Image from Google Earth; Data SIO, NOAA, U.S. Navy, NGA, GEBCO; Image © 2011 Digital Globe; © 2011 Cnes/Spot Image; Image U.S. Geological Survey.)

1983/08/04 – Richmond. Dahinden Disputes Blue Mountain Footprints: René Dahinden went on record on August 8, 1983 stating that he believed footprints found in the Blue Mountains, Washington by Paul Freeman were faked (see 1982/10/00 – Vancouver. Scientific Proof Presented). He came to this conclusion after interviewing Dr. Krantz in early 1983, then going to Walla Walla and interviewing Paul Freeman, forest service officials, and Joel Hardin, an experienced tracker with the US border patrol. Dahinden said that pine needles had been brushed away from inside the tracks; the prints of forestry staff on the scene sunk more deeply into the ground than the alleged sasquatch prints; and a dog and horses brought to area soon after the alleged sighting showed no reaction to the smell. He said that Joel Hardin had analyzed the prints (I believe the prints in the ground) and pronounced them fakes.

In discussing this issue with Dahinden in the early 1990s, he said he was sure that Paul Freeman had faked all of the prints. When I asked Dahinden how Freeman did this, he said, "simply carved them into the ground with his hands." This is a tough call, especially for 21 prints.

Source: "Alleged Sasquatch print are fakes, says Dahinden." The *Sun*, Vancouver. BC August 4, 1983. Also Jeffrey Meldrum, 2006, *Sasquatch: Legend Meets Science*. Tom Doherty Associates, New York, New York, USA, pp. 23–27. Also author discussion with René Dahinden.

1983/08/00 – Nanaimo (VI). Up A Tree: Gary Hurtig reported that he saw an unusual creature near Nanaimo in August 1983. He said he and some friends were shooting at apples with pellet guns in an area that was once a park/zoo of some sort. Hurtig walked away from the others by himself and was leaning against a tree when he saw something move about 12 feet up in another tree. He watched in that direction and a creature of some sort turned and looked toward him. It was a dark, hairy thing with big eyes (no hair around them) and a flat face. Hurtig estimated the oddity was 50 to 100 feet away. As he continued watching, the creature jumped down from the tree, knocking off a branch as it fell, and ran off into the woods.

Hurtig was frightened and also left the spot. He said that the creature's size was too big for a bear, and its face totally unbear-like. He thinks the creature had been watching the other boys as he approached from behind.

John Green interviewed Hurtig and investigated the sighting location. However, the site he was taken to did not seem to match adequately the area described.

Source: John Green, Sasquatch Incidents Data Base, Incident Number 1000219.

1983/AU/00 – Parksville (VI). Would Have Fired if Sure Not a Man: Ray Norbert, a hunter, reported that he saw a sasquatch near Parksville in the fall of 1983. Norbert was hunting late in the season at dusk, but still with good visibility. He was in a meadow east of the Island Highway when he heard a crashing in the forest on the far side of the meadow. He then saw the creature come out and walk along the fence line, about 100 yards away. The creature traveled the length of the field, walking like a man, and re-entered the forest at the end of field. It was dark in color and twice as high as the fence posts—about 8 to 9 feet tall. He said he would have fired if he were sure the creature was not a man.

Source: John Green, Sasquatch Incidents Data Base, Incident Number 1000218.

1984/03/00 – Kelowna. Two Sets of Tracks: A boy stated that he found 2 sets of human-like, bare footprints in the snow near Kelowna on Dilworth Mountain during March 1984. He was out hiking with his dog at the time. One set of prints was very large, around 17 inches long. The other set was about 10 inches long.

Both sets of prints appeared to have come from the direction of a residential area. They crossed an open field and then proceeded uphill towards a more forested area. The boy stated that from where he first saw the tracks, they continued on for about 150 yards before going into the tree line. His dog reacted to the prints and followed them. The boy called his dog back and left the area.

Source: BFRO website. The witness was interviewed by Blaine McMillan.

1984/04/12 – Vancouver. Vietnam Vet Hunts Sasquatch: On April 12, 1984 Vietnam Veteran Mark Keller, from California, said that he and 2 companions (one a former U.S. serviceman) planned to hunt down and kill a sasquatch. Keller said they were preparing for a 3-man, military-style expedition into a remote part of Washington, Oregon, and BC. He stated, "I could take photos, but all they would do is cause more arguments. To kill one is the only way to prove that Sasquatch exist." He said he does not have "one cubic centimeter of doubt" that sasquatch are real.

Source: " Sasquatch beware: hunt is on: Secret mission for Vietnam veteran: Special to *The Globe and Mail,* Toronto, Ontario, April 13, 1984.

1984/08/20 – Agassiz. Photos Snapped; Indistinct: Bill Bedry of Burnaby and friend Gordon Fland reported that they saw and photographed a sasquatch in the Agassiz area on August 20, 1984. The pair first saw the creature at a distance of about 115 to 145 feet in a gravel pit. The creature was trying to climb out of the pit and get into the woods. They got within about 98 feet of the oddity.

Bedry stated, "It looked hairy and everything else and made one hell of a racket." He said it was well over 6 feet tall, and looked to weigh between 300 and 400 pounds. It stayed on its "hind legs" at all times. He described the sound as "a big howling scream" that scared the hell out of them. He said it was definitely not a bear.

As the creature kept sliding downward on loose gravel, Bedry snapped three photographs. It eventually got out of the pit and dis-

Bill Bedry

appeared from view. Some men and several dogs went to the site on August 26 and found a footprint and a broken branch nearby.

Bedry said one photograph shows the creature's face, the other 2 are profiles. Steve Bosch, a *Sun* reporter, saw the photographs and said the shots are of an indistinct dark object against a dark background. A sasquatch researcher (probably René Dahinden) also looked at the photographs and said they were not conclusive.

Bedry said, "I kind of believe in sasquatches now because what I saw is some kind of creature. Like I said, it wasn't any bear, and I've never seen anything like that before."

Source: Chris Rose, 1984. "Did Bill Bedry photograph a sasquatch?" The *Sun,* Vancouver, BC, August 28, 1984. (Photo: Bill Bedry, the *Sun*, Vancouver, BC.)

1985/07/30 – Burns Lake. Car Struck 2 or 3 Times by Sasquatch: Glen Stewart reported to the RCMP that an unusual animal had attacked his car near Burns Lake (road from Pendleton Bay to Burns Lake) on July 30, 1985. Stewart's wife was with him at the time. Stewart told Conservation Officer Peter Stent that a large, very tall, hairy creature came out of the woods and ran alongside his station wagon for a short distance at 30 miles an hour. It was screaming and at one point struck the car 2 or 3 times. The officer did not find any marks on the car that appeared to have been made by an animal.

Source: John Green's Sasquatch Incidents Data Base Incident Number 1000221, from the *Burns Lake News* about August 7, 1985.

1985/07/00 – Revelstoke. Appeared To Look Both Ways: A motorist traveling to Kelowna reported that he saw a large, man-like animal cross the highway near Revelstoke between 6:00 and 7:00 p.m. in July 1985. He saw the creature at a distance of about 300 to

400 feet. It came up the side of the road ditch and appeared to look both ways on the highway to ensure no cars were coming. As it crossed, it looked straight at the motorist, who estimated that it was 8 to 9 feet tall. It had very big shoulders, long arms and was covered in orange-brown hair. Its weight was estimated at between 700 and 800 pounds. It took long steps (6 or 7) in crossing the highway, and walked with a stooped posture.

The motorist stopped and inspected the area where it was last seen. He could still see the creature moving away through the trees. He said he was going to wave some people down, but the oddity was moving too fast and, "just too many trees. It was gone."

Source: BFRO website; Witness interviewed by Blaine McMillan

1985/YR/00 – Queen Charlotte City (QCI). Deer Struck Down By Sasquatch: A man who was fishing near Queen Charlotte City on the Honna River reported that he saw a deer struck down and carried off by a sasquatch in 1985. The man saw what appeared to be a very tame deer come out of the woods and drink from the river right by where he was fishing. Suddenly a rock flew out from the trees, striking and knocking down the deer. A tall, hair-covered, human-like creature immediately emerged from the woods, threw the deer over its shoulder, and dashed back into the woods. The man quickly left the area.

Source: BFRO website, from "The Sasquatch in Northern British Columbia, Canada," HBCC UFO Research, April 07, 2001.

1986/04/00 – Lone Butte. Trail Led To a Swampy Lake: John Duthie and his wife saw an unusual creature near Lone Butte in April 1986. The couple had ascended a hill to look at the view. Down below they could see a small section of curved road, on which a large, dark (brown/black, mottled), very heavy creature was walking erect. It did not have a snout. It turned off the road and into the trees. Later Duthie returned with a friend who knew what part of the road was in sight from the hill. The friend stayed on the road and Duthie went back on the hill, enabling him to make a size comparison between the creature he saw and his friend. He estimated that the oddity was about 8 feet tall. They had a dog that followed a scent through the trees, leading to a trail of crushed old vegetation (reeds and so forth). This trail led to a small swampy lake (a distance of about 200 yards). At this point whatever had made the trail apparently swam off.

Source: John Green, Sasquatch Incidents Data Base, Incident Number 1000104.

1986/SP/00 – Golden. A Gorilla, Dad! It Was a Gorilla!: An 8-year-old boy said he saw "a gorilla" near Golden in the spring of 1986. The boy was traveling with his father in their truck on the Trans-Canada Highway. At about 1:30 a.m. the father felt fatigued, so pulled off the highway and parked in a quiet spot to get a little rest.

He had no sooner turned off the truck's motor when his son grabbed him in a state of panic and begged him to get out of there. The boy was in such a state that his father thought it best to simply drive away.

A little later after the boy had cooled down, his father asked him what was the problem. The boy said, "A gorilla, Dad; it was a gorilla. It was coming up to the truck." The boy had looked out the window and saw the creature approaching. He thinks it actually came up and leaned on the truck. He was absolutely positive that it was walking on 2 legs.

It is, of course, easy to dismiss a sighting of this nature, writing it off as probably just a man coming over to the parked truck. However, the eyes of an 8-year-old are generally very good, and for him to say that he saw a gorilla deserves some thought.

Thomas Steenburg interviewed the boy about 6 years after the incident and was somewhat impressed with his story. At no time did the boy mention the word "sasquatch" or "bigfoot." He simply confirmed that what he saw looked like a gorilla.

Source: Thomas Steenburg, 2000. *In Search of Giants: Bigfoot/Sasquatch Encounters.* Hancock House Publishers Ltd., Surrey, BC, pp. 29–31.

1986/05/21 – Harrison Hot Springs. Hair Looked Stiff: Adam Harvey reported seeing a sasquatch near Harrison Hot Springs at Deer Lake on May 21, 1986 (Sasquatch Provincial Park). He and Lucas Groenfeld had camped next to a water pump that was by a big hill and a meadow. At 3:00 a.m. Groenfeld saw someone (thought it was a person) look into their tent. Harvey got up, went outside, and shone a flashlight downhill from their position. He saw a brown creature 7 to 8 feet tall, standing upright. He held the flashlight on it and watched it run off, jumping over a 4-foot stump and disappearing from view. He said it was running very fast and noticed that its hair looked stiff. Its nose was flat and its eyes appeared greenish-yellow.

Deer Lake in Sasquatch Provincial Park.

Source: John Green, Sasquatch Incidents Data Base, Incident Number 1000204. (Photo: C. Murphy.)

1986/06/15 – Gibsons. What Looked Like a Bed: Two boys hiking up a mountain trail near Gibsons reported they saw a strange creature at about 11:00 a.m. on June 15, 1986. They described it as black in color and noted that its back touched the branch of a tree. It walked off into the bush and the boys went to where it had been. They sensed an odor like a wet dog, and judging by the height of the tree branch the creature had brushed they said it must have been 9 to 10 feet tall.

The boys saw a trail that the creature had made in the bush and followed it to a point under some cedar trees. They saw huckleberry bush leaves on the ground that were shaped into what they think was a bed.

As huckleberry bush leaves are very small, it would take a great many of them to make a "bed." The leaves were likely a mixture containing huckleberry leaves.

Source: BFRO website.

1986/07/00 – Coalmont. Appeared To Be Female: A man (resident) reported he saw an unusual creature near Coalmont in early July 1986. On his way down to a local creek, he saw something on the roof of an abandoned cabin about 100 feet away. It was hunched over or squatting. The resident stated:

> It was big, and it was light brown, about the color of a cinnamon bear. That's what I thought it was at first. Then it stood up and it wasn't a bear.

As the creature was on the far side of the roof, the lower part of its legs were not visible, thus its full height was difficult to determine. The resident was able to see its face and described it as follows:

> The face was sort of a blackish color, and there wasn't much hair on the face. I remember it had a large mouth, but the mouth was closed. I didn't see the color of the eyes. It had no nose that I could see, looked like to holes where the nostrils should be.

He noted that the hair on its head was not much longer that that on the rest of its body, and was the same color as the other hair. The head itself was:

> ...sort of pointed, not real pointed like a gorilla, but it had more of a point than a human. The point was sort of in the middle of the skull, not at the rear like a gorilla or ape.

He also noticed breasts that were very long and droopy, indicating to him that the creature was without doubt a female. The breasts were covered in hair the same as on the rest of its body.

The creature's arms were not overly long for the body, "They

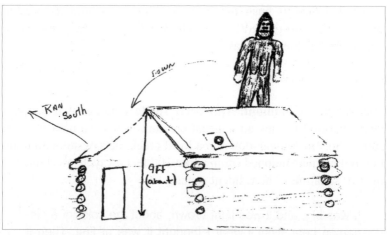

Drawing of the creature seen by the resident created under his direction.

looked pretty much like a human." However, he pointed out that they were very muscular. As to its hands, "The hands were about level with the knees, not hanging below the knees like a gorilla or orang."

The full time the creature was observed was about 5 to 10 seconds. It simply stood there, looked toward the resident and then jumped off the roof and ran into the bush. He checked behind the cabin to see if there were any footprints where the creature had jumped down, but the ground was very hard so nothing was seen.

Remarkably, a relative of the resident stated that he saw what was likely the same creature down by the same old cabin on July 26, 1986. He provided a sketch of the creature as seen here. We do not have details on this incident.

Drawing provided by a second Coalmont resident of the creature he saw near the old cabin in late July 1986.

Source: Thomas Steenburg, 2000. *In Search of Giants: Bigfoot/Sasquatch Encounters.* Hancock House Publishers Ltd., Surrey, BC, pp. 94–98. (Photos/Artwork: Provided by the witnesses.

1986/08/00 – Chilliwack. Sasquatch Snatches Fish Catch: Bobby and Cathy Harris, a vacationing elderly couple, reported seeing a sasquatch while fishing near Chilliwack along the Chilliwack River in August 1986. Bobby had just returned to the river after hanging his catch of about 8 fish on a tree near their camper. He had not finished his first cast when Cathy cried out in alarm and pointed back to the tree. Bobby looked in stunned silence, still holding his fishing pole halfway through a cast.

The Harris's campsite. The fish catch was hung on one of the trees.

To the couple's astonishment, a dark gray, hair-covered creature, walking on 2 legs, grabbed the fish from the tree and walked off through the bush. It crossed a nearby road and disappeared into the forest on the other side. Bobby and Cathy stood and stared like statues. Finally, Cathy broke the silence, "What the hell was that?" she exclaimed. "That was bigfoot, and I still don't believe it," Bobbie replied.

The Harris's weren't the only witnesses to the unusual theft. At least 2 other campers elsewhere in the

Thomas Steenburg at the location where large footprints were found.

One of the footprints found by Steenburg. It was 18 inches long.

campsite saw what happened and rushed up to the Harris's camper. The creature was not seen again, and the hardy Harris couple recovered from the incident. They stayed on for another 2 days and happily replaced the stolen stash with a fresh catch.

After hearing of the incident from one of the other witnesses (who had left the day after the encounter), researcher Thomas Steenburg arrived at the site just as the Harris couple was pulling out. He was able to talk to them briefly and got a first-hand account of the incident.

Steenburg had a good look around the campsite and then decided to venture into the forest across the road where the creature disappeared (directly behind the campsite). Along a small creek that flows into the river he found a line of footprints. The best prints, which were right beside the creek, were up to 18 inches long. The line of prints headed north through a grassy clearing and disappeared at a rock slide area. In all, 112 footprint impressions were counted (many just recognizable because of the hard ground). The best prints were photographed and a cast was made of one of them.

Thomas Steenburg with a cast he made of one of the footprints he found near the Chilliwack River.

Source: Thomas Steenburg, 1990. *Sasquatch/Bigfoot: The Continuing Mystery.* House Publishers, Surrey, BC, p. 60–62. Account provided by T. Steenburg (more information has been provided here than in the book referenced). (Photos: campsite, Steenburg on location, C. Murphy; footprint in soil, T. Steenburg; Steenburg with cast, C. Murphy.)

1986/10/00 – Bella Coola. Ripped the Screen Door Off: William Milligan and his wife, caretakers at Whonnock Industries logging camp near Bella Coola at South Bentinck Arm, reported a sasquatch encounter during October 1986. Mrs. Milligan woke her husband at 3:00 a.m., having heard something at the door. Mr. Milligan pounded on the wall, loaded a shotgun and turned on the outside light. Right under the light he saw (through the window) a tan colored flat face with dark brown hair at the same level as his face. The creature was standing on the porch which was 16 inches lower than Milligan's floor level (he is 6 feet tall). During this time, the creature had ripped the screen door off and was tearing wood off the window frame in the door.

Milligan has very poor eyesight, but could clearly see the creature's face about 18 inches away and well lit. At this moment, the creature turned, jumped off the porch, and ran out of range of his vision. Wide claw marks were found on the door.

Milligan remarked that he had been at the camp for 3 years, and has seen huge footprints in the area 2 or 3 times. They were deep in the sand of side channels of the nearby river, and were like a man's print but with short toes.

Source: John Green, Sasquatch Incidents Data Base, Incident Number 1000223.

1987/03/14 – Dawson Creek. Oil Rig Crew Sees Sasquatch: Workers on an oil rig near Dawson Creek reported that they saw an unusual creature on the night of March 14, 1987. Myles Jack first saw the creature at between 50 and 75 yards. It was in a kneeling position and then suddenly stood upright. Other men then saw it and all said that it looked about 7 to 8 feet tall and weighed between 350 to 400 pounds. Jack said that it looked, "more like a man than an animal."

Jack then called to his buddy, Brian Mestdagh, to have a look, and as he did he clearly saw the creature run off, still upright, into the bush. Jack said it took huge steps—twice those of the average man. It then circled the clearing around the oil rig, peering intermittently at the men as it did so. It twice crossed the unpaved access road that ran through the camp.

Four of the 7 crew members who got a good look at the creature were outdoorsmen and hunters. They all dismissed any chance of it being a bear.

Province newspaper reporter Ann Rees and photographer Les Bazso were in Dawson Creek at the time. Upon hearing of the sighting, they went out the next day and investigated the incident. The creature's footprints were clearly visible in the powdery snow. They measured about 15 inches long and were about 54 inches apart (measurement process not known).

Where the creature had been seen kneeling there was a V-shaped marking, which could have been where its knees touched the ground. Behind this marking, there were 2 indents, possibly made by its toes.

When interviewed by the press, Myles Jack stated, "I never imagined I'd see anything like that in my lifetime." Brian Mestdagh added, "I don't care whether people believe me or not, I saw what I saw." He tried to follow the tracks and stated that sometimes they would head into the bush and then just stop; no more tracks. He never did find a trail he could follow for any distance.

Myles Jacks with his hands in 2 of the footprints.

John Green and Bob Titmus went to the site and investigated the incident. More details are found on Green's data base. Commenting on the photographs taken, Green stated:

From the description and the pictures, each leg was about a foot wide. The knees measured about 3 feet apart. A human can't get his knees 3 feet apart no matter how big he is."

That there happened to be newspaper people in town at the time was looked into and ruled out as a possible "hoax" factor.

Source: Don Hunter with René Dahinden, 1993. *Sasquatch/Bigfoot: The Search for North America's Incredible Creature.* McClelland & Stewart Inc., Toronto, Ontario, pp. 5–7, from the *Province*, Vancouver, B.C., March 16, 1987. Also, Terry Gilbert, 1987. "Sasquatch Watch: Sighting claim sparks bid deal over Bigfoot." The *Sun, Vancouver,* BC, April 29, 1987. Also John Green, Sasquatch Incidents Data Base, Incident Number 1000245. (Photo: Les Bazso, *Province* photographer.)

1987/04/13 – Fort Nelson. It Was a "Nahganee" A motorist traveling with a friend and his small daughter reported that they spotted a man-like creature in the Fort Nelson region (Highway 77 to Fort Laid, NWT) on April 13, 1987. The oddity was spotted beside the highway between 10:00 and 11:00 a.m, at a distance of between 200 and 300 yards. The motorist said the creature was dark, appeared to be about 8 feet tall, and looked to weigh over 500 pounds. It took a few large steps and quickly disappeared into the bush. He said, "We call it *Nahganee,* which means bushman in the Dene language."

Source: BFRO website. Witness interviewed by Stan Courtenay.

1987/08/00 – Golden. It Stayed On 2 Legs the Whole Time: Agnes Perkins and Charlotte White are sure they saw a sasquatch near Golden at Rogers Pass during August 1987. The women saw what they thought was a man on the roadside about 800 yards ahead. As they got closer, they saw that it was a 7-foot tall creature covered in black hair.

As they continued to approach the oddity, it suddenly turned right and climbed up the steep hillside. "It stayed on 2 legs all the time," Perkins stated.

The women did not stop. Perkins kept her eyes on the

Rogers Pass is a narrow valley surrounded by a number of mountains. It is formed by the headwaters of the Illecillewaet River to the west and by the Beaver River to the east. Both of these rivers are tributaries of the Columbia River, which loops about 150 miles around to the north of the pass. (Wikipedia)

297

road and White watched the creature until it entered a thick growth of lodgepole pines high up on the hill. Perkins pointed out that she was impressed with distance the creature covered as the car went by.

Source: Thomas Steenburg, 2000. *In Search of Giants: Bigfoot/Sasquatch Encounters*. Hancock House Publishers Ltd., Surrey, BC, pp. 26. Also John Green, Sasquatch Incidents Data Base, Incident Number 1000224. Also Wikipedia. (Photo: Rogers Pass sign, Wikipedia Commons, please see the Wikipedia site for details.)

1988/07/05 – Harrison Hot Springs. Showed Intelligence: Joseph Verhovany reported that he saw a sasquatch picking berries in the Harrison Hot Springs area, mouth of the Silver River (east side of Harrison Lake) at 4:00 or 5:00 p.m. on July 5, 1988.

Verhovany had gone into the area to get berries, and it was in this process that he saw the creature. He stated he was about 100 feet from the oddity and observed it eating berries. It drew a branch to its mouth and ate just the ripe berries. As it did not destroy

Joseph Verhovany creating a sketch of the creature he saw.

the berry bush in this process, Verhovany got the impression that the creature was very intelligent.

He estimated the creature was over 6 feet tall. It stood on 2 feet and was covered in hair. He observed it for about 3 minutes. The creature turned and looked at Verhovany, stepped backwards for a few feet, and then turned and ran away. After it had gone, Verhovany said he sensed a sulphurish odor, like something rotten. He did check for footprints, but they were only marginally visible in ground debris.

Source: Thomas Steenburg, 2000. *In Search of Giants: Bigfoot/Sasquatch Encounters*. Hancock House Publishers Ltd., Surrey, BC, pp. 39–42. Also John Green, Sasquatch Incidents Data Base, Incident Number 1001365. (Photo: J. Green.)

1988/08/24 – Wycliffe. Too Tall To Be a Man: A motorist and his friend reported seeing a strange, upright walking creature near Wycliffe on Highway 95 at about 10:30 p.m. on August 24, 1988. They saw the oddity come out of the trees on the left side of the road and cross the road in front of their car. It simply crossed at a fast walking pace and disappeared into the trees on the other side of the road. One of the men described the creature as covered in black hair and being very large. He said it was too tall to have been a man.

This report was given to Thomas Steenburg by one of the men. Steenburg asked him if he would submit to a detailed interview, but he declined as he had a reputation as a "no-nonsense" sort of guy and didn't want it "getting out" that he saw a sasquatch.

Source: Thomas Steenburg, 2000. *In Search of Giants: Bigfoot/Sasquatch Encounters.* Hancock House Publishers Ltd., Surrey, BC, pp. 27.

1988/09/08 – Woss (VI). Thought It Was His Son: A motorist reported that he saw a strange creature in the Woss area along the Nimkish River on September 8, 1988. He was traveling with his 3 sons and had stopped at a rest facility. They heard the rustling of grouse behind the outhouse and decided to flush them out to get a look at them.

The father sent his sons into the bush to flush the birds toward him, and then proceeded down a trail and took a position to view the birds. As he waited, he noticed what he thought was one of his sons standing and leaning against a tree with his back to him, but he was in the wrong place. He shouted at the boy asking why he was there and got the surprise of his life. What was standing by the tree turned around and revealed itself to be a hair-covered creature, dark brown or black in color, with wide shoulders. The long hair on its neck and shoulders flew straight outward. It simply walked off with a graceful stride, "as if it were very fit."

The boy, assumed to be in the wrong place, shouted back to his father from exactly where he should have been. The father quickly ordered his boys back to their vehicle and left the area immediately.

Source: Dr. John Bindernagel, 1998. *North America's Great Ape: The Sasquatch.* Beachcomber Books, Courtenay, BC, p. 10.

1988/10/00 – Strathcona Provincial Park (VI). Wildlife Biologist Finds Sasquatch Tracks: Dr. John Bindernagel, a wildlife biologist, and his wife Joan, found sasquatch tracks in Strathcona Provincial Park during October 1988. The tracks were found in 2 muddy places on a hiking trail. They measured about 15 inches long. The best track was partially overprinted by a hiker's boot sole. Dr. Bindernagel made a cast of this track (see 2000/SU/00 – Courtenay [VI]. Wildlife Biologist Says It All). He has been interested in the sasquatch issue since 1963, and started research in BC in 1975.

Dr. John Bindernagel

Source: Dr. John Bindernagel, 1998. *North America's Great Ape: The Sasquatch.* Beachcomber Books, Courtenay, BC, p. 57, 213. (Photo: C. Murphy.)

1989/SP/00 – Harrison Hot Springs. Flexed Its Back: Betty Unger reported a sasquatch sighting near Harrison Hot Springs on a road in Hemlock Valley at about 4:00 p.m. during the spring of 1989. She was driving with her 9-year-old grandson, Dallas Yellowfly, and they both saw the creature in the roadside ditch. It was ascending the steep bank below the road. Unger thought it was a small black bear. She stopped her car to watch as it reached the road, stood erect, and flexed its back. She said that it looked very human-like, but covered in dark brown hair. She estimated that it was less than 6 feet tall. The creature crossed the road diagonally towards the car, coming within about 30 feet. It paused in its passage, looked at both Unger and her grandson, then jumped up onto the bank above the road on the other side. It looked back down at them for a moment and then continued up the bank and disappeared in the trees. Unger said that its movements were faster than a human could move. She described its face as dark with little hair. She said she could clearly see its facial expression, muscle movement and eye movement.

Hemlock Valley with Harrison Lake in the background on the right.

Source: John Green Sasquatch Incidents Data Base, Incident Number 1000225. (Photo: Image from Google Earth; Image © 2011 Province of British Columbia; Image © 2011 DigitalGlobe; © 2011 Cnes/Spot Image.)

1989/09/20 – Fort St. James. Strange Spur: Walter Patrick reported finding strange footprints near Fort St. James along the Stuart River in September 1989. Patrick, a foreman of a salmon enhancement program, found 8 prints about 11 inches long, and slightly wider than a human foot in one location, then 2 sets of the same type of prints, but smaller, behind a clump of bushes.

The prints at the first location sank about 2 inches into the ground and had pointed toes. The smaller toes inclined sharply from the big toe. The arch area was curved and the inside heel portion seemed to have a strange bump with a type of spur jutted from the back of the heel. There were no signs of claws.

Patrick speculated that a large creature had 2 smaller ones along but kept them at a distance.

Source: "Strange footprints found near northern B.C. river said to be from Sasquatch." *Canadian Press* release (newspaper not known), September 20, 1989.

1989/11/00 – Bella Coola. Running Like a Human: Glen Clellamin, his mother, and Jimmy Nelson, reported an encounter with an erect-walking creature near Bella Coola on November 11, 1989. The three were sitting around the kitchen table talking at the Clellamins' home. They suddenly sensed a terrible odor, "like a dead dog." Glen left the table to get a drink of water, and looking out the window saw some sort of animal. He exclaimed, "There's an animal out there!" thinking it was a bear that was after deer meat they had hanging. Glen and his mother immediately came to the window. They saw a large, hairy creature with wide shoulders and about 7 or 8 feet tall, running away like a human. The got a good look because the porch light was on and it illuminated the yard.

The following night, Clellamin's dog started barking at something and when the boys looked outside, they spotted the creature in the yard next door. When it ran off, the boys followed it down the valley toward a creek. They heard a growl and then a high-pitched scream that made the hair stand up on the back of their necks. Their curiosity was greater than their fear, so they kept moving (although much slower) toward the creature, closing in on it. When they were about 30 feet away, it stopped moving and looked at them. The boys yelled at it and the creature started to walk towards them. The pair then ran in the opposite direction.

The following morning, they went to look for tracks in the area where the creature had last been. They found 3 sets: one large, one medium, and one small. It appears a family of the creatures had been there.

Several people later came forward and said that they had seen the creatures, some the large one, and others the medium or small one. Footprints were photographed and provided to a local newspaper (*Coast Mountain Courier*).

Source: Thomas Steenburg, 2000. *In Search of Giants: Bigfoot/Sasquatch Encounters*. Hancock House Publishers Ltd., Surrey, BC, p. 79, from the *Coast Mountain Courier*, Hagensborg (Bella Coola Valley) BC, November 1989. Also John Green's Sasquatch Incidents data base. Incident Number 1000247.

1989/YR/00 – Burnaby. BCSCC Formed. In 1989 the British Columbia Scientific Cryptozoology Club (BCSCC) was formed in Burnaby by writer James A. Clark, scientist Dr. Paul LeBlond and journalist John Kirk. The organization comprises a broadly-based membership of enthusiasts dedicated to the investigation of various animals as yet unidentified by science—in particular, BC cryptids such as sasquatch, Ogopogo, and Cadborosaurus.

BCSCC logo.

John Kirk (left) and Dr. Paul LeBlond, 2 founders of the BCSCC.

Source: BCSCC website: http://www.bcscc.ca/membership.htm. (Photos: BCSCC logo and Kirk portrait, J. Kirk; LeBlond portrait, C. Murphy.)

1989/DE/00 – Sechelt. Students Attacked With Barrage of Rocks & Debris: A group of high school students sleeping out on a beach near Sechelt said something attacked them in the night during the 1980s. Around midnight they were awakened by loud animal sounds and the sound of breaking branches. This was immediately followed by a barrage of sticks and small logs from the adjacent forest. The terrified students ran off, abandoning their sleeping bags and other equipment.

The next day they went back to the spot to retrieve their belongings. Their sleeping bags had been ripped apart and strewn across the beach. They investigated the forest area and found a wide path of torn-up vegetation and churned up soil just inside a shrubbery border.

Source: Dr. John Bindernagel, 1998. *North America's Great Ape: The Sasquatch.* Beachcomber Books, Courtenay, BC., pp. 118, 225, from information provided to Graham Suther of Denman Island by fellow high school students from the Sechelt Peninsula.

1990/10/00 – Ottawa, Ontario. Canadian "Sasquatch" Postage Stamp Issued: In October 1990 the Canadian Postal Service, Ottawa, released a postage stamp depicting a sasquatch. The stamp was part of the Postal Service's Folklore Series: Canada's Legendary Creatures. The Patterson/Gimlin film, which was taken in California, was instrumental in the stamp issue.

This was the first, and currently only, government postage stamp that depicts a sasquatch. The Postal Service newsletter that announced the stamp had several stories of sasquatch sightings, plus an image from the Patterson/Gimlin film (frame 352).

Canada's "The Sasquatch" postage stamp.

Source: "Sasquatch: Wild Man of the Western Forests." *Heritage Post,* published by the CRB Foundation and Canada Post Corporation, Antigonish, Nova Scotia, 1990, (Photo: Sasquatch Stamp, © Canada Post Corp., 1990.)

1991/05/00. Radium Hot Springs. Long Blond Hair: Courtenay Huggins reported seeing an odd bipedal creature near Radium Hot Springs on the Stanley Glacier Trail in May 1991. Huggins was one of 6 in a group walking single file, with between 10 and 30 yards between them. The creature ran across the trail between the first and second hiker (Courtenay Huggins), who said it was about 6 feet tall with long blond hair. No clothing could be seen, just hair.

A small dog belonging to one of the hikers ran after the creature, then the first and second hiker (now together) heard a tree creaking as it about to fall some 75 yards away. They could see the top of a dead burnt tree moving with the creaking sound, and could hear the dog barking in the distance. All of a sudden the tree came crashing down and the dog stopped barking. The dog's owner frantically called his dog, which came back about 15 seconds later.

The group reasoned that the creature had pushed the tree over, and thought it best to leave the area right away.

Source: T. Steenburg Sasquatch Incidents File BC No. 10125. Steenburg interviewed Courtenay Huggins and his brother, Tyler, who was also with the group. Also BFRO website. One of the witnesses was interviewed by Blaine McMillan and later by Matt Moneymaker.

1991/09/18 – Golden. Well Built: Voytek Tertell reported that while driving near Golden through Rogers Pass he saw a large, hair-covered, bi-pedal creature on the roadside on the night of September 18, 1991. His account is as follows:

All of a sudden I saw it in the headlights—what I thought was a grizzly bear. I was going about 75 miles per hour, so I didn't see it for a very long time. I'm positive it was brown. I'm positive it was walking upright—well built, but not fat, slim sort of thing, and it was traveling in the same direction so I didn't see any facial features. Just as I was passing it, it kind of wandered off to the right towards the woods.

Source: Thomas Steenburg, 2000. *In Search of Giants: Bigfoot/Sasquatch Encounters.* Hancock House Publishers Ltd., Surrey, BC. pp. 39–42. Also John Green's Sasquatch Incidents Data Base. Incident Number 1000107.

1992/06/00. Port Townsend, Washington, USA. British Columbia Could Be "Last Stand" For Sasquatch: Biologist and nature writer David George Gordon stated in Port Townsend, Washington in June 1992 that BC could be the last stand for the sasquatch. Gordon, who authored the book, *Field Guide to the Sasquatch,* provided his views as follows, based on forest clear cutting and the population explosion pushing northward from California:

> I think we're greatly impacting the range of the sasquatch. We are quite likely seeing the last of a dying breed. It's highly likely B.C. would be the last great stronghold of these creatures."

Gordon, who refuses to take a firm position on the sasquatch issue, is nonetheless appalled that scientists choose to ignore the creature.

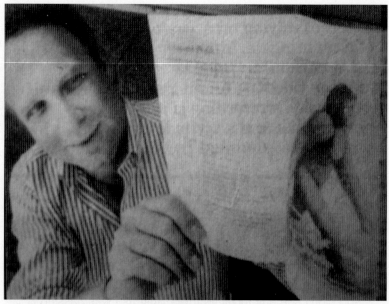

David George Gordon

Source: Larry Pynn, 1992. Sasquatch sage suggests man-beast may make last stand in B.C. The *Sun*, Vancouver, BC June 4, 1992. (Photo: Mark Van Manen, the *Sun*, Vancouver, BC.)

1992/SU/00 – Comox (VI). Night Sounds Frighten Campers: Mary Strussi and her friend had a frightening night while camped near Comox at Comox Lake in the summer of 1992. The lake is just outside Strathcona Provincial Park.

During the night they heard frightening sounds, which included a mournful bellowing accompanied by the sounds of an animal running back and forth and breaking branches in the nearby forest. The pair abandoned their camp for the security of their vehicle. The sounds continued intermittently all night long.

Source: John Bindernagel, 1998. *North America's Great Ape: The Sasquatch.* Beachcomber Books, Courtenay, BC, pp. 122, 227, from information provided by Mary Strussi.

1992/10/00 – Nelson. Half As High As the Trees: Mark Hourie reported that he saw a sasquatch through his binoculars near Nelson in the Blueberry–Paulson Pass in October 1992. Hourie was helping friends spot deer. The creature was half as high as the 17-foot trees. He knew it had to be a sasquatch because it was walking on 2 legs and covering too much ground. It was black, and walked with its arm swinging.

Source: John Green Sasquatch Incidents Data Base, Incident Number 1000226, from Suzy Hamilton, 1993. "Ape-man alert: Kootenays awash in Sasquatch sightings. The *Province,* Vancouver, BC, January 12, 1993.

1992/10/00 – Castlegar. Mystery For the Rest of My Life: Hunters believe they saw a sasquatch near Castlegar in the Christina Lake area during October 1992. They watched the creature through binoculars. "It was running bare-naked through three feet of snow," they said. It ran effortlessly without snowshoes. They described it as being black, and a minimum of 8 feet tall. One of the men remarked, "I'll have a mystery in my mind for the rest of my life."

Source: Clare Ogilvie, 1993. "Believers spellbound by bigfoot: Meeting draws experts, laymen." The *Province,* Vancouver, BC, May 23, 1993, from a talk provided by one of the hunters at a sasquatch conference at Harrison Hot Springs in May 1993.

1992/AU/00 – Fort Nelson. In the Middle of Nowhere: K.R. Tubbs reported seeing an unusual creature while driving near Fort Nelson in the Laird Hot Springs Provincial Park in the fall of 1992. He saw what he thought was a person on the right side of the road up ahead, and wondered what someone would be doing out there "in the middle of nowhere."

As Tubbs got closer to the the oddity, he could see that it was very tall, perhaps over 7 feet, very wide at the shoulders, and it was all the same color—brownish-gray with a lighter tone on the outside. It appeared to be pulling on tree branches on the hillside next to the road.

When Tubbs got to within one-half mile of the creature, it turned around and walked on 2 legs across the road where there was a steep cliff. The creature simply went "over the edge."

Tubbs stopped and inspected where the creature had first been, and saw that the snow was all disturbed in one place. He could not see any definite footprints. He went to the other side of the road and saw that the snow was disturbed down to where the drop-off hit the slope on the bottom land, which had many alder trees and a stream. He said that the trail looked like brooms had swept a path, although there were deep impressions spaced at about the distance of those that would be made if a person were running.

Tubbs waited for awhile to see if he could again see the creature, and then heard what he said sounded like a cross between a howl and a scream in the woods below. Then, about 300 yards to his right, he saw the alders flattening, like something huge was moving through them—like watching a car slowly driving through a cornfield. He said the alders had to be a couple of inches thick.

Source: Linda Coil Suchy, 2009. *Who's Watching You? An Exploration of the Bigfoot Phenomenon in the Pacific Northwest.* Hancock House Publishers, Surrey, BC, pp.155–158, from information provided by K.R. Tubbs, CEO of a packaging company.

1992/YR/00 – Port McNeill (VI). All On 2 Legs:

Dennis Richards reported that he saw a sasquatch near Port McNeill on Crease Island in 1992. Richards was in his boat at the time, traveling to Knight Inlet. He said the creature was walking along the island beach. It stepped over some large beach logs and then disappeared into the forest. It traveled a distance of over 200 feet, all on 2 legs. He mentioned that the size of the logs eliminated a bear; even if it could walk that distance on 2 legs, its legs would have been far too short.

Source: John Bindernagel, 1998. *North America's Great Ape: The Sasquatch.* Beachcomber Books, Courtenay, BC, pp. 29, 209, from personal communication.

1992/YR/00 – Pullman, Washington, USA. First Scientific Book On Sasquatch. Dr. Grover Krantz, a physical anthropologist at Washington State University, has authored the first truly scientific book dealing with the sasquatch. The book, titled, *Big Footprints: A Scientific Inquiry into the Reality of Sasquatch,* released in 1992, presents his findings on the creature since his involvement in the sasquatch issue that began in 1963. The book painstakingly argues the case for the existence of the creature. It is like a university text, with exhaustive detail, diagrams, and photographs of footprint and hand print casts.

Source: Stewart Bell, 1993. "On a myth and a footprint." The *Sun,* Vancouver, BC, October 23, 1993. Also, Grover Krantz, 1992. *Big Footprints: A Scientific Inquiry into the Reality of Sasquatch.* Johnson Printing Company, Boulder, Colorado, USA. (Photo: Johnson Printing.)

1992/CI/00 – Remote Island. Clam Diggers Attacked With Rocks & Driftwood: Samson Cecil and several other men said they were attacked by ape-like creatures on a remote BC island (off Vancouver Island) in the summer of about 1992. The group had beached their boat in a bay and began walking along the beach to dig clams. One of the men spotted an ape-like face watching them from behind a large stump. The group was then subjected to a barrage of rocks and driftwood from the area where the face was seen. "A large rock just missed my head!" recalled Cecil. As the group retreated back to their boat, a loud whistling noise was heard.

Source: John Bindernagel, 1998. *North America's Great Ape: The Sasquatch.* Beachcomber Books, Courtenay, BC, p. 117, 225, from a report provided by Samson Cecil to Jerry Markus; also information provided by Henry George, the father of one of the men in the group.

1992/CI/00 – Port Alberni (VI). Camper Lifted: A woman reported that she had a frightening experience while sleeping in her camper near Port Alberni at Sproat Lake in 1992 or 1993. In the middle of the night the front end of her vehicle was lifted about a foot off the ground and then dropped. She rushed outside to see what had happened, but did not see anything.

Although Sproat Lake is most known for being a scenic vacation spot, it is known to archaeologists as a site occupied by Archaic people dating back to 11,000 BC. There are numerous petroglyphs in the area.

Source: John Bindernagel, 1998. *North America's Great Ape: The Sasquatch.* Beachcomber Books, Courtenay, BC, p. 114, 224. Also, Wikipedia.

1993/01/00 – Nakusp. Footprints In the Middle of Nowhere: A

forestry technician reported that he found unusual footprints in the Nakusp area in January 1993. He had been flown by helicopter to a remote area about 60 miles from Nakusp. The purpose of the trip was to look at road layout. There was snow on the ground, and as the technician went about his survey he noticed 6 footprints of bare, human-like feet about 10 inches long, but wider than a human foot. They were definitely not made by a bear. He later asked the helicopter pilot if he had made the prints, and he responded that he had not done so. The engineer said that he knew of no reason why someone would be out in the middle of nowhere at 10 to 20 degrees below zero making tracks in snow with bare feet.

Source: BFRO website.

1993/03/30 – Harrison Hot Springs. Eating Devil's Club Roots:

Joseph Verhovany, who reported seeing a sasquatch eating berries in the Harrison Hot Springs area in 1988, had another sighting in the same area on March 30, 1993. He was near Hornet Creek Road getting cedar wood to make roof shingles. He noticed that some devil's club bushes had been pulled out of the ground and their roots had been chewed by something.

At a distance of between 200 and 300 feet Verhovany then saw a brownish-red sasquatch squatted down eating devil's club roots, like a person would eat corn on the cob. As he watched, the creature stood up and, walking on 2 legs, went to another plant. It then proceeded to pull out the plant and eat the roots. Verhovany estimated that the creature was over 6 feet tall. In describing the oddity, he said

Devil's club. This extremely spiny shrub often grows into entangled masses up to 8 feet high. First Nations people used the plant both as food and a medicine.

it looked "more like a monkey, something between a monkey and a human."

The creature was observed for between 4 and 5 minutes, then melting snow under Verhovany's feet caused him to fall over a tree trunk. The creature heard him fall and immediately looked in the direction of the noise. It saw Verhovany and moved away, "first four-legged, then after, 2-legged."

As in his previous sighting, Verhovany sensed an odor when the creature left, but not as strong. He reasoned that the creature was probably not sweating as much, as the temperature was much colder.

Verhovany went to the spot where the creature had been and found footprints in snow patches. He said they looked like human footprints, but much wider and much longer.

Source: Thomas Steenburg, 2000. *In Search of Giants: Bigfoot/Sasquatch Encounters.* Hancock House Publishers Ltd., Surrey, BC, pp. 59–66. Also John Green's Sasquatch Incidents Data Base. Incident Number 1001384. (Photo: Dr. H. Fahrenbach.)

1993/04//00 – Richmond. Murphy Teams Up With Dahinden: After meeting René Dahinden in April 1993 at Dahinden's home in Richmond, Chris Murphy became good friends with René, and they worked together on resolving the sasquatch issue. They had met as a result of a college presentation on the sasquatch being prepared by Murphy's son, Daniel.

Murphy retired from a 36-year career with the BC Telephone Company (now Telus) in September of the following year (1994) and soon started devoting most of his time to sasquatch research.

Up to 1993, and his meeting with Dahinden, Murphy had not given the sasquatch issue a second thought. All he had read about the creature was a newspaper article concerning Rant Mullens and the faking of footprints with wooden feet. Dahinden presented the case for the sasquatch in his highly characteristic way, and thus commenced the friendship.

Murphy went on to enter into a partnership with Dahinden in the marketing of sasquatch-related books and other items under the company names Pyramid Publications and Progressive Research.

Chris Murphy, left, and René Dahinden in 1993.

Source: Author's experience and files. (Photo: C. Murphy,)

1993/08/01 – Ainsworth. Cycling Encounter: Matt Gagnon reported that he encountered a sasquatch while cycling up a dirt road about a mile and a half west of Ainsworth on August 1, 1993. At about midnight, as Gagnon rounded a corner, he saw by moonlight (full moon) a creature standing on its hind legs beside the road 300 or 400 yards ahead. It ran away up the road on 2 legs, arms swinging, going for a distance of about 30 yards. It then stopped, looked at Gagnon, turned left and ran into the woods.

Gagnon also fled. He estimated that the creature was 7 to 8 feet tall. He said it was black and had arms down to its knees. It did not make any noise. All that was heard was the sounds it made crashing through the brush. He searched for footprints the next day. Only a few depressions were found.

Source: John Green, Sasquatch Incidents Data Base, Incident Number 100889.

1993/08/00 – Fort Nelson. A Big Red Monkey: Brian Fertuck and another man encountered a sasquatch near Fort Nelson, about 35 miles north of the Alaska Highway on road 317, in August 1993. The men were doing a "check cruise" for the BC Forest Service. In an area of over-mature timber they heard footfalls around them. Fertuck looked back and saw "a big red monkey," 8 to 10 feet tall with a dark face. His partner, who had poor eyesight, could not see the creature. It stepped smoothly out of sight behind the roots of a fallen tree. The men found 2 sets of tracks in the mud of the root-pad hollow, one huge and one smaller.

Source: John Green, Sasquatch Incidents Data Base, Incident Number 101302.

1993/08/00 – Fort St. James. Somewhat Athletic Gait: Fishermen reported that they saw a large hominoid north of Fort St. James beside Inzana Lake in August 1993. The creature was seen walking near a tree line on a clear-cut. Tree planters had been at the clear-cut for 5 days, and the fishermen said they were able to judge the size of the creature based on the workers who had previously seen it. One of the witnesses stated:

> ...it [the creature] had a fluid somewhat athletic gait. It's speed was deceiving because it looked like a person out for a leisurely stroll yet when I looked back to the point I first

saw it there was an amazing amount of ground covered in a short time.

Source: BFRO website.

1993/09/00 – Cold Fish Lake. Flailing Its Arms: A hunter reported seeing a strange ape-like creature near Cold Fish Lake, Spatsizi Plateau Wilderness Park, in September 1993. He was walking down a trail when he saw a large, reddish-brown, ape-like animal standing on the lower branch of a tree situated atop a rock outcrop. The animal appeared to be unaware of the hunter's presence. The hunter retained his composure and even whistled at the creature to get a reaction. The animal bobbed its head vigorously and jumped from the branch, flailing its arms and landing upright. It ran away, remaining upright, turning to look at the hunter before disappearing into the forest. The hunter estimated that it was just over 6 feet tall and "really wide." Its hair was "a good shiny coat," not matted.

Source: Dr. John Bindernagel, 1998. *North America's Great Ape: The Sasquatch.* Beachcomber Books, Courtenay, BC, pp. 9, 10 from information provided by the hunter.

1993/10/00 – Nakusp. Incident at Hot Springs: Two couples had a concerning experience near Nakusp at St. Leon Hot Springs in mid October 1993. The group had been bathing in the undeveloped hot springs. As they made their way back up the trail single file to their vehicle at about 11:30 p.m., 2 of the group heard something in the bushes keeping pace with them. Later the other 2 also heard twigs snapping. They all thought it was probably a bear, whereupon an extremely loud vocalization was heard. One of the group said, " The closest thing I can compare it to is a peacock's call, but it was at such a high decibel it actually made our ears vibrate." Then whatever the creature was, turned and ran very quickly down the hill through the bush away from the group. From what they heard, it was obviously on 2 legs not 4.

The group got back to their vehicle and left the area. The next day, one of the men went back to the hot springs and found very large, human-like footprints in the mud around the springs. They were much larger than those left by the group the previous night.

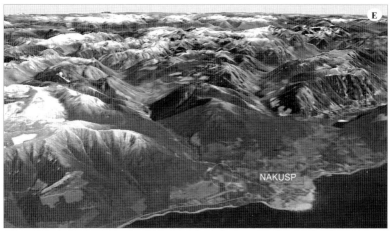

The Nakusp area facing east with Upper Arrow Lake in the foreground. The developed hot springs are in the area indicated with a circle. The water is piped in from natural springs in the mountains.

Source: BFRO website. One of the witnesses was interviewed by Blaine McMillan. (Photo: Image from Google Earth; Image © 2011 Province of British Columbia; © 2011 Cnes/Spot Image; Image © 2011 DigitalGlobe; Image USDA Farm Service Agency.)

1993/YR/00 Pocatello, Idaho, USA. ISU Scientist Joins Search: Dr. Jeffrey Meldrum, an anthropologist with the Department of Biological Sciences at Idaho State University, Pocatello, took up an interest in sasquatch studies in about 1993. He worked closely with Dr. Grover Krantz of Washington State University. After Dr. Krantz died in 2002, Dr. Meldrum became the main scientist in the sasquatch/bigfoot field.

Dr. Jeffrey Meldrum

Source: Christopher L. Murphy, 2010. *Know the Sasquatch: Sequel and Update to Meet the Sasquatch.* Hancock House Publishers, Surrey, BC., p.153. (Photo: J. Meldrum.)

1994/05/00 – Remote mainland inlet. Loud Howling Concerns Workers: Road-building workers reported a strange incident that occurred at a remote mainland inlet in May 1994. A sudden loud howling or wailing sound was heard that continued at intervals for several days, causing a great deal of uneasiness in the camp. The unknown and exceedingly loud calls emanated from valley walls.

Source: John Bindernagel, 1998. *North America's Great Ape: The Sasquatch.* Beachcomber Books, Courtenay, BC, pp. 122, 227, from information provided by 2 Campbell River residents.

1994/05/31 – Castlegar. On All Fours, But Odd: Nellie Campbell reported seeing a very strange animal near Castlegar on May 31, 1994. While driving east on Highway 3, about 14 miles from Castlegar, she saw what she thought was a bear coming up the bank about 375 yards away, and slowed to watch it. It was a different creature—on all fours, but placing very little weight on its front limbs. It traveled very fast. Its gait was peculiar—one side and then the other. Its limbs were long, especially in front. Its body was thin, covered with brown hair streaked with dark gray. If it stood erect, she estimated it would have been 6 to 7 feet tall. Campbell reported her sighting to Professor Paul Tennant (believed to be at the University of British Columbia).

Source: John Green, Sasquatch Incidents Data Base, Incident Number 1001517.

1994/08/00 – Harrison Hot Springs. Footprint Mystery: Mr. M. Jasmin found 4 unusual footprints near Harrison Hot Springs on a trail in the Deer Lake campsite area (Sasquatch Provincial Park) on August 17, 1994. Five toes were clearly visible. He made careful measurements of one of the prints (left foot) and "staked" it off with small twigs. He also measured the distance between the prints. He recorded the following information:

-Length from big toe to heel	15.5 inches
-Length from little toe to heel	14.5 inches
-Overall width at toes	9.5 inches
-Width of heel	5 inches
-Depth of heel	1.5 inches
-Depth of small toes	0.5 inches (approx.)

-Depth of big toe	1 inch (approx.)
-Stride	42 inches (approx.)
-Step	24 inches (approx.)

-Soil quality on August 16: Top soil dark, very fertile, no rain for 42 days.

-Estimated weight of animal: between 400 and 600 pounds

-Tracks found on a slight upgrade

Note: The stride and step measurements, if correct, indicate that the creature took very short steps for a 15.5-inch foot length.

Mr. Jasmin waved down the campsite fee collector, Sylvia Pool, and showed her and her daughter, Veronica, the footprints (Veronica was accompanying her mother on her rounds). Sylvia subsequently informed her boss of the finding. He went and looked at the prints and said he was going to inform his superiors.

During this time, both Mr. Jasmin, and Sylvia Pool independently contacted sasquatch researcher Thomas Steenburg. Both told Steenburg that the prints were clear and distinct, showing all 5 toes clearly. However, when Steenburg talked with the daughter, Veronica, she said that only the big toe and heel were clearly indicated.

Steenburg contacted John Green, who in turn contacted Bob Titmus, and the 2 researchers went directly (same date) to the location of the prints. By the time they got there, the prints had all been tramped upon by people and were now nothing more than shapeless depressions in the ground, including the print that appeared to have been "staked" off.

Mr. Jasmin provided his measurements in a letter to Thomas Steenburg (received about August 24) and mentioned, "The park supervisor contacted the BC Parks Department which dispatched someone to arrive today, August 17" (evidently quoting from his notes).

I think a little speculation is in order here. Is it likely the prints were seen and photographed by parks board officials and then destroyed to eliminate concern to campers? I have been told that park rangers in the US are directed not to talk about sasquatch-related incidents. If they do, then their job could be jeopardy. Although this is very upsetting for sasquatch researchers, I can see justification for it. One might muse that the parks people in both the US and Canada have far more information on the sasquatch than we realize.

It needs to be mentioned that prior to this incident there was a report of a strong odor in Sasquatch Provincial Park. There was also an unconfirmed report of a sasquatch sighting. We don't have any details on the incidents, however, we are told that some campers in the area folded up and left their campsites.

Source: Thomas Steenburg, 2000. *In Search of Giants: Bigfoot/Sasquatch Encounters.* Hancock House Publishers Ltd., Surrey, BC, pp. 103–109.

1994/09/11 – Vernon. Rifle Scope Crosshairs On Eyes: Hunter Randy Rudney reported that he had a sasquatch in his rifle scope while hunting deer north of Vernon (Okanagan Valley) on September 11, 1994. Rudney said he watched the creature for almost 15 minutes—never thinking of killing it, although he was close enough to move his scope crosshairs from eye to eye.

Source: John Bindernagel, 2000. "Sasquatches in Our Woods." *Beautiful British Columbia* magazine, summer 2000.

1994/09/00 – Alert Bay. Makes You Gag…: Tom Sewid reported that he saw 2 sasquatch near Alert Bay on Village Island in September 1994. The incident occurred at night and he saw the creatures in the spotlight of his seine-boat, *Skidegate.* He described one of the creatures as having a chest "like one-and-a half 45-gallon drums." As Sewid watched, one of the creatures rushed into the forest and he heard the sound of snags crashing to the ground.

Jo Jo Christianson, a crew member on the boat, caught wind of the creatures and said that the smell "makes you gag or want to vomit."

Source: John Bindernagel, 1998. *North America's Great Ape: The Sasquatch.* Beachcomber Books, Courtenay, BC, pp. 114, 224. Also same author, 2010. *The Discovery of the Sasquatch,* same publisher, p. 90. Also John Bindernagel, 2000. "Sasquatches in Our Woods." *Beautiful British Columbia* magazine, summer 2000.

1994/AU/00 – Tofino (VI). Insistence of Tree Rapping: Adrian Dorst reported that he heard repeated tree rapping near Tofino in the Ursus River Valley area during the fall of 1994. The sounds occurred sometime after midnight.

It needs to be noted that aboriginal elders from some BC coastal villages accept slow, rhythmic tree-striking as evidence for the proximity of *Buck'was,* or sasquatch.

Source: John Bindernagel, 1998. *North America's Great Ape: The Sasquatch.* Beachcomber Books, Courtenay, BC, pp. 197, 240.

1994/11/00 – Balfour. Four Paces To Cross Highway: Ken Starchuck reported seeing an odd creature near Balfour in November 1994. While he was driving to work at 7:40 a.m. (just getting light) he saw a black figure on the road. It took 4 paces to cross the highway. One step took it from the side of the pavement almost to the center. When on the other side of the highway, it went up the bank into the trees. Before stepping out of sight, it turned and looked at Starchuck's car.

Source: John Green, Sasquatch Incidents Data Base, Incident Number 1001329.

1994/YR/00 – Pullman, Washington, USA. Jacko Disappearance Theory: In 1994 Dr. Grover Krantz of Washington State University, Pullman, stated there might be an answer to the mysterious disappearance of Jacko, an alleged ape-boy captured in 1884 (see 1884/06/30 – Yale. Jacko the Ape-boy).

The last we know of Jacko is that he was shipped in a cage destined for England to be used in a sideshow, but apparently he never arrived there. However, later that year, across the continent in New York City, the Barnum & Bailey Circus presented Jo-Jo the Dog-Faced Boy, a 16-year-old youth covered in long hair. Jo-Jo, seen here, whose actual name was Fedor Jeftichew (b.1868), was alleged to have been found in Russia along with his father, who was also covered in hair.

Fedor Jeftichew, who became known as Jo-Jo the Dog-Faced boy.

319

Dr. Krantz has coincidentally connected Jo-Jo with Jacko. He reasons that Jacko could have been purchased by U.S. circus man P. T. Barnum and billed for a sideshow, but died before he could be exhibited. Barnum thereupon quickly found a replacement, Jo-Jo.

This speculation is given some credibility, Krantz stated, because the original advertising image (dated 1884) of the creature to be exhibited did not match an advertising photograph of Jo-Jo taken in 1885. A photograph of Jo-Jo taken in 1884 could not be found.

I have not been able to find the original advertising image Krantz refers to (I don't know if it was a drawing or a photograph). Krantz apparently had it, but I don't know of anyone else who saw it.

The Barnum & Bailey collection (documents and photographs) was purchased by an Eastern university and subsequently sorted into museum "drawers." I noticed on their website that there was one marked "Jacko," so I contacted the university and asked what was in it. A woman got back to me and said she had checked it and the drawer was empty. It would certainly appear that there had been some sort of documentation for the drawer to have been created, but nothing has come to light, to my knowledge.

Source: Dr. G. Krantz, 1999. *Bigfoot/Sasquatch Evidence*. Hancock House Publishers, Surrey, BC, p. 204. Also, "What is it? A Strange Creature Captured Above Yale. A British Columbia Gorilla". Correspondence to the *Daily Colonist*, Victoria, BC, July 3, 1884. Also John Green, Sasquatch Incidents Data Base, Incident Number 1000128. (Photo: Public Domain.)

1994/CI/00 – Gold River (VI). Straight Gray Hair. Adrian Dorst reported seeing a strange creature in the Gold River area around 1994. Dorst and a friend were canoeing the west coast of Vancouver Island. They had come ashore to rest, and at dusk Dorst was lying under his canoe. He smelled something like a dead animal and came out to investigate. Looking around he saw a large, human-like, hair-covered creature crouched or sitting nearby. It had straight gray hair. When it saw Dorst it "took off" and was in the woods in a few strides. Dorst also then heard a sound like blowing across the opening of a bottle.

Source: John Green, Sasquatch Incidents Data Base, Incident Number 1001358.

1995/07/00 – Bella Coola. Talks With First Nations: In late July 1995 the author went to Bella Coola to discuss the sasquatch with some of the First Nations people there. This was not difficult. Immediately I arrived in a Bella Coola motel parking lot, the town's "good will ambassador," Darren Edgar, came up to me, introduced himself and asked if I would like him to show me around. It was getting late, so I asked him to come back in the morning.

Darren was waiting outside at about 9:00 a.m. I told him I was interested in talking to First Nations people about the sasquatch. He took me to meet a couple of his relatives who told me that people in their family had seen the creature. One of them said that fishermen on the Bella Coola River know that a sasquatch is around when they see a large rock fall from the mountain side and splash down near their fishing boat. To them, this is a signal that the sasquatch wants one of their fish. The fish (salmon) they net in the river are very large, and in a word, "beautiful." The fishermen answer the call by throwing a fish up onto the shore and then leaving. The fish is soon taken by the sasquatch.

Darren then took me up to the local petroglyphs. On the way down we met a group of First Nations youths heading up the trail. Darren stopped them and told them I was interested in the sasquatch. One of the youths told me he had seen one when he was about 12. Some of the others said they had seen sasquatch footprints, and others said they often hear the creatures; they apparently call to each other.

I asked the youths about carrying a camera and taking photographs of footprints, or perhaps the creature if they saw it. They all sort of laughed in a way that told me they did not need proof of the creature—it was I needing proof.

Source: Author's personal experience.

1995/08/03 – Harrison Mills (Chehalis Reservation). Footprints Found & Later a Sighting: Brad Tombe found unusual footprints near Harrison Mills (Chehalis Reservation) along the Chehalis River at about 3:30 p.m. on August 3, 1995. He photographed the prints and made casts of some of them. He took the casts to show René Dahinden; I happened to be visiting René at the time. Brad showed us the casts and provided the following information in a written report.

Location: Chehalis River
Weather: Rainy/Overcast
Date: August 6, 1995
Time: 3:30 p.m.

I walked down river and fished the runs as I went. The river bank was quite rocky and in places sand occasionally appeared in small stretches. When I walked past one of the patches it appeared to have some sort of tracks through it. As I began to look at them the tracks appeared to be footprints. At the time I was not sure what they were, but it looked like a large human foot. One could clearly see a heel and large mound of sand that had been pushed up. The front of the foot could also be seen and it appeared that a large toe was present. I decided that I would practice my plaster skills and poured out a few casts. When I proceeded to do so it was then that I could see the shape of the foot-

Brad Tombe with 2 casts taken of footprints he found along the Chehalis River.

print. Another angler stopped and helped me measure what appeared to be the stride [pace] of the person and it was 50 inches. There were four tracks and all were 12 inches in length and I photographed them all with the measuring tape beside them for comparison.

A short time later, René received a telephone call from Chief Alexander Paul of the Chehalis Band. Chief Paul told René that one of his people had seen a sasquatch. René, my son Dan, and I went

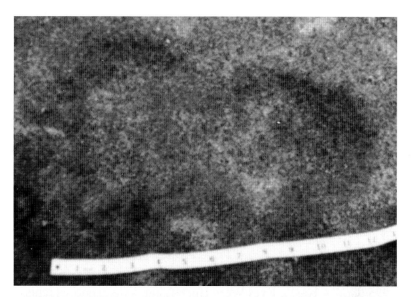

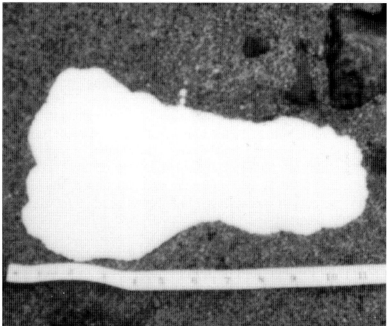

(Top) A footprint in the ground, one of four found by Brad Tombe along the Chehalis River. (Lower) A print filled with plaster. I am not certain that the 2 prints are the same.

up to the reservation to talk with the witness. He told us that he saw the creature, black in color, standing in the Chehalis River, up above its knees, and bending over. He thought it was a bear at first, but then it straightened up and walked out of the river.

I cannot, of course, verify that the 2 events were connected; it just appears logical. In talking with Chief Paul, he said that there had been 3 sasquatch sightings reported to him in the last 3 years.

Chief Alexander Paul

Source: Report provided by Brad Tombe and author's files.

1995/08/27 – Kaslo. Eyes Were Horrifying: Two canoeists reported seeing a huge strange creature near Kaslo at Kootenay Lake on August 27, 1995. They went ashore to have lunch at a remote wooded area on the lake. They found a small clearing and after about 5 minutes heard strange sounds coming from the woods behind them at about 200 feet. They pushed through the bush to see what was making the sounds, which were like a "deep murmured growl." As they got closer, they sensed a disgusting odor, then at about 20 feet they saw some sort of huge animal that appeared to be stooped down and tearing apart a small animal and eating it.

They backed off, keeping their eyes on the oddity. When they were about 50 feet away, the creature stood up, turned, and looked right at the men. It was at least 7 to 8 feet tall, and in the words of one of the men:

> It was covered in dark, thick fur similar to bear fur; its face was for the most part furry as well, with just a bit of black leathery skin on its forehead and cheeks showing through. Its eyes were horrifying, we both thought it was going to kill us.

Both men froze on the spot. They didn't want to turn and run because they were sure the creature could chase and catch them very quickly. Then suddenly it grabbed its meal off the ground, turned, and ran off into the brush in the opposite direction. The men ran back to their canoe and left.

Source: BFRO website.

1995/08/00 – Tofino. (VI). As If Walking On Air: Jennifer Little reported that she and a friend watched a sasquatch from their Zodiac (inflatable boat) near Tofino on Vargas Island in early August 1995. They watched the creature stride across the island beach. Little said it moved "as if it was walking on air."

Source: John Bindernagel, 1998. *North America's Great Ape: The Sasquatch*. Beachcomber Books, Courtenay, BC, pp. 41, 211. Incident initially reported to Adrian Dorst.

1995/AU/00 – Prince Rupert. Walked With Bent Knees: A female forestry technician reported seeing an unusual animal while working near Prince Rupert on Spirit Island in the autumn of 1995. Having completed her project, she was walking down an old road at about 4:00 p.m. to a float plane pick-up location. She noticed that something was watching her and turned to get a better view. She saw a tall, hairy, dark brown animal with long arms that took 2 steps and hid behind a large boulder. She noted that when it walked, it had bent knees. The lady took out her bear mace and proceeded to her destination without incident.

Source: BFRO website. Also John Green Sasquatch Incidents Data Base, Incident Number 1001396

1995/11/00 – Ainsworth. Crossed Road In 2 Paces: A motorist reported that he saw a strange creature near Ainsworth Hot Springs (4 miles south) at about 7:40 a.m. in November 1995. He said it was crawling on the bank on the left side of the highway. When it got down to the highway, it stood up and appeared to be between 7 and 8 feet tall. The motorist was about one-quarter mile away at this point. The creature looked at the him and then crossed the road in 2 paces, slightly slouched over. It took a third pace over the roadside ditch, and then a fourth to the edge of the forest. Here it stopped and turned to look again at the motorist who was now about one-eighth of a mile away, and could see that it appeared to be covered in dark hair. The motorist did not stop at this time, however, he went back to the site later but could not see any footprints.

1995/YR/00 – Harrison Hot Springs. Sasquatch Again On Welcome Sign: In 1995 the town of Harrison Hot Springs again put up a "welcome" sign showing a sasquatch. They had a sign of this nature in 1966 (see 1966/05/00 – Harrison Hot Springs. Sasquatch Not Welcome) but removed it as a result of residents not wanting to be associated with the creature. However, it appears things changed over the last 29 years and the creature again greeted visitors as they drove into the town. This lasted until 2009 when again the sasquatch vanished, just as it does in reality.

The last "on again, off again" Harrison Hot Springs sasquatch greeting sign. I was so used to seeing it that I failed to notice it had vanished until going to Harrison in April 2011. It has been replaced with a sailing ship, evidently promoting sailing on Harrison Lake. I have reasoned that the possibility of meeting up with one of the creatures might be of concern to some people, so can appreciate that not all merchants in Harrison would hold the sasquatch "near and dear."

Source: Thomas Steenburg, 2000. *In Search of Giants: Bigfoot/Sasquatch Encounters.* Hancock House Publishers Ltd., Surrey, BC, p. 49. (Photo: T. Steenburg.)

1996/01/21 – McBride. Freezer Opened & Strange Footprints: A woman who lives on a forestry road about 5 miles from McBride reported a strange incident that occurred on January 21, 1996. The family had a large freezer outside the house and between 6:00 a.m. and 6:30 a.m., she heard the freezer lid slam. Her husband had left for work and only the woman and her 15-year-old daughter were in the house. The woman was too afraid to go outside as it was still dark.

When day broke she inspected the freezer, but nothing had been disturbed. She looked for footprints in the snow-covered yard and found 2 near the front driveway about 50 feet from the house. One of the prints was fairly clear and showed that it was human-like and made with a naked foot. The other print was not clear, and was likely a partial print. She measured the clear print with a ruler and it was 13.5 inches long. The distance between the prints was 22 inches. She later took photographs of the prints. The woman expressed she was concerned that only 2 prints were found because her own footprints were quite visible in the wet snow.

Source: BFRO website.

1996/SP/00 – Port Renfrew. (VI). Small Stones Thrown: Otto Winnig, a mechanic who repaired logging equipment on location at night, reported that he had a sasquatch encounter near Port Renfrew at Harris Creek in the spring of 1996. While he was on the job, 2 small stones (under 2 inches in diameter) were thrown separately and landed on the hood of his truck. After the second stone hit, the mechanic walked around his vehicle to investigate. He saw a "big, hairy ape man" stride on 2 legs up a steep bank.

Source: John Bindernagel, 1998. *North America's Great Ape: The Sasquatch.* Beachcomber Books, Courtenay, BC, pp. 116, 225, from 41, 211 (?) from information provided by a friend of Otto Winnig.

1996/YR/00 – Burnaby. Patty's Portrait: In 1996, Chris Murphy of Burnaby took an image from the Patterson/Gimlin film and enhanced it with pastels. Chris had been closely associated with René Dahinden for about 3 years and was provided with film material for the production of posters. Patty, as the creature is affectionately called, was imaged onto ordinary paper and methodically colored.

Details that were missing due to film resolution were "guesstimated" based on details that were present. Some adjustments were made to give the creature a more natural appearance.

The resulting image, shown on the right, was a little surprising, and met with Dahinden's immediate approval. It was sent to a photo library and subsequently found its way into many books (even front covers) and other publications. It is likely the most publicized artwork of a sasquatch portrait. It was featured for a number of years on the Canadian National Library website in a short write-up on the creature. (Additional images provided in the color section.)

Patty's Portrait by Chris Murphy.

Source: Christopher L. Murphy, 2010. *Know the Sasquatch: Sequel and Update to Meet the Sasquatch.* Hancock House Publishers Ltd., Surrey, BC, frontispiece and p. 76. (Photo/Artwork: C. Murphy.)

1997/03/25 – Spuzzum. One Shape All the Way Down: Mike McDonald, a hunter, reported that he saw a sasquatch near Spuzzum, along the Anderson River, on March 25, 1997. He had gone into this area to hunt bear. He sat down against a tree to rest for awhile, and dozed off for a short time. When he awoke, he took his 10-power binoculars and scanned the area. He saw what appeared to be a brown bear below his position, about 30 yards beyond the other side of the Anderson River.

Mike McDonald

McDonald now picked up his rifle and trained the cross-hairs of his rifle scope on the back of the creature. At that moment, to the hunter's amazement, the creature stood up on 2 legs. He then realized that it was not a bear, but one of the elusive forest creatures he had heard of—the sasquatch. He went back to using his 10-power binoculars and observed as the creature pulled down tree branches and ate leaves. He estimated it to be about 7 feet tall, covered with chocolate-brown hair. He described its facial features as follows:

The head was big, that's one thing for sure; covered with fur or hair. And under the eyes and the cheeks and nose was not covered with hair. It was pretty much open, dark skin...the teeth were yellowish or brown. I saw it eating one branch and it opened its mouth at one point, and as it was chewing I could see the teeth that were... not white, that's for sure. They were definitely yellowish or brownish... maybe they could have been white but stained. I don't know. The eyes were dark. Dark...I could see it with the binoculars, I was looking at...even at 150 yards...with 10-power binoculars it looked close enough. I could see its eyes were dark. But at one point I could see it look to one side and I don't know if it was reddish or pinkish on the inside of the one eye. I was focusing on his face. I mean...I could hardly take my eyes off it you know, off the face, when it was eating, when it was facing me.

When asked to describe the creature's arms, McDonald stated:

[They were] like a human's but a lot longer...they were...and the fingers; I could definitely see the hands of it. The fingers were long. The arms were long, again covered with hair...bent at the elbows but...just like ours do. But compared to the rest of its body, its arms looked thin. I don't know if that's because of the length of them or what. They didn't seem to match its body.

When he was asked what stands out most in his mind about the creature he stated:

The view across the Anderson River taken in the spot McDonald was sitting when he saw the creature.

> The size of it. The size of its torso…it was big and thick. I don't know how to describe it, like…like a big man with…with just a blocky body that was kind of like one shape all the way down in its torso. And the color of its hair, and its face. Those three things I think, stand out in my mind.

He observed the creature for about 8 to 10 minutes, and then decided to get his camera which was in his truck, parked about one mile away. The round trip took him about 30 or 40 minutes, and by this time the creature had gone.

As it was starting to get dark, McDonald quickly left the area. Two days later he went back and inspected the spot where the creature had been. He did not find any footprints or hair on the trees from which the creature had been eating leaves.

He contemplated calling the police and telling them what he had seen, but, "I kept thinking they're going to think I'm nuts, you know."

He told his girlfriend and his mother of his experience and later learned of a sasquatch symposium that was being held (actually in session) in Vancouver at the Vancouver Museum. He went to the museum and asked at the desk if there was someone he could talk to about a sighting. Thomas Steenburg was asked to talk with him, and after being interviewed McDonald was asked, and agreed, to relate his experience to the symposium audience.

Thomas Steenburg and Barbara Butler later went to the sighting location. Unfortunately, McDonald was unable to go with them. However, he had provided a map and everything he explained about the location was found to be essentially correct. Although nothing was found, Steenburg is highly convinced that McDonald was telling the truth.

Source: Thomas Steenburg, 2000. *In Search of Giants: Bigfoot/Sasquatch Encounters.* Hancock House Publishers Ltd., Surrey, BC, pp. 9–22. Also John Green, Sasquatch Incidents Data Base, Incident Number 1001398. (Photos: T. Steenburg.)

1997/05/00 – Chilliwack. Something On Tape:

Wayne Oliver and Julie Ellif, both of Somerset, England, videotaped something large walking erect on a dry flood plane near Chilliwack (off the Trans Canada Highway) in early May 1997.

The figure, circled, as seen from the car.

The couple was vacationing in BC and Ellif was taking video of scenery as they traveled. Oliver noticed the figure off to the left, pointed to it, and said "What's that?" Ellif pointed her video in that direction and exclaimed "sasquatch!" When they later viewed their video, they saw that they had captured the figure and subsequently contacted sasquatch researcher Thomas Steenburg.

Steenburg had the video professionally examined at the University of Calgary. Unfortunately, the resolution of the figure was not sufficient to make a definite judgement as to what it was. All that could be concluded was that the figure was not a stationary object (tree stump, rock). It was was seen to be a large, upright, moving object that was dark in color.

Source: Thomas Steenburg, 2000. *In Search of Giants: Bigfoot/Sasquatch Encounters.* Hancock House Publishers Ltd., Surrey, BC, pp. 43–47. (Photo: Wayne Oliver and Julie Ellif.)

1997/07/01 – Chilliwack. Bob Titmus Passes On: The sasquatch/bigfoot "fraternity" was saddened with the death of fellow researcher, Bob Titmus, at Chilliwack in July 1997. The following summary of his sasquatch-related accomplishments was provided by John Green, a close friend of Bob Titmus:

Bob Titmus, a key figure in Sasquatch/Bigfoot investigation for almost 40 years, died at Chilliwack, British Columbia, on July 1st, following a heart attack suffered a few days before at his home in Harrison Hot Springs. He was 78.

Bob Titmus inspecting a tree that had been snapped off. Such occurrences are often associated with sasquatch. Bob dedicated the last 20 years of his life to resolving the sasquatch issue.

Not widely known to the public because he never sought publicity, Bob contributed more solid evidence for the existence of the creature than any other individual, and had the most extensive collection of original footprint casts, most of them from tracks he found himself.

He had a role in the public debut of "Bigfoot" in California in October, 1958. At that time he had a taxidermy shop near Redding, and he supplied his old friend Jerry Crew with the plaster-of-paris and the instructions for using it that enabled Jerry to make his famous cast of one of the 16-inch prints that kept showing up on the dirt road where he was working above Bluff Creek in the northwest corner of the state.

A few weeks later Bob and a friend, Ed Patrick, blew away the notion that "Bigfoot" was a freak individual, by finding and casting distinctly different 15-inch tracks on a sand bar beside Bluff Creek.

Those tracks were not in soft dirt like those on the road,

but in hard-packed wet sand, yet they averaged an inch in depth, making it impossible to dismiss them as being easy to fake. The casts Bob made then, and a year later on the same sand bar, are still possibly the best ever obtained anywhere.

Bob was one of the men who brought Tom Slick into the investigation in 1959 and was the original field leader of the "Pacific Northwest Expedition" that Tom financed in California. Later, again with Tom's backing, he shifted his search to a new area on the central British Columbia coast in the early '60s. Bob's move to British Columbia proved to be permanent. He spent several years operating from a boat among the islands and inlets between Bella Coola and Prince Rupert, and found Sasquatch footprints on several beaches, one a fresh set coming out of the water and into the woods on a small island which the creature could only have reached by swimming through a storm-whipped sea.

After Tom's death Bob continued the search until he ran out of money, then started a taxi business in Kitimat so that he could continue searching along the coastal streams and beaches on a part-time basis.

The casts he made during

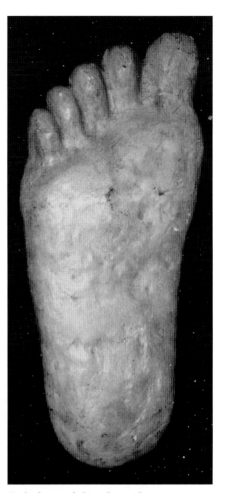

Bob donated this cleaned cast to the Vancouver Museum. On the back he wrote:

"This is an actual cast of Bigfoot imprint made Oct. 2. 1958 in Bluff Creek in Humboldt County, California. 'Bigfoot' is not a hoax."

He signed his name and showed: "Taxidermist, Anderson, Calif."

that period were all lost when his boat was destroyed by fire, but in 1976, while he was clearing a homestead near Hazelton, he was notified that some boys had found tracks close to the Skeena River at Terrace, and was able to get two superb casts of 15-inch tracks there.

When Roger Patterson and Bob Gimlin claimed to have filmed a female Bigfoot at Bluff Creek in October, 1967, Bob came from Kitimat to the first public showing, in Vancouver, B.C., and then went on down to California, where he made casts of ten tracks at the film site. He was the first investigator to go there and what he found left him totally convinced that the movie was genuine. Most of his investigations continued to be in British Columbia, and he became a Canadian citizen, but he returned to Bluff Creek for several months in the fall on a number of occasions, and was successful in finding tracks there several more times, once getting casts of both knee and hand prints.

A lifelong hunter, he was firmly in the camp of those who considered Sasquatch to be normal animals, and he hunted with a gun to collect one for study, but never had the opportunity to use it.

Another disappointment concerned some brown hairs that he collected one by one from bushes and branches near Bluff Creek while following an apparent Sasquatch trail. Years later it became possible to identify the hairs by immunological reaction. They proved to be from a higher primate; the eminent scientist who did the tests limited the possibilities to human, gorilla or chimpanzee.

As a veteran taxidermist, Bob was sure that the hairs could not be from any of the three, as could easily have been established with a comparison microscope—but the scientist had ground up all of them.

Bob's achievements were recognized by the *International Society of Cryptozoology*, which made him an honorary member in 1987, the first person from the Untied States to achieve that distinction.

Because of a back injury sustained in 1962 during a storm on the B.C. Coast, he lived with constant severe pain

Bob saw a hand print at the bottom of a pond in California. He drained the pond to make this cast. The length of the print is about 12 inches.

for more than 30 years, when other health problems limited his activities.

His last trips to Bluff Creek were in 1994 and 1996, but by then he was unable to go far from his car. On his last search along the sandbars of Bluff Creek, in the late 1980's, he was again able to cast several tracks, but because it was getting dark he did not attempt to take them out at that time, and left them buried under a tree at the film site. He has never been back, so if anyone can find the right tree the casts may still be there.

The photograph on the front cover of this book shows Bob Titmus holding casts he made of the prints found near Terrace along a Skeena River slough in 1976 (see 1976/07/17 – Terrace. Major Track Finding). Bob is recognized as the preeminent sasquatch field researcher.

Source: John Green, Harrison Hot Springs, BC, July 1997. (Photos: Titmus with tree, G. Krantz; Hand cast, C. Murphy.)

1997/YR/00 Vancouver. Dahinden & the Kokanee Commercials: During 1997, Columbia Brewing Company contracted René Dahinden for a number of television commercials for their Kokanee beer. The sasquatch had long been used for Kokanee commercials, so Dahinden was a prime candidate. The commercials were very humorous and extremely well done. René was a "natural" for the part he played. He was so good, in fact, that he won an actor's award.

In addition to the television commercials, large advertising placards showing René with an actual plaster footprint cast were displayed in government liquor stores. Columbia had 400 of the casts made and they were used as prizes in a contest.

René Dahinden on a Kokanee placard. A large color version of this image is in the color section.

A picture (framed) presented to René by the Kokanee people. It shows René carved into a mountain side along with the dog "Brew," and the sasquatch (played by William Reiter) seen in Kokanee beer commercials. The allusion, of course, is to Mount Rushmore, which shows the heads of four US presidents carved in the mountainside.

Although I worked with René and the brewing company people to bring about the commercials, in retrospect, having René appear in them was not a good idea as they "made light" of the sasquatch issue. Nevertheless, on the other hand...

Source: Christopher L. Murphy, 2010. *Know the Sasquatch: Sequel and Update to Meet the Sasquatch.* Hancock House Publishers Ltd., Surrey, BC, p.237. (Photos: C. Murphy.)

1998/06/00 – Hood River, Oregon, USA. NASI Report Released: In June 1998 a report by forensic examiner, Jeff Glickman (director of the North American Science Institute [NASI], Hood River, Oregon) on his analysis of the Patterson/Gimlin film was released.

Glickman's report, titled "Toward a Resolution of the Bigfoot Phenomenon," closes with the following statement:

> Despite three years of rigorous examination by the author, the Patterson-Gimlin film cannot be demonstrated to be a forgery at this time.

Glickman actually found that the film showed a natural creature. He determined its walking height to be about 7 feet 3.5 inches, and that its weight was about 1,957 pounds.

It was felt, at least by this author, that the ramifications of the NASI report would be significant throughout North America. Unfortunately, proper distribution of the report did not take place. It

Jeff Glickman with his state-of-the-art electronic equipment. His analysis of the Patterson/Gimlin film was highly detailed and authoritative.

was not accepted for publication in a scientific journal, nor was it allowed to be published in a booklet. It eventually found its way onto the web; however, it was never given the prominence it deserved in scientific circles. To this day, I consider it the best, most authoritative, and most detailed analysis of the Patterson/Gimlin film ever undertaken.

Source: Christopher L. Murphy. *Bigfoot Film Journal.* Hancock House Publishers Ltd., Surrey, BC. also J. Glickman, 1998. "Toward a Resolution of the Bigfoot Phenomenon," North American Science Institute, Hood River, Oregon, p. 13 (height and weight). (Photo: J. Semlor.)

1998/09/00 – West Kootenays. Defies Conventional Description:

A resident reported seeing 4 sasquatch in the West Kootenays region (exact location withheld) in September 1998. When he first saw one of the creatures, he thought it was his neighbor walking among the shrubs, but says he soon realized the animal was not human. As he continued looking at the creature, others appeared.

The resident provided a full report on the incident in a letter to Dr. John Bindernagel, which opened with: "I have seen an animal that defies conventional description."

Dr. Bindernagel provided the complete story in a article he wrote for *Beautiful British Columbia* magazine, summer 2000 edition. The full article is included in this work (see 2000/SU/00 – Courtenay (VI). Wildlife Biologist Says It All).

Source: John Bindernagel, 2000. "Sasquatches in Our Woods." *Beautiful British Columbia* magazine, summer, 2000, p. 28.

1998/YR/00 – Kemano. Standing On 2 Legs On Rocky Slope:

A woman and 2 Haisla fisheries men reported seeing a strange creature near Kemano in 1998. The 3 were in a boat going down Gardner Canal on their way to Kowesas. After passing Kemano Bay they observed on the east side of the steep-walled fiord a large brown creature standing on 2 legs on an open rocky slope. It watched them as they went by. The creature was not there when the group returned.

Source: John Green, Sasquatch Incidents Data Base, Incident Number 1001518.

1998/YR/00 – Courtenay (VI). North America's Great Ape: Wildlife Biologist, Dr. John Bindernagel, released his book, *North America's Great Ape: The Sasquatch,* at Courtenay in 1998. The book provides highly convincing evidence of sasquatch existence from a scientific perspective. It was the first book to fully address how the sasquatch behaves and sustains itself in North America. Dr. Bindernagel went on to publish a second book, *The Discovery of the Sasquatch,* in 2010, which reinforced his belief in the creature.

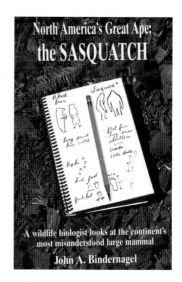

North America's Great Ape: the SASQUATCH

A wildlife biologist looks at the continent's most misunderstood large mammal

John A. Bindernagel

Source: John Bindernagel, 1998. *North America's Great Ape: The Sasquatch.* Beachcomber Books, Courtenay, BC.

1999/01/00 – Hood River, Oregon, USA. NASI Folds: In about January 1999 the North American Science Institute (NASI) in Hood River, Oregon closed its operations. Despite the favorable report on sasquatch existence prepared by Jeff Glickman, a forensic examiner, the backers of NASI withdrew support (funding) for the operation. Unfortunately, during the last 6 months NASI was unable to get its report, "Toward a Resolution of the Bigfoot Phenomenon," published in a scientific magazine or journal. Although during the previous year plans were made to at least have the report available for sale in a printed booklet, certain legalities prevented the publication. The report was, however, eventually published online.

Source: Christopher L. Murphy, 2008. *Bigfoot Film Journal.* Hancock House Publishers, Ltd., Surrey, BC, p. 84.

1999/06/00 – Chilliwack. Hair Over Feet Like Bell-bottom Pants: A man camping with his wife and children near Chilliwack at Cultus Lake reported seeing a possible sasquatch while the family was walking along a trail in June 1999. The man spotted what he thought was a stump, but suddenly it stood up and ran away parallel to the trail they were on and in the direction they were going. The man caught glimpses of it between the trees and bushes at about 50 yards. He noticed that its feet came up to its buttocks as it ran, "I saw a long thigh come up level as it ran. It was very muscular in the back." The color was "like that of a grizzly bear."

He also noticed that the hair around the bottom of its legs was thick and looked like bell-bottom pants. The hair on its back was, "matted down and very neat, like it was brushed down into a 'V.'" The man's wife and children were not looking in that direction so did not see anything.

A loud crack was heard, like a tree splitting, so the man decided it was best to leave right away. He picked up the smallest child, and they all ran out to the road. They met a parks employee emptying garbage bins and the man said to him, "I just saw something big and brown in the bush." To his surprise the employee replied quite calmly, "Was it on 2 legs or 4 legs?" The man responded "2," and the employee said, "Okay, I'll tell the rangers."

Later, the man went back to the area alone. He found a rotted log that had been torn up, and a dead-fall tree broken in the middle. Upon further inspection he found 14- to 15-inch-long tracks deeply imbedded in the soil. They were 5 to 6 inches wide, but toes were not visible. The distance between the tracks was estimated at 6 feet.

Source: BFRO website. Witness interviewed by Kevin Withers:

1999/09/14 – Hope. Like a Roll of Fur at Its Wrists: Frank Simons reported that he saw an unusual creature in the Hope area on September 14, 1999. He was getting firewood near an abandoned gravel pit and noticed a large, dark, reddish-brown animal at the bottom of a gully about 50 yards away. It seemed to be standing on its hind feet scratching or digging at something in some small bushes.

Simons saw it from the back; its head was obscured by overhanging foliage. He noted that its fur/hair seemed to be longer on its back, and that the undersides of its hands (or paws) were light colored. There was "like a roll of fur" at its wrists. The fur/hair on its back was shiny but matted, as if wet. The creature was big and bulky, solid all the way down, with no waist. Its arms seemed longer than its legs.

Simons decided to quickly return to his vehicle, and at that point the creature turned in his direction. However, he was still unable to see its head or face because of tree branches. As Simons ran to his vehicle he glanced back and saw the creature climbing up the steep slope on the opposite side of the gully on 2 legs. It seemed to be pulling itself up by grasping the bushes and trees with its forearms.

At this point, Simons got the impression that its head was small for its bulk. It moved rapidly up the slope and disappeared in the trees. He commented, "I had the impression it moved in a humanoid fashion."

Source: BFRO website. Witness was interviewed by John Green who also inspected the sighting location.

1999/DE/00 – Vancouver. Woman Followed By Sasquatch: A woman reported that she was followed and then apparently chased by a sasquatch in the Vancouver lower mainland area during the 1990s. She remarked on the "slapping" sound of the creature's feet as it ran behind her along a paved road.

Source: John Bindernagel, 1998. *North America's Great Ape: The Sasquatch.* Beachcomber Books, Courtenay, BC., pp. 160, 232.

1999/DE/00 – Sayward (VI). Dragging a Deer: A couple driving south near Sayward had an unusual sasquatch sighting in the 1990s. They said they saw the creature dragging a deer up a mountainside beside the highway.

Source: John Bindernagel, 2000. "Sasquatches in Our Woods." *Beautiful British Columbia* magazine, summer 2000, from information provided by Audrey Wilson of Alert Bay.

1999/DE/00 – Harrison Hot Springs. Ape-like Face In Window: Two men reported a frightening incident in the Harrison Hot Springs area during the 1990s. The men were parked in their truck camper on the west side of Harrison Lake and noticed an ape-like face in their window. Immediately afterwards, the men felt their vehicle being shaken, "as if that thing wanted to turn us over."

Source: John Bindernagel, 2000. "Sasquatches in Our Woods." *Beautiful British Columbia* magazine, summer 2000.

2000/WI/00 – Salt Spring Island. Bateman Paints a Sasquatch: The highly acclaimed naturalist artist Robert Bateman of Salt Spring Island, created a painting of a sasquatch in early 2000. Bateman is known worldwide for his astounding artwork of Canadian wildlife.

The painting was commissioned by the editor of *Beautiful British Columbia* magazine to accompany an article about the sasquatch by Dr. John Bindernagel. The article entitled "Sasquatches in Our Woods," was published in the summer of 2000. Binder-

Author with Bateman's painting. A color version of the painting itself is in the color section.

nagel, a noted sasquatch researcher and personal friend of Bateman, sent the artist a portfolio of eyewitness drawings and a list of anatomical features that recurred frequently. Bateman subsequently sent Bindernagel a near-final version of his painting for an opinion; no changes were suggested. Bindernagel's article won an award for regional magazine articles (see the following entry).

Source: Personal communications with Dr. John Bindernagel. (Photo: C. Murphy.)

2000/SU/00 – Courtenay (VI). Wildlife Biologist Says It All: In the summer of 2000, *Beautiful British Columbia* magazine featured an article on the sasquatch by wildlife biologist Dr. John Bindernagel of Courtenay. The article is, in my opinion, one of the most important in the annals of sasquatch studies. It is so important that I asked permissions, and was accorded such, to reprint it "as is" in this book to give it a more permanent place in sasquatch-related literature.

Dr. Bindernagel has been involved in sasquatch studies since 1965, and has authored 2 books on the subject. He is a highly qualified wildlife biologist who has personally done sasquatch-related field research. He is totally convinced that the creature is a natural species. His latest book, *The Discovery of the Sasquatch: Reconciling Culture, History, and Science in the Discovery Process,* is an urgent appeal to the scientific community in general to recognize the overwhelming evidence of sasquatch existence.

Please note that the incidents related in the following article have been presented in this work under their appropriate dates. It is important, however, that they be repeated in the context of Dr. Bindernagel's message.

Sasquatches In Our Woods
John Bindernagel

Cockamamie myth? Pie-eyed bear sightings? Not so, says this B.C. wildlife biologist—these creatures of folklore really exist.

> "I have seen an animal that defies conventional description."

So began a letter to me from a resident of British Columbia's West Kootenay region, describing what he saw on his rural acreage one morning in September 1998. He agreed to provide details if I would assure him complete confidentiality.

Initially, he thought it was his neighbour walking among the shrubs, but says he soon realized the animal was not human. As he caught increasingly good views, he mentally discarded bears, then cougar, as possible explanations.

> One minute this creature was able to squat and be no more than three of four feet in height. The next, it seemed to be able to draw itself up to six feet, or perhaps more in length. Three others appeared, and one of the smaller ones climbed into the branches of a tree. It was then that I first had the thought that I was looking at monkeys ...The profile was like that of an orangutan. It seemed to have hair not fur.
>
> The one that...had first drawn my attention drew up behind slightly denser bush and walked off on its hind legs...This really floored me...The movement seemed so naturally bipedal.

After about 35 minutes, all but one of the animals had disappeared into the forest.

It seemed that this would be the last chance to determine...whether I had really been watching bears, I was not disappointed. It, too, finally rose up. I was extremely struck with its barrel chest and seemingly long, skinny legs and arms. This, though, was only relative to its large chest and upper body. It walked off! Maybe two or three strides and it was gone.

It most definitely walked erect with shoulders sloped forward. The gait was not like gorillas that I have seen either in the zoos or on TV. Nor was it that peculiar walk of the orangutan. This was a full easy stride.

I suppose for me, I must accept that I probably saw sasquatches that day. I keep waiting for a news report that tells me that a private zoo has lost some apes or something similar."

Author and biologist John Bindernagel, seen here on a field trip near Courtenay on Vancouver Island, does not doubt the existence of the sasquatch.

National political columnist Allan Fotheringham once characterized B.C. as the land of "Social Credit and the sasquatch." B.C. journalist Stephen Franklin called the sasquatch, "the godsend of newspaper cartoonists in the silly season when politicians are on vacation." The creature is, as Canadian folklorist Carole Carpenter once said, part of B.C.'s "cultural identity." Long before our time, the Kwakwaka'wakw people depicted Tsonoqua, "Wild Woman of the Woods," on West Coast totem poles and masks as a giantess with pursed lips, along with her male counterpart, Bukwuss. The words "sasquatch" itself comes from the Coast Salish term Sasqits, or "hairy man."

It is my conviction that the sasquatch, or bigfoot, is real, and that its very existence has generated the myths, aboriginal legends, newspaper reports—and yes, the jokes and

World-celebrated wildlife painter Robert Bateman of Saltspring Is-
land created this sasquatch illustration especially for this story,
carefully portraying the creature's face as described in eyewitness
sightings.

hoaxes. As a wildlife biologist, I have examined the issues for
more than 25 years, and not only do I believe the sasquatch
is a real animal but one about which we know a great deal.
We have samples of its tracks and scat, and reports of its
"eye-watering" odor. We know details of its physical appearance, its diet, and its behaviour in feeding, nesting, and de-
fending its territory.

There have been more than 200 reported sasquatch
sightings in B.C. alone. Eyewitnesses are remarkably consistent in their basic description of the animal encountered or observed, reporting a large, upright, human-shaped animal
standing or walking on its hind legs. The shoulders are prominent, like those of a human, but the neck is short and thick.
The animal is covered with hair, usually dark in colour. The
arms are proportionately longer than those of a human. Witnesses are often impressed by the animal's huge size.

In some cases more detail has been observed: a barrel
chest ("its chest was as deep as it was wide"), a flat, broad

nose ("like a gorilla"), an area of bare skin on the face around the eyes, and the apparent absence of ears, which, if present, were hidden by long hair. Some witnesses say they saw breasts on what appeared to be adult females.

Despite the compelling consistency of eyewitness reports, most wildlife biologists do not recognize the sasquatch as a valid wildlife species in B.C. or elsewhere. Skeptics remain the majority, and their arguments merit examination. How, they say, could such a large animal exist here without being seen more often? Why do we have no good photographs of a living sasquatch, no bones or teeth of a dead one? How is it no sasquatch has ever been roadkill, or fallen into the sights of a backwoods hunter?

The media's treatment of sasquatch reports may provide part of the answer. A sasquatch sighting, when reported, normally takes the form of a light-hearted news item, often recounted with a knowing grin and the suggestion that the eyewitness was on the way home from a pub. While a healthy skepticism regarding the existence of an ape-like animal in B.C. is justified, such cynicism and ridicule tend to inhibit other witnesses, such as the West Kootenay man, from coming forward publicly.

John Green of Harrison Hot Springs published *The Sasquatch File* in 1973, summarizing 110 reported sightings in B.C. He concluded then that animals answering the description of sasquatch are seen far more often than we realize. To date, Green has collected almost 350 B.C. sightings of sasquatches or their tracks—but witnesses remain as reluctant, as fearful of ridicule, as they were 27 years ago.

Nevertheless, those who have seen a sasquatch—especially those employed in outdoor occupations who are familiar with bears—know when they have seen an unusual animal and wish to report it. Balanced media treatment of sasquatch evidence inevitably results in an outpouring of fresh reports.

As for the absence of sasquatch remains—bones, teeth, hair, a body—this is understandable in the case of an uncommon or rare mammal. Wildlife biologists and archaeologists recognize how quickly animal remains are scattered, eaten, and broken down in nature, especially in our acidic B.C. soils. The possibility remains that well-preserved sasquatch bones may yet be found in a limestone cave or similar site.

That sasquatches have escaped roadkill may attest simply to their intelligence in avoiding moving automobiles, and their preference for backwoods habitat. None have been shot, though a few hunters have had opportunity. While hunting deer in the Okanagan Valley north of Vernon on September 11, 1994, Randy Rudnyk says he watched a sasquatch in his rifle scope for almost 15 minutes—never thinking of killing it, though he was close enough to move the crosshairs from eye to eye. In October 1955, William Roe says he levelled a gun at a female sasquatch on B.C.'s Mica Mountain. "If I shot it, I would possibly have a specimen of great interest to scientists the world over," said Roe. "I lowered the rifle. Although I had called the creature 'it,' I felt now that it was a human being and I knew I would never forgive myself if I killed it."

While we have no body of a sasquatch, we are not without physical evidence of its existence. Scat has occasionally been reported where sasquatches have been sighted. William Roe backtracked the female sasquatch he observed on Mica Mountain and dissected several scats, finding only vegetable matter. John Christman of Bremerton, Washington, found a deposit on the Olympic Peninsula "the diameter of a pop can and enough to fill a large bucket." Unfortunately, it may be that sasquatch scat is not sufficiently different from that of large bears to provide definitive evidence. But then there are the footprints.

Sasquatch tracks are characterized by their elongated human-like shape, the absence of claw marks, and the occurrence of "hind" feet only. Bear tracks, by comparison, show claw marks, the pointed heel of the bear's hind foot, and the presence of alternating forefoot and hind foot impressions. Compared to human footprints, sasquatch tracks differ mainly in size: those recorded vary from seven to 22 inches (17 to 56 centimeters), with an average in John Green's records of 16 inches (40 centimeters). Physical anthropologist Dr. Grover Krantz, author of *Big Footprints: A*

Scientific Inquiry into the Reality of Sasquatch (1992), states that sasquatch footprints are normally one-third wider than human footprints of the same length.

Most mammals are nocturnal or crepuscular (active mainly at night, or at dawn and dusk), and so are observed infrequently. Wildlife biologists routinely accept mammal tracks as evidence that an elusive species is present, without need of a sighting. This is common practice for bears, wolves, and caribou—yet most scientists will not accord the sasquatch the same benefit of the doubt.

Nevertheless, laypersons have observed sasquatches engaged in feeding and nesting as other animals do. In some cases the creatures have left physical evidence of their activi-

Photograph shows prints of a 38-centimetre-long sasquatch track observed in Strathcona Provincial Park, compared with a print of a size 11 human footprint. Sasquatch tracks are not only longer but proportionately wider as well. Because of the large size and weight of the animal, it normally leaves a deep impression in the soil.

ties. In spring 1997, bear hunter Mike McDonald observed a sasquatch eating willow buds and young leaves in the Fraser Canyon. To reach buds above its head, it stood upright and pulled the branch tips down to its mouth with one extended arm and hand. In October 1967, in a remote part of Oregon's Cascade Range, Glen Thomas observed a male sasquatch dig a pit, from which it obtained and ate ground squirrels. Ed-

Strathcona Provincial Park's rugged wilderness is one of several places on Vancouver Island where creatures resembling the legendary sasquatch have been observed.

win James, along with a group of friends and family members, surprised a sasquatch digging for clams on Gilford Island in the 1950s. A friend of Audrey Wilson of Alert Bay was driving south with her husband near Sayward, Vancouver Island, when she saw a sasquatch dragging a deer up a mountainside beside the highway. Ten minutes later she asked her husband, "Did you see that?" His answer: "Yes. Did you?"

Prospectors, hunters, and biologists from the B.C. coast and Interior have reported sasquatch nests, which differ from bear beds and dens in having a woven rim and sometimes a roof. Dr. Fred Bunnell, a University of British Columbia wildlife biologist described a "sasquatch bower" he discovered with unusual bent and broken branches overarching a nest against a rock face. "No bear makes a day bed like that," Bunnell concluded.

Most biologists assume that eyewitnesses reporting sasquatches are misidentifying upright bears, since a bear on its hind legs is the closest acceptable approximation. Others argue that claimants were duped by tricksters wearing fur suits. Unfortunately, witnesses have had little with which to compare their own observations. The sasquatch is notably absent from our field guides, the authoritative reference books on which we depend to identify the animals of our natural world.

In my recent book, *North America's Great Ape: the Sasquatch,* I have included field guide-style illustrations that compare the squarish, human-like shoulders of the sasquatch with the tapered shoulders of the bear. In profile, the sasquatch's flat face contrasts with a bear's prominent snout. As well, the distinctive sasquatch gait—a graceful, ground-eating stride with arms swinging—is very unlike the bear's briefly maintained upright shuffle on its short hind legs. The fluid sasquatch gait also differs significantly from the stiff-legged human gait. The West Kootenay eyewitness quoted above recalled seeing the animals "moving in the trees in a way no human could have imitated without a lot of noise and superb acrobatic training. They would...move rapidly from one place to another with a gracefulness and ease."

I have studied evidence of the sasquatch since I first found tracks on Vancouver Island in 1988, and my research suggests the sasquatch is, in fact, a great ape. Dr. Krantz proposed that the sasquatch may be the descendant of *Gigantopithecus,* a giant fossil ape of Asia—or, more likely in his view, actually may be *Gigantopithecus,* "still with us." Unlike the great apes of Africa and Asia, however, the sasquatch has a more human-like stride and foot shape.

Occasionally, sasquatches in remote areas have revealed themselves, and their ape-like nature, in their attempt to repel humans. While most sightings involve a sasquatch walking or running away, or watching curiously from a distance, there have been reports of apparent intimidation. While walking up a beach in B.C.'s remote Deserters Group of islands in the early 1990s, some clam diggers spied an animal with an "ape

Field guide-style drawings by Wendy Dyck from North America's Great Ape: the Sasquatch *contrast the shape of an upright bear with that of a sasquatch. In profile, the bear's prominent snout is markedly different from the sasquatch's flat face. In front view, the sasquatch's squarish shoulders contrast with the bear's tapered shoulders. The sasquatch has relatively long legs that allow for a graceful stride, in contrast with the short-legged shuffles of a bear when it walks on its hind legs. A bear's ears are usually visible, while the sasquatch's apparently are hidden under long hair.*

face" watching them from behind a large stump. It then began to throw rocks and driftwood. "A large rock just missed my head!" recalls Samson Cecil, adding that his group then set a possible world speed record for dragging a skiff down a beach into the water.

Campers on beaches and riverbanks have been threatened by the sound of a large animal running back and forth just inside the adjacent forest, vocalizing, breaking large branches, and making a great deal of noise without showing itself fully. When this happened to Mary Strussi and her friend on Vancouver Island's Cruickshank River in the summer of 1992, they abandoned their tent for the security of their truck.

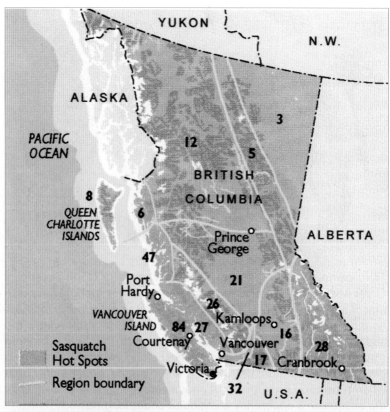

Sasquatch sightings have been reported from all regions of BC. Numbers on the accompanying map indicate the number of reports from various regions, as outlined by the author. Most reports, however, come from the coast, the Fraser Valley near Vancouver, and the Kootenays.

Some years earlier, two men in a truck camper on the west side of Harrison Lake noticed an ape-like face in their window, and immediately afterwards felt their vehicle being shaken "as if that thing wanted to turn us over."

In general, the sasquatch seems quite benign, with little or no interest in harming people. Its uncommon intimidation displays—breaking branches, running back and forth, throwing stones—are remarkably similar to those of chimpanzees. The result is to drive intruders from the animal's range, away from its young or a rich food source, without actual physical interaction.

Another interesting parallel to the ape world is the reported persistence of sasquatch odor, described by observers as a "strong stench of rotten eggs" or "rotten meat." The smell "makes you gag or want to vomit," said Jo Jo Christianson, a crew member on the seiner *Skidegate* who caught wind of two shoreside sasquatches near Alert Bay in September 1994. The overpowering odor is likely of glandular origin, as with mountain gorillas, which, according to noted researcher Dian Fossey, emit "a gagging fear odor" when threatened.

My interest in the sasquatch often raises eyebrows, but I am not alone in my convictions. I have been encouraged in my research by professional colleagues in the international wildlife research community—researchers not subject to our local and regional conditioning. Jane Goodall, widely known for her pio-

Sasquatch
(Gigantopithecus)

Size: Reported average height: males 2.3 metres (7.5 feet); females 2.1 metres (7 feet). Shoulder width approximately 0.9 metres (3 feet)

Weight: 225 to 365+ kilos (500 to 800 pounds, estimated on footprint size and depth.

Diet: Leaves (especially willow), berries, plant roots (including lilly roots), shellfish (especially cockles), fish, ground squirrels, waterfowl, deer, elk. Occasionally cultivated vegetables and domestic livestock.

Breeding: Unknown. Solitary individuals are most common but small family groups are reported occasionally.

Longevity: Perhaps 33 years or more, based on three sightings of one uniquely coloured individual.

Physical characteristics: Typical primate shape, with prominent shoulders and flat face. Stands and walks upright, though often slouched. Resembles an upright gorilla. Hair colour variable, but most often dark (black, dark brown, reddish brown).

Behaviour: Normally shy and elusive, but occasionally indifferent to human observers when not threatened. Rare exhibitions of intimidation.

Range: Most sightings in B.C., Alberta, and the western United States (where it is known as bigfoot), but credibly reported in lower numbers from other parts of North America.

Status: Rare, but perhaps not endangered. The paucity of reported sasquatch sightings may result from media skepticism and official perception as an invalid wildlife report.

neering field studies of chimpanzees in Tanzania, admits to a long-time interest in "'wild men,' the yeti, bigfoot, and the sasquatch." She called my book "exciting" and acknowledged my efforts to "carefully describe the behavioural characteristics that have been recorded for the sasquatch." Dr. Vernon Reynolds, an Oxford University primatologist and author of *The Apes*, was impressed with "the many points of similarity between sasquatch anatomy and behaviour and that of the great apes." Biologist Dr. George Schaller, author of *Year of the Gorilla*, thought I showed "a lot of insight" and made "sensible deductions" in developing my hypothesis that the sasquatch could be North America's great ape.

What appears to be a female sasquatch strides along a sand bar in frame 352 from the controversial 16-mm movie film taken by Roger Patterson and Bob Gimlin in 1967 in the Bluff Creek area of northern California. Although repeatedly debunked, this film may be authentic and does illustrate agree-upon aspects of sasquatch anatomy and locomotion, such as its hunched posture and smooth gait.

I am now convinced that sasquatches are seen far more often than we know. Indeed, some of my professional colleagues in the provincial Wildlife Branch have admitted to being less than diligent in filing sasquatch reports. Yet some witnesses continue to come forward—challenging our collective skepticism, braving public ridicule. As recently as September 1999, a Vancouver youth said he observed a sasquatch near Bradley Lagoon on the B.C. coast.

I fully expect that wildlife biologists will recognize the sasquatch as a species in the not-too-distant future, and subsequent researchers will find B.C.'s shellfish-rich West Coast to be one of its prime habitats. But until other North American biologists are willing to look beyond their own continent for possible explanations—and until we finally have irrefutable evidence that sasquatch exist—doubt will continue to linger, and witnesses will remain reluctant to speak out.

As the West Kootenay eyewitness mused at the end of his letter: "I think if anybody asks me I might deny the whole thing and cover it up by saying I saw bears and it was a joke."

Editors Note: John A. Bindernagel Ph.D., is a registered professional biologist in B.C, with more than 30 years of field expertise. He is the author of North America's Great Ape: The Sasquatch *(Courtenay: Beachcomber Books, 1998. Box 3286, Courtenay, B.C.), a field guide to the anatomy, ecology, food habits, and behaviour of the sasquatch.*

— 0 —

Cast of one of the footprints found by Dr. Bindernagel in Strathcona Provincial park in 1988. Close examination will reveal the treads of a hiker's boot in the upper portion of the cast.

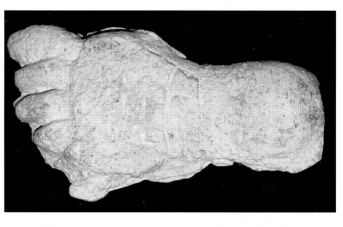

Source: John Bindernagel, 2000. "Sasquatches In Our Woods." *Beautiful British Columbia* Magazine, Victoria, B.C., Volume 42, Number 2, Summer 2000. (Photos: J. Bindernagel kneeling, J. Bindernagel; Bear/sasquatch drawings, Wendy Dyck; Footprints–sasquatch and boot print, Strathcona Provincial Park, Map, J. Bindernagel; Frame 352 of the P/G film, R. Patterson, Public Domain, Cast, C. Murphy.)

2000/08/15 – Hope. Realized It Wasn't a Bear: Kevin Pringle reported that he saw a sasquatch in the woods behind his camp near Hope at Silver Lake Campground on August 15, 2000. Pringle stated:

> I was mucking around the camp. I looked up, had the feeling something was watching me; thought it was a person at first, realized it wasn't a bear; tried to get my wits about me and before I could do that, it was gone.

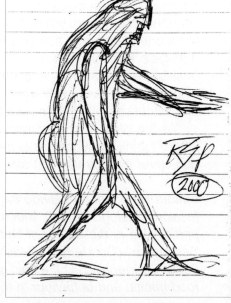

A sketch by Kevin Pringle of the creature he saw.

Pringle was about 120 feet from the oddity, which he said was standing upright, covered in black hair, and had very long arms, He estimated its height at between 7 and 8 feet. He did not discern any facial features as the creature immediately turned and left when he looked at it. He stated, "I got the impression it was running away from me seeing it."

Pringle immediately reported the sighting to park ranger Darlene Reed who was doing chores only yards away from where Pringle was camped.

Ironically, Thomas Steenburg happened to go to the campground 6 days after the sighting to get rid of a bag of garbage. Pringle, upon seeing the "Sasquatch Research" sign on Steenburg's vehicle thought that he had come to interview him. He thought that the park ranger had reported the incident to Steenburg. Upon clearing up this issue, Steenburg set about to investigate the incident.

The area where the creature was seen was inspected, but nothing was to found to indicate its presence or passage. Steenburg then interviewed Darlene Reed who said that Pringle was very excited when he told her of the incident. Other than that, she had nothing to offer. When asked about an official record being kept, she said she

Kevin Pringle at the spot where the creature was sighted.

Park ranger Darlean Reed.

did not think such would be recorded. It was determined that sasquatch sightings are not given the same attention by park officials as bear sightings.

Source: Thomas Steenburg Sasquatch Incidents File, Number BC 10136. (Photos/artwork: Sketch, K. Pringle; Pringle at spot, and Darlean Reed, T. Steenburg.)

2000/09/00 – Harrison Hot Springs. Meandering Tracks: John Miles reported that he found possible sasquatch tracks in the Harrison Hot Springs area during September 2000. Bill Miller investigated the report and provided the following account:

On September 29th of 2000, I was taken by the logging camp watchman (John Myles) and shown a feeder stream coming off of Lake Harrison where tracks had been discovered just prior to my arrival. Myles stated that a cougar had been coming near his camp, so he had taken some chicken parts he had soaked in a bucket of water for a few days and had spot-poured it down that stream [on the ground] and away from his home where he then tied the bag carrying the chicken parts to a small tree by the lake.

Upon arriving at the site, I counted 22 tracks in total. These tracks were along what looked to be an old stream that had dried, but the overhanging forest around it kept it shaded and well protected from the elements. The prints were not in a typical striding out track-way, but were instead in what I would call a meandering walk. In other words the foot stepped here and there as if it was looking around for something. Myles had told me that it was on the third day after pouring the chicken-gut water mixture onto the ground that the prints showed up overnight and that the bag of chicken parts had been removed from the tree.

The best way to describe the tracks and how they were laid is to possibly imagine someone looking for their keys that they may have lost. It was if the subject who made the tracks was looking for something on both sides of the narrow dried up stream. Those tracks turned many times as if stepping here and there without a destination in mind at that time. I followed and counted the tracks out to the rocky area that surrounded the tree where the chicken parts had been hung. I recall the prints being around 11 inches long and showing no signs of an arch. On the inner step of the foot's ball [on one print] there was a stone that was sticking up out of the soil about 3/8th of an inch. It looked like the weight of the individual who had stepped on the stone had managed to press it slightly into the soil leaving a gap around it while still pressing the soil flat all around the stone. It appeared to me that a heavily padded foot had absorbed the stone so that bottom of the foot was still able to make contact with the ground.

One of the tracks found by John Miles. The prints were about 11 inches long.

This was not the first time that tracks had been reported found near the 20-mile bay area.

Source: Bill Miller, Bigfoot Field Research. (Photo: Footprint, Bill Miller.)

2000/09/00 – Gifford Pinchot National Forest, Washington, USA. The Skookum Cast: Unusual body prints were found in Gifford Pinchot National Forest, Skookum Meadows area, Washington, USA, during September 2000. A large plaster cast of the prints was made and presented for scientific examination during that month and later on that year.

The prints were found by Thom Powell, Rick Noll, Dr. Leroy Fish, Derek Randles and others (BFRO members) in an area of soft earth and mud which they had baited with fruit, hoping to get sasquatch footprints. It was speculated that the creature reclined in the soft earth/mud and reached over to take the fruit.

The cast was examined by Canadian researchers John Green and Dr. John Bindernagel, along with US professionals, Dr. Grover Krantz, Dr. Daris Swindler, Dr. Jeffrey Meldrum, and Dr. Esteban Sarmiento. It was concluded that the prints could not be attributed to any known animal species.

John Green had the cast duplicated in a plastic material and it is now a part of his collection of sasquatch-related artifacts.

(Left to right) Scientists Dr. Jeff Meldrum, Dr. Esteban Sarmiento, and Dr. Darris Swindler examining the Skookum cast.

John Green (left) with Tony Healy and John's copy of the Skookum cast at his home in Harrison Hot Springs.

Source: Christopher L. Murphy, 2010. *Know the Sasquatch: Sequel and Update to Meet the Sasquatch*. Hancock House Publishers, Surrey, BC, pp. 171–175. (Photos: Scientists, Whitewolf Entertainment; Green and Healy, C. Murphy.)

2001/03/30 – Radium Hot Springs. Running Like a Man: A motorist reported seeing a very tall, strange creature on the roadside near Radium Hot Springs at about 12:30 p.m. on March 30, 2001. The creature was seen through the trees running "like a man" down a small embankment to the edge of the road (motorist's right hand side). At this time it was one-quarter to one-half mile ahead. It did not cross the road, but retreated back into the trees as the motorist approached.

The motorist said the creature appeared to be between 8 feet and 10 feet tall. It was dark brown or black, and clearly had hands, not paws. Its arms were very long and its overall build was "massive."

Source: BFRO website. The witness was interviewed by Blaine McMillan.

2001/04/18 – Richmond. Farewell René—Fondly Remembered: Renowned sasquatch searcher René Dahinden passed away on April 18, 2001 after losing his battle with prostate cancer. As they took him to the operating room, he told the attendants, "If I don't come outa this, I'm gonna be damn mad." He was a very colorful person, well-known to many people, so I have deemed it appropriate to provide a short biography of his life as follows:

René Dahinden (1930–2001)

Born illegitimately in Lucerne, Switzerland on August 23, 1930, René Dahinden was placed in a Catholic orphanage at the age of one month. About a year later, he was adopted by a middle-aged couple who ran a stationery wholesale business. He pointed out, however, that the legal requirements for the adoption were never completed. Nevertheless, his new parents were fairly well off, so René's infancy and early childhood got off to a good start. René enjoyed skiing, was taken on trips, and had the luxury of spending his vacations at a summer home.

At about age nine, his foster mother died. His foster father took a new, younger wife within a year or so who unfortunately did not take a liking to the boy. So at age 11 René was sent to a boarding school. In René's words, "Neither of them wanted me, so I was put in a boys' institution in Lucerne." Life at the school was fairly good. He attended regular school, worked on the school farm, and occasionally was sent out to work for outside farmers. Nevertheless, the last 2 years had hardened him to the ways of the world, and this was reflected in his character.

Remarkably, a little over a year after René entered the school, his natural mother showed up and claimed him. By this time she had married and had 2 children. René joined the family, but his homecoming was a catastrophe. He lasted about 4 months before he was fostered out to a farming family. While the recent years had been trying for René, past hardships paled to those he would now face. "Life was hard," he once exclaimed, "There was absolutely no time allowed for play. As soon as I got home from school, I had to start the chores, and I worked at them until bedtime. It wasn't that these people were cruel. They just had no time for affection."

He remained at the farm for 3 years and then tried again to live with his mother and her family. Now 15 years old, he stayed about 2 weeks and then struck out on his own. Finding work when and wherever he could, he survived the next 3 years with little trouble. When he turned 18, he had his mother sign the necessary forms for a passport. For the next 5 years, he wandered all over Europe, working long enough at one job to get enough money to move on to another one.

In Sweden, during September 1952, he met his future wife, Wanja Twan. The following year, René decided to immigrate to Canada, and shipped out in October 1953. He went to work on the farm of William Willick near Calgary, Alberta. While there, he and Willick heard a CBC radio program about a *Daily Mail* expedition to find the yeti. This interested René and he remarked to Willick, "Now wouldn't that be something—to be on the hunt for that thing?" Willick responded, "Hell, you don't have to go that far; they got them things in British Columbia." René pressed his boss for more information and found out he was not kidding—René learned about the sasquatch.

René moved to Williams Lake, BC the following spring, where he found work at a sawmill. He spent his spare time doing sasquatch research. Wanja came to Canada in 1955 and the couple married the following year. Their first son, Erik, was born later that year.

In 1958, René operated a boat rental service on Harrison Lake. Wanja worked for a local bank. In those years, skeet shooting was held on the muddy fringe of Harrison Lake by a traveling skeet club. Wanja commented on the lead pollution caused by the spent shot, stating that the shot should be retrieved. Her comment gave René an idea and he thereupon recovered several tons of it for a profit of about $2,500.

The couple's second son, Martin, was born in 1963. During this time, René became totally consumed with the search for the sasquatch. Indeed, over the last 9 years he had made a name for himself in the field—very little happened in the West, sasquatch-wise, without René's involvement. He left the job in Harrison and concentrated on lead shot salvaging at gun clubs, eventually working full time at the Vancouver Gun Club in Richmond. This club soon provided him with living facilities on club property.

By 1967, René was spending very little time with his family. He had made a conscious decision that nothing else mattered except the

René hard at work in the early 1990s. I convinced him to make copies of his footprint casts to sell to sasquatch enthusiasts. He did not like the idea at first because he said people might use them to hoax footprints. His casts were the best as they were all done by hand using river sand.

In Loving Memory of

Renè Dahinden

Born
August 23rd, 1930
Weggis, Switzerland

*There are mysteries unexplained within the forest deep
And legend lures many a man. In search of a giant
big footed creature, too elusive to keep.*

Passed Away
April 18th, 2001
Richmond, British Columbia
Age 70 years

*Among these folks there was a man who stood out well,
as a one of a kind rare breed. For 45 years he traveled
the land, following the Sasquatch's lead.*

Memorial Service
Saturday, April 28th, 2001
at 1:00 p.m.
The Vancouver Gun Club
7340 Sidaway Road
Richmond, British Columbia

*Throwing his backpack over his shoulder, his camera
in hand, Sasquatch researcher Renè Dahinden
forged his life in this wild land.*

*Renè's well worn journey has moved us all deeply
as a father, a grandfather and a friend dear.
And in the quietest moments you would swear, that you
can hear, his footsteps winding over some mountain path
so very near.*
...Michelle Beauregard

Cremation

Memorial card for René Dahinden. A "Celebration of Life" gathering was held at the Vancouver Gun Club. René's sons, ex-wife and many friends were there to share memories. It was a very nice event and truly fitting for my old friend.

sasquatch, including his family life. Erik was then 11 and Martin 4. Fortunately, Wanja was a responsible mother and able to carry on reasonably well without René. The couple formalized the situation with a divorce in August 1967. While Wanja was not happy with the situation, she understood that René's passion would give him no peace until he solved the sasquatch mystery. They remained good friends to the end.

René's relentless search for the creature took him all over California, Washington, Oregon, and BC, where he spent weeks at time in the wilderness. In an effort to get a scientific analysis of the Patterson/Gimlin film, he traveled to England, Finland, Sweden,

Switzerland, and Russia. In 1973, a full account of his life and findings to that date was published in his book, *Sasquatch-Bigfoot: The Search for North America's Incredible Creature*, by Don Hunter with René Dahinden.

While René's research convinced him that the creature existed, he never actually saw a sasquatch. His many attempts to get assistance from governments or major research organizations met with bitter disappointment. Nevertheless, René's burning drive was to either find the sasquatch or die in that quest. He took full responsibility for his actions and never looked back.

Source: Don Hunter and René Dahinden, 1993. *Sasquatch-Bigfoot, The Search for North America's Incredible Creature.* McClelland & Stewart Inc., pp. 75–78. Also personal communication with René Dahinden in previous years and with Wanja Twan (ex-wife of Rene Dahinden). (Photos: René portrait and making casts, C. Murphy; Memorial card, Dahinden family.)

2001/05/00 – Hope. Near the Toll Gate: Gordon Bomersback reported that he saw a strange creature near Hope along the Coquihalla Highway in May 2001. Just after passing through the toll gate, he noticed a large, upright figure walking some distance up the hillside on the right (west) side of the road. At first he thought it was a bear, but changed his mind after seeing that the creature walked upright the whole time. He described it as being covered in what appeared to be dark brown or black hair. He said the oddity was in sight for about 10 seconds.

Thomas Steenburg was given this report on September 26, 2001 after pulling into a highway rest stop near the sighting location. Bomersback saw the "Sasquatch Research" sign on Steenburg's vehicle and approached him. Bomersback was with his wife reflecting on his sighting experience in the area when he saw Steenburg's sign.

Source: Thomas Steenburg Sasquatch Incidents File Number BC 10138.

2001/09/00 – Mission. Chattering Like a Bear: Mikel Crowther reported that he saw an unusual creature near Mission at Stave Lake in September 2001. He was out bear hunting. and while sitting on the top of a tree stump at about 10:00 a.m., he heard possible footsteps behind him. He turned and saw a strange creature stepping over (like a human) a large fallen tree 87 feet from his position. Crowther said he could hear it "chattering" to itself, much like a bear does when it's angry—low grunting noises. When it was straddling the log, it saw Crowther and froze. The 2 gazed at each other for a few seconds, and then the creature turned and walked away. During this time, Crowther sensed an unpleasant odor, like a wet dog or rotten meat.

He described the creature as having speckly brown hair (salt and pepper). However, the hair seemed a little thinner around the shoulders and a lighter color around the stomach area, "Like [the hair on] the top of a dog is thicker than the bottom of its belly." It had a lot of thick hair down the back of its neck and around its lower back. He commented that the hair was not like human hair, "It was kind of shaggy—thicker than ours. It was a lot thicker than ours in certain parts."

Mikel Crowther atop the tree stump from which he observed the strange creature.

He estimated its height was at least 7.5 feet. It had sloping shoulders, and was thick around the collar bone area. "Looked like a guy wearing a sweater with football gear underneath." He could not distinguish the length of its arms, but noticed that its forearms were quite large.

Of its facial features, Crowther said it had a heavy forehead, low structure nose, and a thick jaw line. The skin on its face was dark.

Crowther provided the following account leading up to his sighting:

> When I walked into the area which I have quite often hunted, when I got into the area, I was already uncomfortable. I don't know why, I couldn't explain it. I pay attention to the

ravens and the crows in the area. I've found that they seem to follow bears sometimes and all the predatory animals, and this one particular time there were no ravens around. They were far off somewhere, I could hear them. I was a little uneasy and I don't know why. I thought maybe I had too much coffee that morning. That's where I like to hunt bear from on top of these large stumps [see photo] off different game trails and whatnot. While I was up there, I heard this noise behind me. I heard something walking—thumping. I'm not talking smashing or anything like that, it wasn't that loud. Being there was nobody else out there it was quite clear. It was just a thud, thud, like somebody walking on a peat bog sort of thing. I looked over just as it started to step over this log.

When asked what he thought the creature was, Crowther stated, "I think it was an ape that lives quite comfortably in our environment. Some kind of bipedal primate of some sort."

This report was provided to Thomas Steenburg after Crowther saw the "Sasquatch Research" sign on Steenburg's vehicle. He called out, and then waved Steenburg over to talk. Steenburg later went to the area of the sighting with Crowther and John Green. The distance of 87 feet from Crowther's position to the creature was determined at that time.

Source: Thomas Steenburg Sasquatch Incidents File, Number BC 10141.

2002/02/11 – Windermere. A "Bridge" Going Across the Foot: Two youths reported finding unusual human-like footprints in snow on Windermere Lake on February 11, 2002. The pair went out onto the frozen lake for something to do and saw the tracks about half way across. They said the prints were 14 to 16 inches long with well-defined toes and a "bridge" going across the foot. They followed the tracks for several hundred yards towards the opposite shore from the town of Windermere. They described the trackway as being "direct." The "stride" was said to be 5 feet or greater (this appears to be too short for a stride and too long for a pace.)

Source: BFRO website. One of the witnesses was interviewed by Blaine McMillan.

367

2002/02/14 – Port Angeles, Washington, USA. Renowned Scientist Dr. Grover Krantz Passes On: The world of sasquatch research was shaken upon hearing the news that Dr. Grover S. Krantz had died at his Port Angeles, Washington, home on February 14, 2002.

Dr. Krantz was a physical anthropologist with Washington State University. He became involved in the sasquatch issue in 1963 and spent the next 39 years relentlessly investigating the evidence provided to him. He found what he considered indisputable evidence in dermal ridges that he discovered on some footprint casts. In his own words:

Dr. Krantz in his laboratory.

> When I first realized the potential significance of dermal ridges showing in sasquatch footprints, it seemed to me that scientific acceptance of the existence of the species might be achieved without having to bring in a specimen of the animal itself. It was this hope that drove me to expend so much of my resources on it, and of my scientific reputation as well.

Unfortunately, Dr. Krantz was not able to bring about the scientific acceptance he envisioned. Nevertheless, it was confirmed by Jimmy Chilcutt, a fingerprint expert who had made a special study of the dermatoglyphics on the hands and feet of non-human primates, that dermal ridges discovered by Krantz indicated they were definitely those of a non-human primate.

Dr. Krantz's most notable sasquatch-related accomplishment was his reconstruction (model) of the skull of *Gigantopithecus blac-*

Dr. Krantz giving a talk in Vancouver in 1999. He was always a keynote speaker at conferences, and a very genuine and nice person to talk with.

ki, an extinct primate that lived in southern China more than 300,000 years ago. Krantz theorized that the sasquatch may have descended from this primate. His model is based on a lower jaw fossil of the creature.

Dr. Krantz was also the major supporter for the authenticity of the Bossburg "cripple foot" casts (made from alleged footprints of a crippled sasquatch). He studied these casts intently and provided a proposed bone structure for each cast. Despite skepticism, these casts are very intriguing.

Although highly regarded as an anthropologist, Dr. Krantz' reputation suffered because of his belief in the reality of the sasquatch. While certainly bothered by this development, he did not let it stop him in any way. He spoke at sasquatch symposiums, appeared in many television documentaries, and was continually quoted in newspaper articles.

Source: Christopher L. Murphy, 2009. *Know the Sasquatch: Sequel and Update to Meet the Sasquatch.* Hancock House Publishers Ltd., Surrey, BC. pp. 232, 233. (Photos: Krantz in laboratory, G. Krantz; Krantz at podium, C. Murphy.)

2002/07/00 – Grand Forks. We Know What We Saw: A 15-year-old boy and his friend reported a sasquatch encounter near Grand Forks in July 2002. One of the boys related the incident publically when he was a university student (age 19). The following are his own words.

When I was 15 [2002], my best friend Landon and I went camping with my parents for a weekend in the summer. We live in Vancouver, BC, so we took a long drive to a campground East of Vancouver near a little town called Grand Forks.

My family arrived at the campground with everyone stiff, tired, bored, and cranky. My dad and mom checked-in while Landon and I pitched the tent at the campsite where we were to stay. After several failed attempts we finally hoisted the tent. We built a fire (it was getting dark) and roasted marshmallows that night.

By the time we unpacked, lit the fire, readied the sleeping bags and got the food cooked and eaten, it was around 9:30 p.m. and the sun had just set. Landon suggested that we take a walk around the campsite to see what was there. We grabbed our flashlights and headed out.

We followed a dirt path heading deeper into the forest; we had been walking for some time when all of a sudden we heard heavy breathing to the right of the path. We both immediately stopped, thinking it was a bear (we each lived by a big park where bears had been spotted several times so the two of us were extremely paranoid). We quickly stepped way back putting distance between ourselves and the creature and shone our flashlights into the dark bushes and trees.

What we saw would question the work of most scientists to date about the complex web of the animal kingdom; if they saw what we saw, they would most definitely add the creature "Bigfoot" or "Sasquatch" to the long list of primates.

The creature that we first thought was a bear had been crouching or kneeling and as we shined our lights on the bushes, it grunted and began to slowly rise to stand on two legs. It kept getting taller and taller behind the bushes. I remember thinking, "When is this creature going to stop getting taller?" When it was fully erect on two legs it stood about eight feet tall, I know this because Landon and I were close to six feet and this thing was more than two feet taller than us. It was covered in dark brown hair and had the posture and body structure of a human.

With our mouths open, the two of us could not move; we could only stare while this huge creature, through the dark hair covering most of its face, stared back at us. Then it turned and ran swiftly into the dark forest. We were scared out of our minds and too afraid to follow it.

We headed straight back to the campsite to tell my parents everything that happened, but they didn't believe us. We were so terrified neither of us slept a wink that night, although we secretly hoped it would come back, if only so my parents would believe our story. Landon and I still talk about that night; we know what we saw.

Source: Linda Coil Suchy, 2009. *Who's Watching You: An Exploration of the Bigfoot Phenomenon in the Pacific Northwest.* Hancock House Publishers, Ltd., Surrey, BC., pp. 197, 198, from information provided directly by the student.

2002/08/17 – Tete Jaune Cache. Thought It Was a Moose: A motorist traveling on Highway 16 reported that he saw a sasquatch about 9 miles west of Tete Jaune Cache at about 6:30 p.m on August 17, 2002. It was first seen at a distance of about 600 yards as a dark animal of some sort. Then at about 500 yards he thought it was a moose. When he was at about 300 yards the creature turned and looked at him, and from its profile it was realized that the creature was a sasquatch. It crossed the highway from south to north. At about 200 yards he watched the creature as it stepped off the road and disappeared into thick bush.

He stopped and inspected the ground where the creature had been, but there were no tracks as the ground was very rocky.

Source: BFRO website.

2002/08/00 – Port Alberni (VI). Thought It Was a Human With Down's Syndrome: A man visiting from Florida reported seeing a strange creature while camped near Port Alberni in August 2002. The man was with a group, but was left alone to take care of the campfire.

He went into the forest to gather firewood and in the process of trying to break off a branch, he saw what he thought was a bear. He did not want to make eye contact with it, so turned around, pulling the branch with his back to the creature.

When he turned around to see what the creature was doing, he got the impression that it was a human with Down's syndrome, and tried to talk to him. At that point the creature stood up and the 2 were eye-to-eye for about 3 seconds.

The creature was about 6 feet tall, very wide, and stood with its arms at its side and its palms turned out. The camper noted the palms were covered in black skin. It had a flat face, flat nose, big forehead, and the head appeared pointed. Only the chin stuck out a little. The facial skin was black. It had hair on its cheeks but not over its eyes or mouth. The eyes were dark but they were similar to a human—he could see the sclera (white part of the eye). Its ears were not visible. It had a short neck and a massive chest. Its body was covered in hair that looked black, shiny, and well groomed—2.5 to 3 inches in length.

After the brief mutual stare, the creature then simply turned and walked away on 2 legs. The camper said that it walked leaning forward at an angle that would be impossible for a human. He was surprised by how quietly the creature moved away from him through the bushes. He heard 3 or 4 steps then he could no longer hear anything.

Source: BFRO website. The witness was interviewed by Colleen McDonald:

2002/09/00 Mission. Steenburg Moves to BC: Alberta sasquatch researcher Thomas Steenburg moved to Mission, BC in September 2002, stating that he wished to live and work in Canada's "sasquatch" province. Being an avid outdoorsman with military training, he took up the search, as before, by going into the wilderness and physically looking for evidence. While he was living in Alberta, he authored 3 books on the sasquatch: *The Sasquatch in Alberta* (1990), *Sasquatch/Bigfoot: The Continuing Mystery* (1993) and *In Search of Giants: Bigfoot/ Sasquatch Encounters* (2000).

Thomas Steenburg in 2011.

All of the main BC sightings Steenburg has documented (both in his books and his personal sasquatch incidents file) are included in this work of which he is an associate author.

Source: Christopher L. Murphy, 2010. *Know the Sasquatch: Sequel and Update to Meet the Sasquatch*. Hancock House Publishers, Surrey, BC, p. 246. Steenburg's first book, *The Sasquatch in Alberta*, was published by Western Publishers, Calgary, Alberta. His other books were published by Hancock House Publishers, Surrey, BC. (Photo: C. Murphy.)

2002/10/00 – Little Fort. It Bent Straight from the Hips: A motorist reported that she saw an odd creature near Little Fort in October 2002. While driving, she saw a dark gray, man-like figure across the Thompson River. She thought it looked strange so stopped at a turn-out to get a better look. The figure was by a small creek running into the Thompson. It was bending up and down, but the bending was very strange, not like a human. The body seemed longer than that of a human and, "It bent straight from the hips, without an obvious curve to the back."

She could see that it was a dark gray color all over its body ("softish looking hair") except the face and hands, which were lighter.

At one point the oddity stood straight up and for several moments looked right at her. She thought perhaps that she should flag someone down to also see it, but flashing through her mind was that if it were a sasquatch, would people try to shoot it? She continued watching, and in her own words:

> I stood there and the look it gave me had power in it. I know it saw me, possibly better than I saw it. I felt I was invading somewhere I should not have been. I looked for a while longer, the subject went back to doing what it had been doing—looked something like picking up rocks and throwing them out of the water. I later thought, could it have been fishing, or what? I am sad that I did not stay there and watch to the end, as I left before the subject did.

Source: BFRO website. Witness was interviewed by Dave Bruce.

2002/11/00. Tofino (VI). Real Big Orange Eyes: Arnold Frank and his nephew, Patrick, reported seeing a strange creature near Tofino during early November 2002. They said they saw the same or a similar creature twice. On the first occasion, while walking in the woods near Radar Hill, they heard a crashing sound and then glimpsed it through the bush. "We saw real big, orange eyes, real high off the ground," Arnold Frank said.

A few nights later, they saw a creature of the same nature at a car rest stop. They saw it dashing into the woods. They estimated that the creature was about 8 feet tall.

Also, in this same area and time frame, an elderly lady reported that a gigantic biped crossed the road in front of her car. It stopped, looked directly at her, and then proceeded into the bush on the other side of the road.

Source: Mark Hume, 2002. "To see something that big would scare anybody." The *National Post,* Vancouver, BC, November 14, 2002. Also "Bigfoot has them talking—again." The *Ottawa Citizen*, Ontario, November 15, 2002. Also, Graham Andrews, 2002. "Sasquatch spotted: Reports of apparent local sightings on the rise." The *Alberni Valley Times,* Alberni, Vancouver Island, November 13, 2002. Also Thomas Steenburg Sasquatch Incidents File, Number BC 10139.

2002/12/00 – Lillooet. Looked Like a Monkey Butt: A motorist traveling on Highway 40 near Lillooet reported that he saw an odd creature walking on 2 legs along the highway at about 8:30 p.m. around December 20, 2002. The creature, illuminated by headlights, was on the opposite roadside, going in the same direction as the motorist, so it was seen from the back. Its arms were swinging, and as he went past, it looked over its right shoulder at him. Its face was described as being flat with no snout. He estimated that it was 6 to 7 feet tall. Its body covered in black hair, 2 to 3 inches in length. There was no hair on the palms of it hands, nor on its buttocks, which he said "looked like a monkey butt." It walked in a fluid and graceful manner.

Source: BFRO website. The witness was interviewed by Blaine McMillan.

2002/YR/00 – Harrison Mills (Chehalis Reservation). You Will Never Catch a Sasquatch: Kelsey Charlie, a Chehalis resident, and a female companion reported an unusual sasquatch sighting in the Harrison area in 2002. The pair was walking in the woods and saw a female sasquatch and her child by the edge of a creek. Charlie first thought the figures were a tree stump, but then saw he was looking at 2 sasquatch. The mother was squatted down scooping water from the creek in her hand and feeding it to her child. The child had its arm on the mother's shoulder. Charlie estimated the size of the child sasquatch to be about his height (around 5 feet, 7 inches). Because the mother was squatted down, he could not estimate her height.

Charlie informed that he had seen sasquatch tracks in the past, and that the creatures can occasionally be heard from the Chehalis Reservation—a crazy scream. He said it sometimes sounds like an African jungle in that area.

He said that both his father and grandfather had seen sasquatch. His grandfather was of the belief that the creatures are "shape shifters."

After a filming session with Charlie in June 2005, I asked him about catching one of the creatures. He looked at me very intently and said, "You will never catch a sasquatch."

Kelsey Charlie and his daughter, Angela Raven (2005).

Source: First Story: *Sasquatch & Bloodlines*, June 2005. CTV production, Vancouver, BC. Also from personal communication with Kelsey Charlie from being with him in the same television production. Some of the film was taken at the author's home.

2003/03/00 – Burnaby. Film Site Model Provides Insights: In March 2003 Chris Murphy (author) completed construction of a scale model of the site where a sasquatch was alleged to have been filmed at Bluff Creek, California in 1967.

Working with documents and photographs given to him by the late René Dahinden, and with the assistance of Igor Bourtsev in Russia, Murphy was able to determine the placement of the trees, stumps and forest debris at the actual film site. These were plotted on a sheet of plywood and the site constructed using modeling materials, sand, and tiny pieces of bark and drift wood. The completed model provided a three-dimensional view of the site which facilitated a better understanding of the film.

Top: Frame 352 of the Patterson/Gimlin film on which the model (below) was primarily based. The triangle at the bottom was Roger Patterson's position. It is immediately seen that the low camera height for the film frame gives one a very different perception of the actual scene.

About 2 years later, additional information came to light and the model was revised accordingly. What is shown here is the final model.

Source: Christopher L. Murphy, 2003. "Bigfoot Film Site Insights. *Fate* magazine, Lakeville, Minnesota, USA, March, pp. 26–29. Also Christopher L. Murphy, 2010. *Know the Sasquatch: Sequel and Update to Meet the Sasquatch.* Hancock House Publishers, Surrey, BC., pp. 64–70. (Photos: Frame 352 of film, R. Patterson, public domain; Film site model, C. Murphy.)

2003/06/00 – Duncan (VI). Prints In Strawberry Field: Francis Joe reported finding large, human-like footprints in his strawberry field near Duncan in June 2003. The tracks were 15 inches long running in a straight line and were about 3 feet apart. Francis Joe stated, "That's not the tracks of an ordinary human. You could tell if it had shoes on." His fields are bounded by bush to the north and the Cowichan River to the south. The Cowichan First Nations people call the creature believed to have made the prints *Thumquas.*

Source: Peter Rusland, 2003. "Local family wondering what left the huge tracks." The *News Leader*, Duncan, June 11, 2003.

2003/11/00 – Vancouver. Sasquatch Heads Created: During November 2003 Vancouver artist Penny Birnam completed the sculptures of 4 sasquatch heads for display at an upcoming sasquatch exhibit at the Vancouver Museum. Birnam worked from an image taken from the Patterson/Gimlin film, but made different facial features for each head. She reasoned that the creature would probably vary in this regard.

Penny's reasoning brings into play some intriguing speculation. If the creatures differ widely in their facial features, they would be quite different

Artist Penny Birnam holds one of the sasquatch heads she created for my Vancouver Museum sasquatch exhibit.

from other non-human primates. From my observations, such are highly similar in this regard—so much so that it is hard to tell them apart. Other animals, of course, are for the most part almost identical when they are of the same type or breed.

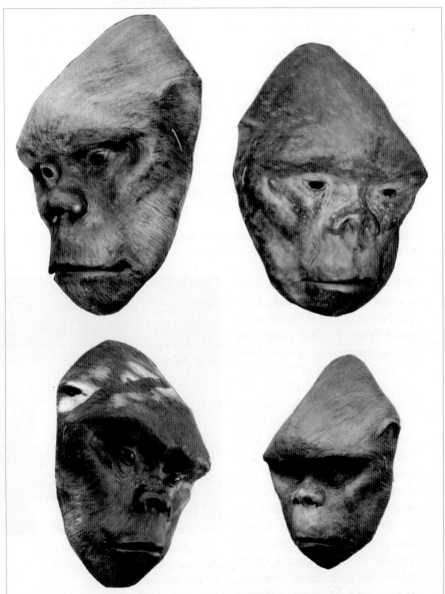

The four heads created by Penny Birnam.

Source: Christopher L. Murphy, 2010. *Know the Sasquatch: Sequel and Update to Meet the Sasquatch.* Hancock House Publishers, Surrey. BC, p. 113. (Photos: C. Murphy.)

2003/12/18 – Vancouver. First Nations Elders Provide Sasquatch Stories: During my research for a sasquatch exhibit at the Vancouver Museum in 2004, 3 council elders of the Nisga'a Lisims First Nations Government (British Columbia) shared their individual stories about the sasquatch in a letter to the museum. The following are their submissions, directly as recorded by Allison Nyce, manager, Ayuukhl Nisga'a Department. I have left the stories verbatim to preserve their integrity and highly unique presentation.

NISGA'A LISIMS GOVERNMENT
COUNCIL OF ELDERS

Horace Stevens: I remember when all the villages used to have public works and when they were talking about the big hairy man. The young men used to go hunting up on the mountain, when they would head out, they did not go for only one afternoon. They would go for a week and before they headed out to their hunting grounds, they would say where the canoes would land. They would stop at Nass Harbor and walk up the mountain until they came upon Ginluuak. They would go up a certain trail and come back down this way again, that is where they would put down what they were carrying. My father, Albert Stevens, was not a hunter; he told me that when he heard that the hunters were going out hunting he would grab his packsaddle. There would be about 8 or 10 of them but they were not all hunters, The would have those who would be the packers for them. They got the wool from the mountain goats; they know how to make the wool. They were very proud of themselves for making wool. They would make woolen socks, woolen pants and shirts so that a hunter would not catch any sickness while out hunting. My father did not tell *adaawak* (traditional stories) but he told us what they did back in those days. He told about the hunting expedition up on the mountain one time during the winter. They seen footprints in the snow at the shady area, they called it the footprints of the hairy man. One time they were looking down and seen a man walking, he would walk way down. They would always spot him at Nass Harbour. One time they went out hunting for about a day and a half just to get enough food and that is when they seen it again. While they were walking, they seen his footprints going across, it used to snow early a long time ago, late September or early October. The little ponds used to be frozen over and they seen the footprints sunken in about one inch deep; they knew it was the hairy man. They called it the *naxnok* (supernatural), because when they spotted him, he would disappear. I believe this story is true, no one has captured the hairy man; they have only seen him going by. We have also read this in the newspaper of other native people seeing the Sasquatch, especially those who live near or in the mountains. Those who have been close to the area where he walked, they said it was very stinky. My second year hunting with my uncle, while we were walking he stopped all of a sudden and started sniffing the air. During the first snowfall, it would snow and then melt. It would not get very deep at that time. In the olden days, the hunters used to build house wherever they would go hunting. When they knew they were not going to make it back home for a while from the mountains, they used to build their houses with the trees on a slant. They used to stay at a place where they could build a fire under a rock cliff, it was just like a cave, and the rock would come over. One morning when they woke up, they seen footprints going past their camp, they did not hear anything during the night. I have seen it one time when my wife and I were returning from Terrace. We have to tell the stories we know about the hairy man. All of our grandfathers and uncles were hunters and they have told about this. This is all I have to say for now, Mr. Chairman.

—Horace Stevens, Council of Elders, December 18, 2003.

Emma Nyce: Thank you Mr. Chairman, I am happy to be able to be in attendance and I agree with what Horace has said that we should tell what we know. I have also heard this a long time ago when I was with my grandmother Annie in Greenville when I was small. She told us that our grandfather was not trying to scare us when he was telling us about this in Greenville. I heard William Stevens and Peter Calder telling that they did see it; there were three of them. I forgot the place they had mentioned where they had seen this and we laughed when we heard that they said that he had no clothes on. My grandmother told us not to laugh because it was *hawahlkw* (taboo). Only those who have cleansed themselves were the ones who seen it; it was not seen by everyone. They seen it at the mouth of the Nass, it was walking along the sand beach. My dad told us that it was across from Mill Bay. He said that it was true and that only those who cleansed themselves were able to see him. He also told us not to laugh at it because it was alive. When I hear children laughing when they hear this story about the Sasquatch, I tell them not to laugh because he will follow them. My dad told us that he (the hairy man) was a living creature of some kind. When I heard what Horace had mentioned, I remember what I heard from my grandmother. I know they have seen it at the mouth of the Nass but I have not heard of anyone seeing it up this way. The hunters seen this a long time ago and they say that it is part *naxnok* (supernatural). My grandfather told this *adaawak* (traditional story), during the public works. This is all I know about this that is all I will say for now.

—Emma Nyce, Council of Elders, December 18, 2003.

Peter Clayton: One of my sons used to go fishing trout, they went up. They went three times, one day as they were coming down the trail, they seen a man standing down below. When they had reached down below, the man had disappeared. I told him not to go up there anymore because I remember what my father had told us about the *sbinaxnok* (supernatural being). I had forgotten some of the story, this is probably the same as what we are discussing now. I will ask my son who was with him at the time so we will know the place where they were. There is another time when I

381

used to *gihl'askw* (translation not shown). I went with Ester Adams and them and they seen it and she told me not be afraid. This is a very strong *adaawak* (traditional story) that Horace has mentioned. A man can die if he breaks the law regarding the cleansing. I know my son does not have any difficulty whenever he goes hunting. We should ask other elders who have seen a Sasquatch or the *naxnok* and where so that it would become more clear. Somebody saw a man on the other side of Sand Lake.

—Peter Clayton, Council of Elders, December 18, 2003.

Source: Correspondence from the Nisga'a Lisims Government Council of Elders to the Vancouver Museum, BC, 2003. (Photo-Logo: Nisga'a Lisims Government.)

2003/CI/00. Bowen Island. Ladies Stared In Disbelief: A woman staying on Bowen Island reported that she and 2 friends saw a sasquatch on the island in about 2003. The three women had gone for a walk on a trail when all of a sudden the creature jumped out of the bush some distance ahead and stood in the middle of the trail looking at them. All three women stared back in disbelief. The creature then ran off. The description of the oddity provided by the woman matched the general description for the creature.

This incident was related to me by the woman, who was a nurse in a rest home which looked after my father. I had provided the home's library with my book, *Meet the Sasquatch*. The nurse approached me while on a visit to my father and related her experience.

Bowen Island is about 1.2 miles off the coast of BC, and has a land area of about 19.3 square miles. For some years in the 1950s it was used for school picnics, and is now a popular vacation home spot. As of 2007, there were 3,551 permanent residents, with about 1,500 visitors each year. Each day about 700 workers and students commute to the island.

With the above facts, one would not think the island would be very high on the list for sasquatch visits—especially being an island. However, we can conclude that there would be a lot of food there, and for a sasquatch, getting to the island or escaping from it by swimming would be easy.

Bowen Island. There are other smaller islands that could be used as "stepping stones" for swimming.

Source: Author's file and Wikipedia. (Photo: Image from Google Earth; Image © 2011 Province of British Columbia, Data Living Oceans Society; Image © 2011 Digital-Globe; © 2011 Cnes/Spot image.)

2004/05/10 – Merritt. Standing By a Pond: Tim Martindale reported seeing what he believes were 2 sasquatch about 38 miles east of Merritt on May 10, 2004. He was on the Coquihalla Highway heading towards Kamloops when he saw a large and smaller figure standing erect by a pond about 300 feet away. He thought the figures were 2 people fishing, but then realized that the pond was "scummy" and that there would probably not be any fish in it.

He then reasoned that what he was seeing were bears, but as he got closer, the larger figure knelt down and he could see that they were not bears or people.

Martindale estimated that the creatures were 7 feet and 5 feet tall. He reasoned that the larger of the 2 had knelt down to get a drink or wash its hands. During this time, the smaller creature watched Martindale (actually glared at him) as he drove past. He estimated that they were in view for about 10 seconds.

Both creatures were covered in dark brown or black hair. The

Tim Martindale standing near the spot by a pond where he saw what he believes were 2 sasquatch.

smaller creature had lighter hair around its face, "It was lighter than its body hair." He could not clearly see the other creature's face. He noticed that the smaller creature, which was standing up the whole time, had arms that were "knee length."

Martindale was so intrigued with his sighting that he immediately sought other people who had seen sasquatch. He placed an ad in a Merritt newspaper asking people with similar experiences to come forward. A subsequent *Daily News* (Kamloops) article on the incident stated that he wanted people to contact him and provided his email and telephone number.

Source: Thomas Steenburg Sasquatch Incidents File Number BC 10144. Also, Mike Cornell, 2004. "Fleeting roadside encounter triggers obsession with the sasquatch." The *Daily News,* Kamloops, BC, July 2004. (Photo: T. Steenburg.)

2004/05/00 – Hope. Looked Like an Ape: Doug Cariou reported that he and his 15-year-old nephew saw an ape-like creature while driving near Hope during May 2004. The creature was seen in a dry creek bed along the Coquihalla Highway. In his own words, Cariou stated (edited for clarity):

My first thought was is it a man, but it was fully black. You could see the entire torso, the arms, the legs, the head. In an instant I had shock waves going through me, without any thoughts in my mind because of what I was actually seeing. My mind had to catch up with what I actually saw. It looked like an ape and this thing, its limbs or whatever, were considerably thicker than a normal man's, and you could tell it had fur or hair on it. It stood completely erect.

Cariou estimated that the creature was 8 feet tall, or taller. Its hair was, on reflection, dark brown.

Source: Thomas Steenburg Sasquatch Incidents file, No. BC 10163.

2004/06/17 – Vancouver. Sasquatch Exhibit at Vancouver Museum: An opening ceremony with some 150 guests was held June 17, 2004 for a sasquatch exhibit at the Vancouver Museum provided by the author with the assistance of John Green and Thomas Steenburg. My book, *Meet the Sasquatch,* written in association with Green and Steenburg, accompanied the exhibit.

The exhibit, which covered 1,800 square feet of floor space, featured numerous sasquatch-related artifacts, artworks, videos, and scientific materials provided by numerous contributors. The exhibit had been one year in planning. Its theme was the life of the noted sasquatch researcher John Green in his long search for the sasquatch. About 28,500 people attended the exhibit which closed on January 31, 2005.

The exhibit and the book essentially got their start in about 2000 when I started assembling sasquatch-related photographs with a book possibly in mind. I became friends with Yvon Leclerc, a researcher in Quebec, and a talented artist/illustrator. Leclerc worked with me on my project and it was decided to definitely write a book, originally called *Meet*

Yvon Leclerc. His assistance with both the book and the museum exhibit was invaluable.

The entrance to the Vancouver Museum "Sasquatch" exhibit. The entire history of the creature was traced with artifacts, artwork, scientific posters, and a selected segment of the video, Sasquatch: Legends Meets Science. *The Patterson/Gimlin film was projected onto a large screen. The video and film were shown continuously.*

Bigfoot. The objective was to provide a highly pictorial work (coffee table book).

In early 2003, I wrote to the Vancouver Museum providing photographs of numerous sasquatch-related artifacts and suggested that the museum might include some sasquatch material in their permanent exhibit, which I could provide or possibly obtain. I reasoned that BC is considered the main habitat of the creature, so such was highly appropriate. The museum offered to provide a one-time sasquatch exhibit, and worked with me as a co-curator to that end. David Hancock of Hancock House suggested that the book in progress could accompany the exhibit. The book was therefore given priority with John Green and Thomas Steenburg as associate authors. Through the diligent efforts of Hancock House, the book did accompany the exhibit. The title was changed to *Meet the Sasquatch* as this was more appropriate for a Canadian book.

I continue to feel that mention of the sasquatch needs to be in

John Green (left), Thomas Steenburg, and author signing books (inset) at Hancock House Publishers.

the museum's permanent exhibit, especially since the museum already has 3 highly noteworthy sasquatch-related artifacts in storage.

Source: File material. (Photos: Museum entrance, C. Murphy; Leclerc portrait, Y. Leclerc; Book signing, D. Hancock.)

2004/07/22 – Canal Flats. Rocks Thrown: A man out fishing after dark in a rowboat near Canal Flats on White Swan Lake reported a strange incident that occurred on July 22, 2004. He was in an area that is highly inaccessible by land, especially during the night. All of a sudden an object (assumed to be a rock) splashed down very close to his boat. He had not seen it coming. This was followed by a definite rock, which he saw with his flashlight as it hit the water. He said it was the size of a large watermelon.

He quickly left the area, and when back at his campground asked if anyone had heard rocks hit the water. He was told that 5 splashes had been heard. He said that his rowing likely muffled the sounds of the other 3 rocks.

Source: BFRO website.

2004/7/00 – Kincolith. As if Swimming, But Not Moving: A fisherman observed a strange creature near Kincolith in the Kincolith River during July 2004. When he arrived at the river, he noticed what he thought was a person swimming in the water about 150 yards away. He thought this was an odd place for someone to go swimming. He whistled, yelled and waved his arms, to get the person's attention, but there was no response. The "person" then stood up and looked at the fisherman, who again waved, but again no response.

The fisherman now thought he was seeing a bear, so proceeded to go fishing but kept an eye on the creature. It again went down into the water and started moving and splashing, as if swimming, but remaining in one spot.

It finally stood up and walked out of the water on 2 legs. The fisherman was now sure it was not a bear. He noted that the creature was about 6 feet tall, of medium build, had very long arms, no neck, and long legs. He determined that the water in which the creature was "swimming," was only 2.5 feet deep. In a later interview, the fisherman said that the creature was much taller than 6 feet.

Source: BFRO website. The witness was interviewed by Tom Yamarone.

2004/08/15 – Sunshine Valley. Four or Five Steps to Cross Road: Jeanne Britt reported that she saw a strange creature walk across the road in Sunshine Valley on August 15, 2004. She was about 200 yards away at the time. She said that it took 4 or 5 steps to cover the distance.

The woman and her husband, Don, were in the process of gathering rocks for their rock garden. She was sitting alone in their parked vehicle when she saw the creature. She said

John Green (left) with Jeanne and Don Britt at the location the sighting.

that it appeared to be about 7 feet tall. It had a flat face and a muscular body covered in auburn-colored hair. Its hands hung down close to its knees.

This sighting was investigated by Thomas Steenburg and John Green. They met the Britts at the sighting location but were unable to find any evidence of the creature's passage.

Source: BFRO Website. Also T. Steenburg Sasquatch Incidents File BC No.10145.

2004/08/00 – Canal Flats. Walking Along a Ridge: Three loggers in the Canal Flats area at White Swan Lake said they observed a man-like creature walking along a ridge on August 20, 2004. It was in view for about 20 minutes, and walked on 2 legs the whole time. The men estimated that it was 11 to 13 feet tall. At one point, it stopped walking and watched the men for about 5 minutes. It then bent over "like a man would do to touch his toes" and then again stood up straight. The men had to start work, so looked away, and during that time the creature left.

Three days later three loggers went up to the ridge and found footprints that were about 16-inches long and 7 inches wide.

When at a new location (same area) on August 31, 2004, the men found human-like footprints on a new muddy road. Four prints were seen that were not much larger than a man's size 12 foot, but were wider and had larger toes than a man.

Source: BFRO website.

2004/11/30 – Tofino (VI). Stood Its Ground: Alice John and her young niece reported seeing a sasquatch at night near Tofino (between Tofino and Port Alberni) on November 30, 2004. The creature stepped out onto the road in front of their car. They stopped to avoid hitting it, and it stopped right in front of them and stood its ground. It did not move until the lights of an approaching car apparently frightened it off into the bush.

The 2 women said that the creature was covered in dark hair and was between 6 and 7 feet tall. They said that it looked "more human…" and was not a bear.

Remarkably, another member of the Frank family (see page 373) saw sasquatch on 2 occasions in the same area in November 2002. Also, in the same area and month, an unidentified elderly lady said a "gigantic biped" stepped out in front of her car and also stopped. (See 2002/11/00 – Tofino (VI). Real Big Orange Eyes.)

Source: Thomas Steenburg Sasquatch Incidents File Number BC 10147. Also, the report was provided on BC TV *Global News*, December 9, 2004.

2005/02/13 – Squamish. Running In the Bushes: Andrew Mac-Gregor, hiking with his wife, Terry, and son, Cole (11), along a service road in the Squamish area at Checakmus Canyon reported that they saw a large, bipedal creature running into the bushes on February 13, 2005. His wife had bent down to tie her shoelace when a deep growl was heard. Andrew then spotted the creature at about 80 yards running off the road into the bush. It was black in color and moved very fast. The group sensed a strong, wet, musky odor.

The 3 proceeded a little further up the road and again sensed the odor. Their son did not want to go any further, so they left the area.

Location of a probable sasquatch sighting by the MacGregor family at Checkamus Canyon.

Source: T. Steenburg, Sasquatch Incidents File BC No. 10148. Also BFRO website. The man was interviewed by T. Steenburg. (Photo: T. Steenburg.)

2005/03/08 - Sechelt. Swinging Arms & Hunched Forward: Andrea Wray reported that she saw a sasquatch near Sechelt at Halfmoon Bay on March 8, 2005. She was driving at about 6:50 p.m. (getting dark) and saw the oddity clearly in her car's high beams. It crossed the road diagonally in front of her car. In her own words (edited for clarity):

> It was walking quickly from one side of the road to the other. I think what struck me also at the time was where it came out of—hilly. At first it looked as though it was a rocky area. Like it was coming [had come] down the hill and I thought, how can anything get down that without sliding, falling. It was walking [erect], its head was forward. It was in a hurry, it was very fast. Its gate [pace] must have been a good 5 to 6 feet in between its walking [footsteps]. It was swinging its arms, hunched forward just a bit; black from head to toe. Never looked over in my direction, not once.

She estimated that the creature was about 8 feet tall, and said it took about 7 paces to cross the road. Commenting on its physical make-up, she said (edited for clarity):

> One of the distinguishing things that I saw was its silhouette. It looked quite squared. The head looked quite squared, long straight back, hunched over. I didn't see any indication of roundness. Its front was quite flat, but it was quite aways away so, and I'm seeing a side view. The front part of it wasn't too elongated through this area; it was more flat and squarish.
>
> Also when it was swinging its arms, like because when it took a step, its arms swung—huge big swings, also with its legs. I had a chance to see the full side view of its chest and back. The chest was straight up and down. I never saw any bumps as though it were a woman. He wasn't, you know, barrel-chested. It was more, how you say, more massive.

Wray went back the spot the next day and looked for footprints, but none were found. Thomas Steenburg investigated the incident 7 days later and determined that it took him 13 paces to cover the same distance as the creature. He also looked for footprints, but again none were found. Nevertheless, he said it did appear that something large had moved through the thick, thorny bushes.

Andrea Wray showing the spot where the creature came out of the bush (left) and were it entered the bush on the other side of the road (right).

Source: Thomas Steenburg Sasquatch Incidents file No. BC 10149. (Photos: Wray, 2 images; T. Steenburg.)

2005/06/10 – Hope. Glaring Eyes: A motorist driving a tractor-trailer reported seeing in his headlights a tall, bipedal creature near Hope at about 5:00 a.m. on June 10, 2005. The creature was walking down the highway on the left side (facing away from the motorist). When the motorist was about 50 to 60 yards away, the creature stepped off the road into the roadside ditch. At that point it turned and looked at him.

He described the creature as being over 6 feet tall and covered in what looked like reddish hair. The face did not have hair, just dark skin. The motorist stated that what really caught his attention was the creature's eyes. "They were just glaring at me as if I did something to offend it." As he got closer, the creature almost casually strolled away.

Source: BFRO website. The witness was interviewed by Blaine McMillan.

2005/07/28 – Victoria (VI). Museum Anthropologist Don Abbott Passes On: Royal Museum anthropologist Don Abbott passed away on July 28, 2005. Abbott played a major role in the early "bigfoot days." It was Abbott who was sent to California in response to John Green's request that the museum send a professional to Blue Creek Mountain in August 1967 to look at the many unusual footprints found there. (See 1967/09/01 – Victoria. Not a Subject for Mirth.)

Abbott was among those scientists who later saw the Patterson/Gimlin film at the University of British Columbia on October 26, 1967. Although he seemed to be receptive to the existence of sasquatch, he refrained from becoming highly involved in the issue beyond what he had done. In my opinion, he "toed the line" and then abandoned the issue to avoid criticism from his peers and superiors.

Source: The *Times Colonist*, Victoria, BC, obituary entry, July 31, 2005. And Christopher L. Murphy, 2008. *Bigfoot Film Journal*, Hancock House Publishers Ltd, Surrey, BC, p. 12.

2005/SU/00 – Cowichan (VI). Three Sightings: People camping or visiting in Cowichan area during the summer of 2005 reported 3 sasquatch sightings.

A group heard something circle their camp at night, and then saw what they believe was a sasquatch in the light of their campfire. It ran off when a loud dirt bike approached.

A man driving saw a sasquatch dart across the road and leap over a fence. The creature then used a tree to launch itself into the bush. It uprooted the tree in this process.

A lady reported that during the night something looked in the window of the truck in which she and a friend were sleeping. It walked around the truck and then wandered off.

Furthermore, many unusual sounds were heard—vocalizations, whistling, and tapping noises (some were sequenced).

Source: BFRO website, The person who reported these incidents was interviewed by Blaine McMillan.

2005/08/14 – Prince George. Primate-like Face: A man driving in a motor home reported seeing a sasquatch near Prince George at about 6:45 a.m. on August 14, 2005. The creature was walking down a hydro cut line that approached the highway. The motorist saw it take 3 or 4 steps and then stop at the edge of the highway on the right hand side as he passed. It was dark brown, with hair about 4 to 5 inches in length on its body and shorter hair on its head. Its face was described as being primate-like, but it did not have a protruding snout. Its arms were long, with its hands reaching below its knees.

Source: BFRO website. The witness was interviewed by Blaine McMillan.

2005/08/26. Harrison Hot Springs. Tracks On Logging Road Appeared Very Fresh: Bill Miller and John Myles reported finding what appeared to be sasquatch tracks in the Harrison Hot Springs region. Miller provided the following detailed report (edited for clarity) which gives an excellent account of their experience as pertaining to field research in BC:

It was on August 26th, 2005, a very bright, sunny, mild day, that I drove my Polaris 6WD Ranger UTV up the forest road on the west side of Lake Harrison in hopes of stopping at the 20-mile logging camp to see if the camp watchman (John Myles) would like to accompany me for a ride exploring forest areas where I had not been before. When I arrived at the logging camp, I found Myles with his rifle and backpack heading for the boat dock, as he was just about to leave in his boat for a trip to Stokke Creek. It wasn't long before I got Myles to change his plans and bring his gear over to my vehicle. He asked me where was it that I wanted to go and I reminded him of a logging road he told me about a couple of weeks earlier whereas to his knowledge it had been completely overgrown with shrubs and not traveled for several years.

The trip was pleasant and scenic, and before long we had turned off the main logging road about 15 kilometers up the lake from the 20 Mile Bay campground. We then traveled several kilometers deeper into the mountains before coming to a spot in the road where what looked like an old degraded side road could be seen disappearing under a thick growth of underbrush. As the road went back and forth up the ever rising switch-back, we stopped on 3 occasions to shake out our jackets and shirts to rid them of the vast amount of caterpillars that had fallen upon us from the bushes and branches that we disturbed as we inched our way to the top of the mountain. Eventually we made it to the very top and found that the road was in excellent condition from that point onward.

As we came around a rock wall outcropping, and were approaching some ponds that were believed to be just ahead of us, I noticed a dark spot in the light gray gravel of the logging road. My first impression as we approached was that a larger animal had relieved itself, but upon closer inspection, we discovered that the gravel had been overturned as if something scuffed the road with its foot. The scuff appeared darker than the other gravel because the underside of the rocks was still moist. As we looked around, there were several such scuffs in a straight line and about 4 feet apart in 2 directions. I opted to follow them to the edge of the hill, while Myles went in the opposite direction.

The scuffs I followed led to a hillside that looked like it had been clear cut some time ago as I could see a long way down to the tree line below, and beyond that was the lake. It looked to me that we were about a mile above Lake Harrison. As I walked back towards Myles, I noticed he was looking at something and as I paused to look at the original scuff again, he called to me to come and see what he was looking at. I found before me in the clay mud (from what appeared to be an area where water had pooled for a period before evaporating) a good footprint about 12 inches long. Just ahead of that print was a partial print in the same clay that was shared by a flat stone where a portion of the foot had landed on both.

Over the next 15 to 20 minutes we photographed and filmed the prints and scuffs leading across the road. The last thing I did before leaving was to remove a shoe and make a footprint beside the good track in order to have something for scale. We then proceeded on and followed the road a short ways before coming to a dead end above some more still pools where we found lots of bear tracks. We then left the area to start heading back down the mountain, but not before stopping at the tracks once again. It was then that I realized that the individual that left the prints behind may have done so in the moments just before our arrival because the dark scuffs that were previously seen were now gone as the moisture on the overturned gravel had dried from the exposure to mild air and sunshine. The footprint in the clay dirt looked to be drying as well as its color looked to be fading.

The track itself had left me with questions that took me 2 more years before understanding how it had been made. It was Gerry Matthews, owner and operator of the site West Coast Sasquatch, who pointed out why 2 of the smaller toes didn't mark the ground as the others. It seemed that there were stones in the soil just under the outer upper ball of the foot. The toes had flattened the clay, but had not reached the depth of the inner portion of the foot and heel. It was also another researcher's mentioning of the subject's foot cupping itself in the Patterson film which explained how the print we had found in the clay dirt had not touched the ground in the mid-tarsal area.

In summary, I felt the track to be authentic because of the way we discovered it. There was no indication that anyone had passed through the thick unmolested forest growth, which had engulfed a portion of this forest road years before. I found that my 285 pounds at the time of the discovery of the print could not press into the clay and achieve the same depth as the individual that left the track. I considered that no one but myself even knew that I was going into the mountains that day...let alone to this area. And that because the track was fresh, that no one would have known to have placed a track there in hopes we would find it, nor could they have gotten there by road as we came. Someone would have had to have climbed the mountain and spent the better part of the day doing it so to have gotten to this area ahead of us. I should also mention

Footprint (A-lower) found in a logging road, possibly made moments before researchers arrived. Above the print (B-upper) is Miller's footprint for size and depth comparison.

that I took Thomas Steenburg to this site so he could see the track as well, for I was certain the clay would hold the track for a very long time. We failed to make the trip as originally planned, but managed to get to the site a couple of weeks or so later. When we arrived, the clay was as smooth as glass as if nothing had ever stepped there.

Source: Bill Miller, Bigfoot Field Research. (Photo: Bill Miller.)

2005/YR/00 – Port Alberni (VI). Two See Sasquatch: Two people reported they saw a sasquatch in the Port Alberni area in 2005. They submitted a sketch of the creature they saw, as shown here.

Source: John Bindernagel, 2010. *The Discovery of the Sasquatch.* Beachcomber Books, Courtenay, BC, p. 94, from information provided by the witnesses. (Photo/artwork, property of the witnesses.)

Sketch of a sasquatch seen by 2 witnesses.

2006/06/20 – Mesachie Lake (VI). Gait Not the Same As a Human: A motorist reported chasing a large bipedal creature down a road near Mesachie Lake at about 4:30 a.m. on June 20, 2006. The creature was actually on the road, and from a distance appeared to be a bear. The motorist accelerated to chase it off the road.

When the vehicle headlights fully illuminated the creature, it was seen to be running on 2 legs. It was covered in dark hair, with a light colored patch across its back just below its shoulder blades— 3 to 4 inches wide. Also, it was light colored across its buttocks.

It ran about 100 feet down the road and then turned abruptly to the left, running off the road edge, down through the road ditch, and then up a bank and into the bushes. The motorist stated, "Its gait while running was not the same as that of a human." He estimated that it was about 7 feet tall, and 3 to 4 feet across the back.

He went back to the sighting location later that day and found what appeared to be footprints. The impressions were roughly 13 inches long. He noted that there were 2 sperate trackways, one that came out onto the road, and a second set where the creature ran off of the road.

BFRO investigator Blaine McMillan found that there is a history of sasquatch sightings by the local First Nations people in this area. They call the creature "Mesachie Man."

The Mesachie Lake area (the town is of the same name). The large lake east of the town is Lake Cowichan.

398

2006/07/16 – Pocatello, Idaho, USA. Sasquatch Exhibit at Museum of Natural History: In June 2006 Chris Murphy's sasquatch exhibit traveled to the Museum of Natural History in Pocatello, Idaho. The exhibit opened on June 16, 2006 and it ran for over 14 months. Generally, the items displayed at the earlier Vancouver, BC exhibit were featured, but with a different theme. It was one of the best attended exhibits held at the museum.

(Left) The exhibit entrance. The large sasquatch cut-out is from artwork by Brenden Bannon. (Right) Museum exhibit director Dave Mead (left) reviewing newspaper reports related to the exhibit with author.

2006/10/29 – Dewdney. Halloween Hoax: On November 2, 2006 a Mission, BC newspaper reported a sasquatch sighting that occurred near Dewdney, along the Fraser River, on October 29, 2006. The article showed a fuzzy picture purported to show a sasquatch walking along the river bank. The articles stated that the creature seen acted aggressively—threw rocks at a boat carrying adults and a group of children (19 people altogether).

Thomas Steenburg investigated the incident. He contacted the tour company that provided the boat. Jo-ann Chadwick, a company official, assured him that the "sasquatch" was a man in a costume.

He was pulling a Halloween prank for the kids on the boat, which was on its way to visit the Halloween haunted house display at Kilby. Since some of the kids were always talking about the sasquatch, some of their parents thought the little prank would be fun for the kids, and Chadwick went along with the idea. The man in the suit was, in fact, the father of one of the children on the boat. Chadwick took the photograph of the creature and gave it to the newspaper. It is believed she and the newspaper reporter, Carol Aun, were friends and decided to take the incident a littler further. Chadwick was surprised that it was subsequently investigated by a sasquatch researcher. For certain, it was all done in good fun. There was no malicious intent involved.

Source: Thomas Steenburg Sasquatch Incidents file, Number BC 10152, from Carol Aun, 2006, "Sasquatch Sighted on River's Edge in Dewdney, the *Record*, Mission, BC, November 2, 2006.

2007/01/00 - Harrison Hot Springs. Walking Along the Tree Line: Mr. Chapman, a hunter, reported that he saw a sasquatch in the Harrison Hot Springs area at Hemlock Valley, in early January 2007. He was hunting on a forestry side road about one mile off the paved Hemlock Valley road when he spotted the creature. The hair-covered oddity was walking along the tree line at the edge of a logging clear-cut. Chapman estimated it was about 10 feet tall.

Stories of this sighting at the time had it that Chapman shot at the creature, but missed and hit a tree. Sasquatch researchers Thomas Steenburg and Bill Miller managed to contact Chapman in August 2007 and he stated that he did not shoot at the creature. He told the researchers that after watching it walk off into the cover of the trees, he simply turned around and left the area at a quickened pace.

Source: Thomas Steenburg Sasquatch Incidents file. Number BC 10153.

2007/02/13 – Harrison Mills (Chehalis Reservation). Large Rock Lands Near Fishermen: Matt Hikson and Dustin Jones reported an odd incident in the Harrison Mills area along the Chehalis River (Chehalis Reservation) on February 13, 2007. A large rock was thrown and landed in the water near where they were fishing. They did not see anyone or anything, but heard movement in the bushes high above them. They said that a sasquatch came to mind because of the size of the rock.

There have been other rock throwing incidents in this area, and other areas in BC. It has been assumed that such are simply a scare tactic. It has been reasoned that if a sasquatch intended to hit a particular target, he would have no trouble doing so.

Source: Thomas Steenburg Sasquatch Incidents file No. BC 10154.

2007/03/30 – Ucluelet (VI). Laying On Log Over River: A motorist stated that he saw a strange creature near Ucluelet along the Kennedy River on March 30, 2007. What he first believed to be a bear was laying on a log overhanging the river. After taking a second look he saw that it was an ape-like creature with its arm hanging down into the river as if in the process of catching fish swimming upstream. He said that he distinctly saw its facial features, which were similar to an ape. He estimated that if the creature were standing, it would have been about 7 feet tall, and at least 500 pounds. It had dark brown, long, human-like hair on its body, and hair on its face, but not over its eyes and mouth. Its head was human-like. In general, he described the creature as having human/caveman-like features.

Source: BFRO website. The witness was interviewed by Colleen McDonald. Also, Thomas Steenburg and Rick Noll interviewed the witness on location in December 2009 for the Monster Quest documentary *Mysterious Ape Island* (Whitewolf Entertainment, Minnesota, USA.)

2007/09/25 – Canal Flats. Odor Did Not Linger: A man hunting near Canal Flats in the White Swan Lake area reported seeing an unusual creature on September 25, 2007. He had sat down on a stump to have a snack when he was overcome by a powerful odor, which he described as worse than that of a skunk—a cross between a wet dog and a dead animal. He looked up-wind and standing about 50 yards away was a tall, bipedal creature completely covered in dark brown hair. It was massively built—well muscled and thick through the body. Although he did not feel threatened, the man glanced down, and reached to steady his rifle which was leaning on the stump beside him. When he looked up in the direction of the creature, it was gone.

The entire event lasted about 1.5 minutes. Remarkably, he said that the odor he had sensed disappeared as if someone had "turned it off." In other words, it did not linger.

Source: BFRO website. The witness was interviewed by Blaine McMillan

2007/YR/00 – Pocatello, Idaho, USA. Footprints Recognized by Science: In 2007 Dr. Jeff Meldrum, Idaho State University, achieved scientific recognition for the existence of unclassified footprints found throughout North America. He prepared and published a scientific paper on this subject. The paper does not validate the existence of sasquatch (believed to be the maker of such prints), but recognizes the existence of large unidentified tracks and gives them an official name. The following is the abstract from Dr. Meldrum's paper:

ICHNOTAXONOMY OF GIANT HOMINOID TRACKS IN NORTH AMERICA

D. JEFFREY MELDRUM
Department of Biological Sciences, Idaho State University, 921 S. 8th Ave., Stop 8007, Pocatello, ID 83209-8007

Abstract—Large bipedal hominoid footprints, commonly attributed to Bigfoot or sasquatch, continue to be discovered and documented, occasionally in correlation with eyewitness sightings, and rarely in concert with photographic record of the trackmaker (gen. et sp. indet.). One of the best-documented instances occurred in 1967, when Roger Patterson and Bob Gimlin filmed an over two meter tall upright striding hominoid figure, at the site of Bluff Creek, in Del Norte County, California, and cast a right and left pair of exceptionally clear footprints in firm moist sand. Additional footprints were filmed, photographed, and cast by multiple witnesses. Molds and casts of a series of these are reposited at the National Museum of Natural History, Smithsonian Institution, while ten original casts are among the Titmus Collection at the Willow Creek – China Flats Museum, Humboldt County, California. These casts have been 3D-scanned and archived as part of a footprint virtualization project and scan images are accessible on-line through the Idaho Museum of Natural History. The initial pair, originally cast by Patterson, and the remaining casts made by Titmus, are designated the holotype of a novel ichnogenus and ichnospecies describing these plantigrade pentadactyl bipedal primate footprints – *Anthropoidipes ameriborealis* ("North American ape foot"). The footprints imply a primitively flat, flexible foot lacking a stiff longitudinal arch, combined with a derived, non-divergent medial digit.

This achievement was an important milestone towards scientific recognition of the "track maker" as a distinct animal species. It is of course, believed by many people that the animal is what is called sasquatch or bigfoot; however, more concrete evidence is needed to connect this creature with the prints.

Dr. Jeff Meldrum (left) is seen here with sasquatch notables (from left to right): Bob Gimlin, who was with Roger Patterson at Bluff Creek when they filmed a sasquatch; Dr. John Bindernagel, wildlife biologist and author; Daniel Perez, author, and editor of the Bigfoot Times, a monthly publication on all aspects of the sasquatch/bigfoot issue; John Green, author of the main early books on the sasquatch; and Dmitri Bayanov, Russian hominologist and author. This photograph was taken near Bluff Creek (Patterson/Gimlin film site area), California in September 2003.

2008/02/00 – Vancouver. "Quatchi" Made Olympics Mascot: In February 2008, the Vancouver 2010 Winter Olympics Games Committee selected "Quatchi," a child's cartoon-like sasquatch image, to be one of the games mascots. Stuffed toys and other novelties showing "Quatchi" were created and later sold at the games. Also, the creature was depicted on a postage samp and on a 50-cent "collectors" coin issued by the Royal Canadian Mint.

An effort was made by the author to have the Olympics Committee and BC media provide proper information on the sasquatch for the benefit of the many visitors who would be coming to the games. A sasquatch exhibit was offered and an article written. All, however, was to no avail.

Nevertheless, I created a "Sasquatch Awareness Center" web presentation that was hosted on the Hancock House Publishers website.

Thomas Steenburg and Bill Miller also provided a little sasquatch exposure during the games. They had a sasquatch display at a venue called the O Zone in Richmond.

The "Quatchi" collection. Although these items are trivial, the games were a first class event and the sasquatch "mascot" was at least amusing.

Source: Author's files and from information provided by Thomas Steenburg. (Photos: Quatchi toy and coin, C. Murphy; postage stamp, © Canada Post Corp, 2008.)

2008/06/18 - Bridal Falls. Hands On Tree Stump: Ryan Kozak reported that he saw a sasquatch on a forest service road near Bridal Falls, on Mount Archibald, at about 10:30 p.m. on June 18, 2008. He was 4x4 motoring and saw the creature in his vehicle headlights at a distance of 30 to 35 feet. It was standing erect and had its hands on a large tree stump. When spotted, it pulled itself up onto the stump and then jumped off down the other side into a gully. Kozak said it was "very hairy," covered in black or dark brown hair, and estimated to be between 7 and 8 feet tall.

Kozak stopped his vehicle immediately the creature disappeared and opened his window. He said he heard the oddity crashing through the bushes beyond the stump.

Upon seeing a Chilliwack *Progress* newspaper article (June 19, 2008) on another sighting in the same area (see next entry), Kozak contacted the reporter, Ashley Wray. She contacted researcher Bill Miller and he and Thomas Steenburg subsequently investigated the incident.

Ryan Kozak (left) and Bill Miller inspecting the area where the creature went down into the gully. It pulled itself up onto, and then jumped off, the large stump seen directly to the right of the men.

Source: Thomas Steenburg Sasquatch Incidents file Number BC 10161. (Photo: T. Steenburg.)

2008/06/19 – Bridal Falls. Two Steps to Cross Road: A man reported he saw a sasquatch near Bridal Falls, on Mount Archibald, at about 2:30 a.m. on June 19, 2008. He was driving down a Forest Service road and saw in his headlights something standing on 2 legs at a distance of 100 to 150 feet. It took 2 steps to cross the road and went up the other side and out of sight. He estimated that it was about 7 feet tall and said it was covered in black hair. The man alerted his wife to what he was seeing, but she was not looking out the front window, and by the time she looked, the oddity was gone.

Bill Miller is seen here pointing to what appeared to be a footfall. The disturbed bushes indicated the passage of a large creature.

This incident was investigated by Thomas Steenburg and Bill Miller. No actual footprints were found, but there was evidence of footfalls and that something large had ran through the bush.

It is possible the creature seen in this sighting report was the same individual reported seen the previous day, just 4.5 hours earlier (see previous entry).

Source: Thomas Steenburg Sasquatch Incidents file Number BC 10160, from Ashley Wray, 2008. "If You Go Down In the Woods Today." The *Progress,* Chilliwack, BC, July 11, 2008. (Photo: T. Steenburg.)

2008/06/19 – Chilliwack. Unusual Sounds & a Footprint:
A camper reported that he heard unusual "whooping" sounds near
Chilliwack, in the Chilliwack Lake area, on the night of June 19,
2008. He counted a total of 10 vocalizations over a span of about 5
minutes. The sounds were coming from a location higher in the
mountains. The next day, the camper went for a hike in the mountains and found on a trail what appeared to be a human-like footprint
that was considerably larger than his 11.5-inch boot. The print was
impressed about one-half inch into the soil. He reasoned that only
one print was seen because the trail was very narrow with dense
brush on either side.

Source: BFRO website. The witness was interviewed by Blaine McMillan

**2008/SU/00 – Enderby. Forest Service Worker Tells of Sas-
quatch Sightings:** A forest service worker said that he had 2
sasquatch encounters during his previous time with the service. I do
not know the dates or the specific locations, but I believe they were
in the Enderby region. The information came to me as follows:

During the summer of 2008, I camped with my son, Chris, and
daughter-in-law, Lisa, near Enderby at Hidden Lake. The area is
very heavily forested and great for camping. Two friends, Liza Jane
and her husband Ken, popped by one afternoon.

As we sat around, 2 forest workers came by to warn us of the
high fire hazard. One fellow appeared in his late 50s, and the other
much younger.

We chatted about the current fire situation and pine beetle in-
festation, when out of the blue Liza asked them, "Have you ever
seen a sasquatch?" To my surprise the older fellow said, "As a matter of fact, I have, on 2 different occasions." He then explained that
he saw the creatures from a fair distance, but could see that they appeared like hair-covered men—brownish in color. As I recall, in
both cases he looked away (once in the direction of a sound) and
when he looked back the creatures were gone. Liza then said, "What
do you think they are." The fellow said, after looking at us quite intently, "You know, I'm inclined to believe what the Natives say.
There are things these creatures can do that we just don't understand." (I believe the inference here was to their ability to apparently "vanish.")

Visit by forest workers at Hidden Lake. The 2 gentlemen at the back were the workers. Liza Jane is next to them. I am on the left with my son, Chris, on the right.

Source: Personal communication. (Photo: L. Jane.)

2008/08/12 – Smithers. Leaning Against Tree:

A young woman reported that she saw a frightening creature in a small forest near her home in Smithers on August 12, 2008. She was out walking in the forest about midnight with 2 friends. They heard what sounded like a tree snapping, and the woman went alone to investigate. After a few minutes walking she saw a very tall creature about 15 to 20 feet in front of her leaning with one arm against a tree. All she could see was a dark outline because of the thick bush and the darkness. She called to her friends, but before they arrived at the spot, the creature turned and walked out of the forest and onto a road.

The next day, the woman went into the forest to where the creature was seen. She noted freshly made paths both at that spot and elsewhere in the forest. She later went frequently into the forest and said there were broken branches and even a tree that appeared to have been broken in half. In one spot, a tree had been uprooted.

Source: BFRO website. The witness was interviewed by Milan Kubik.

2008/09/21 – Agassiz. Something Glimpsed & Rock Thrown: A hunter glimpsed a creature of some sort and had a large rock thrown at him while hunting near Agassiz in the Ruby Creek area on September 21, 2008. The following is a statement (edited for clarity) he provided to Thomas Steenburg:

> I was walking down this road, up along Ruby Creek—this marshy area—moving real slow, keeping an eye on the bush, looking for deer, bear. I caught a glimpse of something black out of the corner of my eye—moving. No sound at that point, just something moving, so I quickly swung my upper body towards it and I saw a rock 12 to 15 feet off the ground, kind of arcing down towards me. It landed in the bog—seemed like right next to me, but I don't know where in that bog it landed—seemed like it was right there. So I paused, sort of going through my head, you know—where did it come from? Lots of things went through my head, but nothing made any sense. I backed up a little bit and was looking at this grove of trees. I saw something black start to move, and it took off like a rocket. I could only see a small amount, like a piece of it through the dense trees that were there. I heard it take off straight up 50 or 60 yards, it then stopped, then took off again, another 60 yards or so I guess; stopped again, then it took off until I could no longer hear it moving any longer.

The hunter left the area immediately, and later contacted Gerry Matthews, a sasquatch researcher. Consequently, Thomas Steenburg, Bill Miller and Christine Schnurr went to the sighting location. Steenburg found rough tracks in the bush, and upon further investigation found a reasonably good footprint about 12 inches long, some 40 feet from where the hunter had been standing. He made a cast of the print.

The bog (a pool of water about 4 feet deep) in which the rock landed was searched, but

Footprint found by Steenburg.

The Ruby Creek area showing the location of the incident (small circle). The inset shows Thomas Steenburg beside the bog where the rock that was thrown at the hunter landed. Every effort was made to retrieve the rock but nothing was found of the size the hunter saw.

Cast made by Thomas Steenburg of the footprint found. The toes are quite prominent and appear to show that they dug into the ground. The large pointed mound of plaster is simply an artifact caused by a hollow that formed in thick forest or bog debris when the liquid plaster was poured.

nothing was found of the size the hunter believes was thrown at him (large rock about 2 feet wide). The bog has soft mud at the bottom, and it is believed the weight of the rock caused it to sink well into the mud.

Source: Thomas Steenburg's Sasquatch Incidents File Number BC 10162. (Photos: Footprint in soil, Bill Miller; Ruby Creek area, Image from Google Earth; Image © 2011 IMTCAN; Image © 2011 DigitalGlobe; Image © 2011 Province of British Columbia; © 2011 Cnes/Spot Image; Steenburg by bog, T. Steenburg; Cast, C. Murphy.)

2008/11/00 - Mission. Hair Like a Horse's Tail: Kevin Francis reported seeing a strange creature in the Mission area near his home at about 1:00 a.m. during November 2008. As he drove down the road to his residence, he saw the head and massive shoulders of something in the ditch on the left side of the road, about 3 properties from his driveway entrance. Ground level fog and darkness obscured the lower portion of the animal. He described it as being covered in light brown coarse looking hair, similar to the hair of a horse's tail. He was sure it would have been over 6 feet tall, most likely close to 7 feet, if it were standing on the road.

Kevin Francis.

The roadside ditch where Kevin Francis saw the unusual creature. Ironically, a sign above the entrance to the property reads "BIGFOOT."

Francis continued driving and parked in his driveway. When he got out of his car he heard an odd howl coming from the direction where he had seen the oddity.

Source: Thomas Steenburg Sasquatch Incidents file, Number BC 10167. (Photos: T. Steenburg.)

2009/04/04 - Langley. I Think Of It As a Gift: A woman reported seeing a sasquatch cross the road in front of her car near Langley at about 12:30 a.m. on April 4, 2009. She was on 200th Street intending to turn on 16th Avenue and saw the oddity in her car headlights just before she reached 18th Avenue. It crossed the road at a sharp angle, so she was not able to see its face. She said it was definitely walking on 2 legs and was covered in sort of silver-white hair with dark gray mixed in it. As to the hair length she said, "the hair was long, you know like a shaggy dog's…wasn't prickly. It draped, and a perfect arch between its legs." She said it was not very tall, only about 5.5 feet, but that it was stocky. The total time it was in view was about 7 seconds.

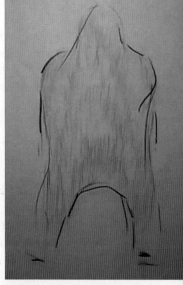

Sketch of the creature seen provided by the witness.

After it crossed the road, it disappeared into the darkness (beyond the area her car headlights illuminated). When questioned as to what she thought the creature was she said that it was not human and not like any animal she had ever seen. When asked how she felt about her experience, she stated, "I feel wonderful. I think of it as a gift."

Thomas Steenburg and Gerry Matthews interviewed the woman. Steenburg summed up his thoughts as follows:

After the interview both Gerry Matthews and I took her over her story again. We came to the conclusion that this charming lady was very truthful with us and was not somebody telling us a made-up story. However, in my opinion we still

The intersection of 200th Street and 18th Avenue in Langley where the creature was seen. Although semi-rural in nature, there are a lot of people in this area. However, in BC frequent heavy cloud cover and a lot of rain greatly decreases visibility at night, so nocturnal creatures of all sorts do wander through both rural and urban neighborhoods.

cannot rule out a possible case of mistaken identity due to the location of the incident. I don't consider it impossible for a sasquatch to be there, but I would think it were very unlikely. On the other hand, this location does border on a large old growth park to the west (Campbell Valley Regional Park), and expensive homes on large lots to the east. Someone who used to live in this area mentioned to me that on occasion, black bears were a problem as they fed on garbage put out by residents. So some wildlife does live in this area. The lady gave us a small drawing based on what she saw (as shown).

Source: Thomas Steenburg Sasquatch Incidents file, Number BC 10164. (Photos: Sketch of creature, witness; Intersection, T. Steenburg.)

2009/04/00 – Merritt. Possible Print—Cell Phone Photo: A man who works as a private contractor for local mills in the Merritt area found a possible sasquatch footprint near Merritt in April 2009. He said he has worked all his life in the forest and started to consider the possibility of the creature's existence when he heard a strange animal cry in the bush during October 2007. He found the footprint about 44 miles east of Merritt and photographed it with his cell phone.

This is the first report I have seen that includes a cell phone photograph. Perhaps we can look forward to more photographs of possible evidence with this new technology.

Photograph of the footprint found taken with a cell phone.

Source: Thomas Steenburg Sasquatch Incidents file Number BC 10165. (Photo: witness image.)

2009/04/00 – Gibsons. Sasquatch Walking Insights: Keith Quinlan reported that he saw a sasquatch near Gibsons at a Peace Camp on Mount Elphinstone in April 2009. He provided a sketch of the creature to me (by way of Dave Collin), but I have not been able to contact him for further details. The sketch, as seen on the next page, is highly intriguing, and differs greatly from other witness sketches.

I believe Quinlan's drawing provides some insights as to how a sasquatch walks when it is in the forest. My impression is that it sort of "slinks" around, taking very long or short paces in accordance with the terrain and forest debris. It would bend its body (which is likely very flexible) and swing its legs forward very quickly. I see this "locomotion" as highly animal-like, although a very athletic human could also move in this way.

We do, of course, have the Patterson/Gimlin film which shows a sasquatch walking along a creek side. The walk is unusual, but not

Keith Quinlan

Sasquath running at Peace Camp
on Mt Elphinstone
seen april 2009 *[signature]* . 18/12/2010

Drawing by Keith Quinlan of the creature he saw near Gibsons.

greatly different to the way a human would walk. However, in this case, the creature is in a clearing so would not need to pick and choose its steps.

We know for certain that sasquatch are very fast when traveling through the forest. They have been known to literally crash through the bush without regard for the thick entanglement of plant life. It appears to me that the way they move accommodates this.

Source: Keith Quinlan plus general information and personal opinion. (Photo/Artwork: Keith Quinlan.)

2010/06/06 – Duncan (VI). An Extremely Loud Grunt: A man motoring (4x4) with his girlfriend near Duncan on Mount Sicker reported that he saw a strange creature at about 9:30 p.m. on June 6, 2010. The couple had motored up onto a look-out near the top of the mountain (clear-cut area) and got out of their vehicle to stretch. On a ridge about 300 to 400 feet away the man saw a tall figure that appeared to be walking on 2 legs. Its shoulders appeared slumped and draped in long black and brown hair. Its arms were very long and its head was a long oval draped in hair. He estimated its height to be between 8 and 9 feet. He said to his girlfriend, "I think I saw a sasquatch," and the 2 joked about it.

They went back to their jeep, and sat talking. Some 10 to 15 minutes later (now getting dark), the girlfriend heard a noise in the distance to the front of their Jeep at the "2 o'clock position." They did not pay any attention to it, but it then repeated and kept repeating, so both now heard it. It sounded like a low deep grunt. They still ignored it for a while and then decided to get out of their jeep to see if they could better hear it.

They advanced towards the bush and the man threw rocks into the bush hoping to get some sort of reaction. He hurled the rocks into both the 2 o'clock and 5 o'clock points, but nothing happened. The girlfriend then yelled several times, but again no reaction. They moved back towards their jeep and while standing near it, an extremely loud grunt was heard. The man said, "It had so much bass that it just cut through the air." It sounded as though it was no more than 20 feet from them at the 5 o'clock position. No other sounds were heard (bush cracking sounds) even though the direction of the grunts had changed.

The pair jumped into their jeep and spun it around to see if anything could be seen in their headlights. Nothing was seen. Now feeling very uneasy, they left the area. Oddly, they said they noticed that there was a small black and white bird that kept squawking around them the whole time they were in the area. When the loud grunt was heard, it was swooping around them.

This incident was investigated by BFRO researcher Cindy Dosen and her husband on June 10, 2010. They went with the witnesses to the sighting location. At the spot where the creature was seen there was a stump, and the distance from the jeep to this stump was 96 yards. Possible footprint impressions were found at this spot and were followed into a draw. Three impressions had some good definition. The prints were 13.5 to 14 inches long and 5.5 to 6 inches wide (widest point). The distance between the tracks (pace) was 53 inches.

On June 13, Cindy Dosen and 2 other BFRO researchers went to the site and from additional details provided by the witnesses it was determined that the creature seen was about 7 feet tall.

This team investigated the whole area around the sighting location and found other footprints about 14 inches long, and 5 or 6 inches wide.

Source: BFRO website. Witness was interviewed by Cindy Dosen and other BFRO researchers.

2010/YR/00 – Port Mellon. Large Hand: A boy about 12 years old reported seeing a large, hair-covered hand in the Port Mellon area during 2010. The circumstances of this incident are as follows.

The author had provided a display of sasquatch-related casts in the shop window of Oscar's Aquatics, Gibsons. During October 2010, the boy went into the shop and told the owner, Lisa Murphy (author's daughter-in-law), that he had seen a big hand like the hand cast displayed in the window (Freeman hand cast from Blue Mountains, Washington). Lisa asked him to write down his story and she later gave it to me.

The boy had gone fishing in the Port Mellon area and took a short-cut home through a forest trail. He heard an unusual tapping sound and when he looked in the direction of the sound, he saw a large, hair-covered hand holding a rock and hitting a tree with it. The boy could not see anything else because the bushes were too thick. He stated, "I got the creeps and ran away home."

The shop window at Oscar's Aquatics with a display of sasquatch casts.
The hand cast is on the left. The inset shows how it appeared.

Source: Author's file. (Photo: C. Murphy.)

2011/02/08 – Hope. Something Walking In the Bush: Liza Lilleste reported that she and her fiancé saw an unusual "figure" in the Hope area while motoring along the Silver Skagit Road at around 10:00 a.m. on February 8, 2011. In her own words (edited for clarity):

Liza Lilleste.

> We were driving. I was the passenger. I looked up straight ahead and I could see a human figure walking. I could see arms and legs—arms swinging—for maybe about 3 seconds, then it was gone. It was among trees to begin with, but it just disappeared. I said, "There's a man walking in the bush up there, but he's all dark brown, that's really weird."

Her companion, did not see anything, but drove to the palace where the creature had been and they looked for footprints, but nothing was found. When they drove further down the road, the companion also glimpsed the creature on a trail.

Lilleste contacted Gerry Matthews, a sasquatch researcher, but failed to mention that the creature was also seen by her companion. Several researchers, including Thomas Steenburg, subsequently went to the location. Again, no tracks were found. Upon interview-

The location where Liza Lilleste saw the creature. At first glance, this photo appears to show simply a forest scene. However, Lilleste's fiancé is seen standing on the right—easily mistaken for a tree stump.

ing Lilleste the second sighting was made known, so the researchers went to this second spot with Lilleste's fiancée. Steenburg noted:

This time, on the trail where the companion saw it we found possible tracks, about 12 inches long. Three tracks were found in total which were closer to the road than the object was when sighted. They also appeared to be traveling toward the road rather than away from it. The witness said that what he saw was heading in the opposite direction. So if these are tracks left by the creature the two witness encountered it is possible that they were made before the second sighting took place as the creature was approaching the road. Of course, this is speculation on my part. It is still possible that these three prints are human in origin and not sasquatch tracks at all. They were not clear—no toes discernable, but the forefoot seemed oddly wide when compared to the heel.

One of the footprints, about 12 inches long, found by researchers. The 3 tracks found had a pace of about 54 inches and 60 inches respectively.

Source: Thomas Steenburg's sasquatch incidents file, No. 10168. (Photos: T. Steenburg.)

2011/04/09 – Harrison Hot Springs. The Sasquatch Summit: A major conference/tribute to John Green was held at Harrison Hot Springs over the weekend of April 8, 2011. Speakers from the US, Russia, and Canada gave presentations. Total attendance was about 240 people.

The event was held in the Harrison Hot Springs Hotel, which has excellent faculties for such events. An extensive exhibit of sasquatch-related artifacts was provided by the author and John Green.

The event organizers, Tom Yamarone, Alex Solunac, and his wife Lesley, provided what the author considers the best sasquatch event he has ever attended.

John Green, who has dedicated over 50 years of his life to sasquatch research, was given due honors and great thanks for his numerous contributions to the field.

Special Sasquatch Summit booklet containing a short biography of John Green and tributes from his many friends.

The tribute dinner was a truly memorable event with highly suitable entertainment provided by Kelsey Charlie of the Chehalis First Nations Band, and tribute talks given by John's friends of many years, including John Allen, local historian and former mayor of Harrison Hot Springs.

John Green (left) with special guests Al Hodgson of the Willow Creek – China Flat Museum, California, and Bob Gimlin, greeting attendees and signing Summit memorabilia.

The Summit speakers (left to right), Loren Coleman, Dr. John Bindernagel, Thomas Steen-burg, Dr. Jeffrey Meldrum, John Green, Igor Bourtsev, and Chris Murphy.

Kelsey Charlie Sr. (right) with his Chehalis dance group (left to right): Darren Charlie, Angela Charlie (Kelsey's daughter), Mikyle Charlie, Kelsey Charlie Jr., Keegan Charlie and Brylee James. The dances and songs reflected the sasquatch in Chehalis history and tradition. Kelsey has himself seen a sasquatch, as did his father and grandfather.

The Summit exhibit included John Green's copy of the Skookum cast, and footprint cast from John's collection (superior original casts) along with others from Murphy's collection. Enlargements from old newspaper articles provided by Scott McClean traced the history of the creature through the last century. Artwork by RobRoy Menzies, Paul Smith and Yvon Leclerc came together with other material to provide a true "sasquatch experience."

The Patterson/Gimlin film site model, camera of the type used by Patterson, and cast copies of those taken by Patterson (along with a human foot for comparison) provided an intriguing display. The images in the back are panels prepared by the Vancouver Museum in 2004.

For certain, the Sasquatch Summit will long be regarded as a major event in the annals of sasquatch history in BC. During the conference, a show of hands was asked of those who had not attended a sasquatch conference in the past. I was taken aback by the number of people who came forward. I am sure many, if not all, will attend future conferences.

The Sasquatch Summit organizers (left to right), Alex Solunac, Thomas Yamarone, and Lesley Solunac. A detailed account of the event is provided at www.bigfootsongs.com/

Source: Author's personal experiences. (Photos: Summit booklet, T. Yamarone; Green, Hodgson, Gimlin at table, speakers line-up, Chehalis group, exhibit general shot, exhibit film site model and related artifacts/images: C. Murphy; Summit organizers, T. Yamarone.)

2011/05/18 – Ottawa, Ontario. Canadian Sasquatch Coin Issued: The Royal Canadian Mint in Ottawa, Ontario, released a special collectors' 25 cent coin depicting a sasquatch on May 18, 2011. The coin, shown here at about actual size (1.37 inches or 35mm in diameter), comes in a presentation case providing a brief account of the creature. The imagery and information does not reflect an in-depth study of the sasquatch. Nevertheless, the coin does call attention to the mystery and is a unique item for collectors. Images of the presentation case are provided in the Photograph Presentation section at the back of this book.

Source: Canadian Mythical Creatures, 2011, Sasquatch, Royal Canadian Mint, Ottawa, Ontario. (Photos: Public Domain.)

Statistics Considerations

The most meaningful statistic that can be gathered from the information presented is that there have been 379 reported sasquatch-related incidents in BC over 110 years.[1]

Basically, all statistics that can be gathered from the incidents provided are biased for several reasons. For an incident to be reported there must be a person to experience the incident, so the "habits" of people become a part of the statistic. For example, people travel more in the non-winter months, so there will be more incidents in these months. Also, people use roads (all types) and park or forest "human-made" trails, so many incidents will be on or near such. That many sasquatch are seen in this circumstance does not imply that they frequently use roads or trails—just that people do.

Another major factor is that the number of incidents reported is directly related to the number of researchers collecting information. Thomas Steenburg has a "Sasquatch Research" sign on his vehicle that has resulted in a number of incidents being reported to him. If there were several hundred researchers with the same sign, then I am sure the number of reported incidents would greatly increase.

The number of incidents reported by decade since 1900 in this work is as follows (last year shown is to December 31 of that year):

1900–1909	13
1910–1919	5
1920–1929	5
1930–1939	13
1940–1949	15
1950–1959	24
1960–1969	80
1970–1979	100
1980–1989	37
1990–1999	40
2000–2009	47
TOTAL:	**379**

1. Only actual BC incidents are included and an adjustment has been made for multiple sightings of what appears to be the same creature.

Most certainly, human population increases, more people going into wilderness areas, and improved communications play a part in the number of reported incidents. However, work by researchers and general public awareness of the sasquatch issue play a greater part.

I believe the great increase in reported incidents from 1960 to 1979 was mostly due to Bob Titmus, John Green, René Dahinden, and other researchers "beating the bushes," as it were. To be sure, the Patterson/Gimlin film (1967) also played a major role. Here we had an actual image of a sasquatch, so interest in the subject was greatly increased with the result that more people "got involved." Although there are certainly many researchers in North America at this time, there is only a handful in BC, none of whom have traveled the remote coastal areas as Bob Titmus did.

I have often been asked the question, "Where in BC do the most sasquatch-related incidents occur?" According to the entries provided in this book, I have arbitrarily set five (5) incidents as a benchmark. The following shows the results by location in order of highest to lowest number (time frame is 110 years):

Location	Count
Bella Coola	19
Harrison Hot Springs	18
Harrison Mills	16
Hope	15
Chilliwack	9
Agassiz	8
Squamish	7
Sechelt	7
Swindle Island	6
Comox	5
Cowichan	5
Golden	5
Nelson	5
Port Alberni	5
Prince George	5
Terrace	5
Total:	**140**

(16 Locations)

In all there are 170 different locations, so the other 239 incidents (i.e., 379 minus 140) are spread out over 154 locations (i.e., 170 minus 16).

It needs to be mentioned that Harrison Hot Springs, Harrison Mills, Hope, Chilliwack and Agassiz are all in the same region. They have a combined incident number of 66. It therefore can be stated that over 110 years more incidents have been reported in this region than any other in BC. However, this is a very "soft" statistic. It does not indicate that there are more sasquatch in this region than elsewhere, just that there are more people to see them or the evidence they leave. If a ratio of human population to sasquatch incidents were established, for certain Bella Coola would come out on top.

BC's Sasquatch Population: Assuming that each sasquatch incident presented involved a different individual, and that sasquatch live for an average of 70 years, then there are *at least* 343 sasquatch wandering around BC (count for the period 1940 to 2009 inclusive).

I believe the true number of incidents is really 8 times the the number reported (by far, most incidents are not reported). This would indicated that the actual number of sasquatch alive and well is *at least* 2,744 individuals (i.e., 8 times 343).

Certainly, there are far more sasquatch than just those who have been sighted by people, or who have left evidence found by people. There is no way I know of to arrive at the difference. However, I think it can be stated that there are enough sasquatch in the province to ensure an on-going population of the creatures. I believe 500 is the minimum number for any creature to sustain itself, and the sasquatch population appears to be much greater than that number.

BC's Sasquatch "Turf": BC has a total land area of 365,900 square miles essentially available for sasquatch "turf." The space taken up by cities, villages, farms and so forth is basically insignificant. The following analysis of square miles per sasquatch is therefore based on the total square miles. I have arbitrarily increased the number of sasquatch in increments of 1,000.

Number of Sasquatch	Area per Individual
2,744	133 square miles
3,744	98 square miles
4,744	77 square miles
5,744	64 square miles
6,744	54 square miles

These figures serve to illustrate the difficulty in trying to find a sasquatch. The creature naturally moves around and generally avoids human contact. This greatly decreases the chances of finding one. Sasquatch are almost always seen by chance. I know of only one case in BC where 2 men went looking for one of the creatures and believe they actually saw one (see 1961/10/00 – Nelson. Silhouetted in the Moonlight).

Hoax Considerations

It is very convenient for professionals to write off the sasquatch issue as one big hoax. This is partly due to the media, which is quick to publicize hoax claims when publicity seekers come forward. It realizes that controversy is "good for business," so play this card at every opportunity.

Claims that footprints found are, or could have been, made by people using carved wooden feet are ridiculous. It is essentially impossible to make a decent foot impression with a wooden foot except in very soft soil. One has simply to cut a piece of wood, or plywood, to 16 inches long by 6 inches wide and attempt to plant it in regular soil. With 96 square inches of surface, there is enough to jack-up a car without leaving a significant impression. Sasquatch leave a good impression because they are very heavy to begin with (500 pounds or greater) and in the process of walking their weight is concentrated on their heels first, and then transferred along the foot to the toes. Even if a person did weigh 500 pounds, this process cannot be duplicated with a wooden foot. Whatever the case, to even consider that scientists would be fooled by a wooden foot impression (or its subsequent plaster cast) is ludicrous. (On this issue, the reader is referred to the entry: 2007/YR/00 – Pocatello, Idaho, USA. Footprints Recognized by Science).

That there has been at least one successful (although short-lived) sasquatch sighting hoax in BC is a fact (see 1977/05/15 – Mission. Misguided Misfits Pull Off Silly Hoax). Although the hoaxers owned-up to this caper, I am certain more research would have revealed the silly prank. Nevertheless, the question must be asked as to the probability of hoaxes with other sightings detailed in this work. In considering the numerous sighting locations, circumstances, and witness credibility, I would have to say the probability is all but non-existent. Some of the early stories (Jacko, Serephine Long, and Albert Ostman) certainly need to be "red flagged." Were they told in modern times, few of us would believe them because hoaxing has now become a way of life in our society.

On this point, there have been a lot of sasquatch evidence claims in BC that have turned out to be blatant hoaxes. In some cases, the hoaxers was caught in the act. Fortunately, diligent investigation has revealed the truth behind such claims. Both Thomas Steenburg and Bill Miller have recently exposed 2 different hoaxing attempts in the BC lower mainland. If the claims had gone unchecked, a good deal of time and money would have been spent in further explorations by both these researchers and others trying to resolve the sasquatch issue. Certainly, a sasquatch researcher must be very vigilant with regard to the possibility of hoaxes. He or she must consider other explanations for alleged sasquatch encounters. Accepting every report at face value simply makes one an advocate rather than a true researcher.

The advent of digital photography has been both a blessing and a curse. While more photographs have become available, this is countered by ease of manipulation. Detecting hoaxes is possible, but it can be difficult and time consuming.

Unfortunately, hoaxes are a fact of life in our society in all disciplines. As they can result in both "fame and fortune," they are not going to go away.

In Conclusion

From the information presented in this volume, I am sure the reader can fully appreciate what a sasquatch in BC is said to look like and draw some conclusions on how it acts and manages to sustain itself.

Although this book concentrates on BC, the creature is essentially the same right the way across North America, for which reported sasquatch-related incidents were 2,557 for the 100 year period ending in 2003. Whether or not this further substantiates the creature's existence, I am not sure. All I can say with my own certainty is that if the sasquatch does exist, then it would exist in BC.

As to other knowledge on the creature which some people believe has been withheld by the government, research institutions, or individual researchers, to my knowledge nothing has been substantiated. I have never been confronted by government people or suffered any interference in the research I have carried out, nor have John Green, Thomas Steenburg, or the late René Dahinden. Also, from my close association with these researchers, I really don't think they are harboring (or harbored) any "secrets."

At this point in time, the sasquatch in BC is essentially what has been provided in this book. If there is more information that would make any significant difference, then such has not been made known to either Thomas Steenburg or myself.

Although, by my count, BC has had more sasquatch-related incidents over the 100-year period ending in 2003 than any other province or state, the tangible evidence is, in a word, marginal. There are only 5 BC footprint casts (including a set of prints) presently known. I am told that Bob Titmus had made a number of BC casts, however they were all destroyed in a fire. Photographs of footprints in BC are also vague (save one). Again, what I have provided in this book is essentially all we have.

As to BC hand casts or casts of other body parts (or photographs of same), there are none that I know of. Furthermore, I do not know of any hair samples collected in BC that indicated they were from a sasquatch or "unknown primate." The same applies to feces or "scat." Nothing collected in BC has been such that it could have been submitted for DNA analysis.

I will guess that 99.99% of the tangible evidence there is in support of sasquatch existence comes from the US (mostly Washington, Oregon, and California). As these states, like BC, border the Pacific Ocean, they and BC are all in the same geographic area. As a result, it can reasonably be taken that US evidence is equally applicable to BC.

I have wondered for years about sasquatch crossing the border into the US. For the most part, there are no substantial man-made barriers between BC and Washington and Idaho. Nevertheless, one would think there is some sort of surveillance. Remarkably, to my knowledge there has never been a sasquatch seen and reported by the border patrol people in this region—nor right across Canada for that matter.

In the last 10 years, there appears to have been a tremendous surge of reported sasquatch-related incidents in the US. However, in BC over the last 30 years, the number has remained very constant (average of 4 per year).

From my own perspective, the likelihood of getting more/better evidence of the creature *increases* in accordance with the number of reported sightings. In other words, if the US sighting numbers are reasonably valid, then there should be much better evidence, especially with regard to photographs/videos.

In many ways, I am inclined to think that the sasquatch issue as presented here for BC is representative of the issue throughout North America. In short, the creature is nowhere near as "available" as sighting records in other regions indicate.

Clearly, assuming sasquatch do exist, they must be very reclusive by nature. People may be in close proximity to them on more occasions than we think. Yet nothing is seen as the creature stands and watches 30 feet back in a tree line. The people in such circumstances go on about their business none the wiser. More than once witnesses have told Thomas Steenburg that if the creature had not moved, or they had not looked right at it, they would have never known it was there. In almost every case, as soon as the creature became aware it had been spotted, it turned and walked away, disappearing back into the deep woods. The witnesses just see something unbelievable for a few moments before it disappears. More often than not, they never even think to raise their camera, if carrying one.

So the sasquatch mystery is one that just keeps going on. In BC, sightings continue at the same rate they always have. It is whether or not incidents are reported to the media or researchers which fluctuates from year to year. As a result, unless hard physical evidence is found in the field (researcher, hunter, hiker) or uncovered in a long-forgotten museum box or drawer, the mystery of the sasquatch will simply carry on.

The authors of this book can only ask readers to keep their eyes open when in the country, and not to be deterred in reporting an incident (footprint finding or sighting) to a researcher. Also, everyone is urged to carry a camera when exploring BC, the Homeland of the Sasquatch.

COLOR PLATES

The following are selected color photographs that supplement the material presented in the main text. Some images have been repeated, but have been greatly enlarged.

ɔb Titmus with casts he made from footprints found along a Skeena River slough in 1976.
ɪe prints measured about 15.5 inches long. Titmus was by trade a taxidermist and a
ghly experienced outdoorsman. He was beyond doubt the preeminent sasquatch field
searcher in his time. He essentially dedicated nearly 40 years of his life to sasquatch
search, and was thoroughly convinced of the creature's existence. Much of his field work
ɪs done in remote BC coastal areas where few other researchers have ventured even up
this time. (Photo: R. Titmus.)

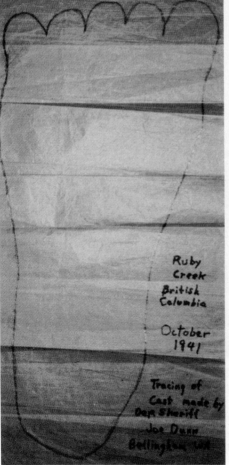

Ruby
Creek
British
Columbia

October
1941

Tracing of
Cast made by
Dep. Sheriff
Joe Dunn
Bellingham WA

Photos from left to right:
—The Chapman house at Ruby Creek where a sasquatch was encountered in 1941. (Photo: J. Green.)
—René Dahinden on a fence where the creature's footprints indicated it simply took in stride. (Photo: J. Green.)
—A tracing of a footprint cast made from one of the creature's prints. (Photo: C. Murphy.)
—Author and Deborah Schneider, a relative of Mrs. Chapman. (Photo: C. Murphy.)

John Green knew the Chapmans, and it was this incident that convinced him of the probable existence of sasquatch, whereupon in 1958 he embarked on lifelong research.

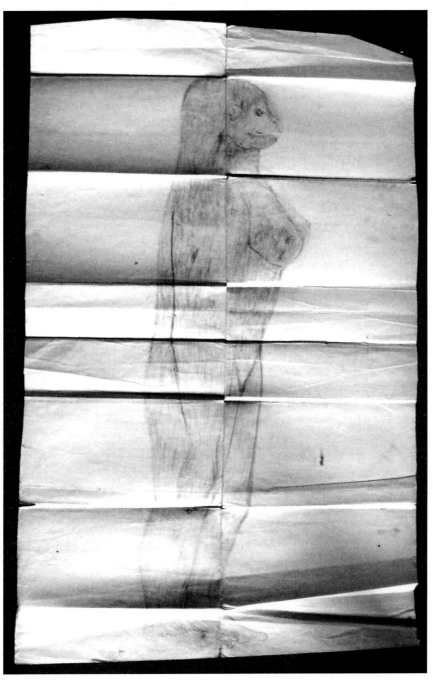

The actual drawing by Myrtle Roe of the creature seen by her father on Mica Mountain in 1955. It had been folded and filed away for some 48 years by the time John Green provided it to me for my museum exhibit. (Photo: C. Murphy.)

A bear head and sasquatch sculpture by the author. The single most frequent explanation given for sasquatch sightings, both in BC and elsewhere in North America, is that what people saw was a bear. Aside from the fact that bears do not walk very far on 2 legs, and the shape of their body is very different from a sasquatch, the most distinguishing features in both creatures are in the head and face.

As can be seen in the above photographs, a bear's head does not rise very far above its eyes. Also, a bear has a long "snout." A sasquatch is "human-like" with its eyes in about the center of its head, resulting in a long receding forehead. Its mouth is a fair distance from its nose, so it has a "muzzle." None of the BC sightings presented state that the sasquatch had a "snout." Many people have referred to a "flat nose." What they are saying here is that it did not have a "snout."

Because this is all so "elementary," it is little wonder that outdoorsmen and hunters get upset when they are told by authorities that they probably saw a bear.
(Photos: C. Murphy.)

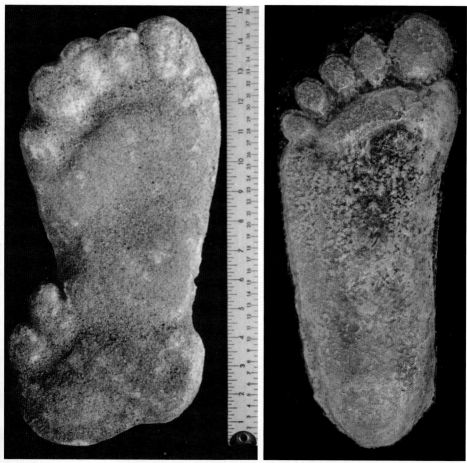

Left, cast of a double-tracked bear print. Right, cast of a sasquatch footprint. I have made the images the same length and matched them for comparison purposes.

A double-tracked bear print is the result of a bear's back foot partially overlapping the print of its front foot. In some cases, the alignment is such that when the print weathers, it can take on the appearance of a single large print—much larger than what would normally be made by a bear. It is often stated that alleged sasquatch footprints are simply double-tracked bear prints.

Nevertheless, there are other differences. The general outline of a bear print (normal or double-tracked) is very different to a sasquatch print. Also, often a bear print will show claws. If more than one print is found in a series, it will be seen that the bear's big toes are on the outside, rather than the inside as with sasquatch and humans.

Naturally, if there is enough detail available, evidence of double tracking is obvious in the bear print. One thing is for certain—anthropologists do not mistake bear prints for sasquatch prints. (Photos: C. Murphy.)

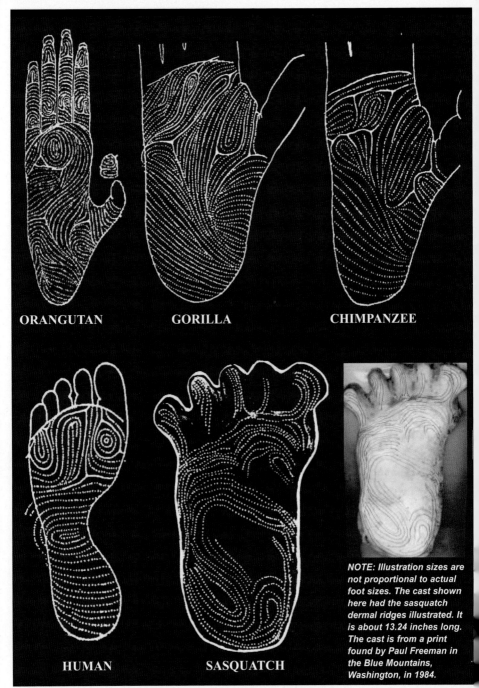

ORANGUTAN

GORILLA

CHIMPANZEE

HUMAN

SASQUATCH

NOTE: Illustration sizes are not proportional to actual foot sizes. The cast shown here had the sasquatch dermal ridges illustrated. It is about 13.24 inches long. The cast is from a print found by Paul Freeman in the Blue Mountains, Washington, in 1984.

Comparison of sasquatch dermal ridges with those of other primates. The presence of ridges on the cast shown was confirmed by an expert in this field. (Photo: Author's collection.)

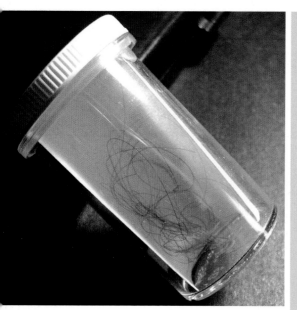

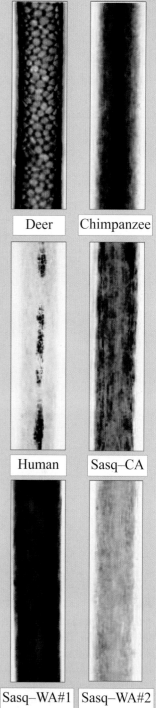

(Above) A sample of what is believed to be sasquatch hair. What you see is about actual size.

(Right) Enlargements of hair samples provided by Dr. Henner Fahrenbach with the following explanation:

Hair micrographs (260x): The deer hair has the cross-section almost entirely occupied by the medulla, an unbroken lattice in hair terminology. It has, of course, a thin cortex and cuticle. The chimpanzee hair, pitch black, has a continuous, mostly amorphous medulla. The human hair has the typical amorphous fragmentary medulla. The three sasquatch hairs (one from California; 2 from Washington) are: (CA) dark brown; (WA#1) very dark (observed as black on the animal); and (WA#2) reddish brown (called buckskin by the observers of the animal). A medulla is uniformly absent in these hairs.

Dr. Fahrenbach stated: Generally, sasquatch hair has the same diameter range as human hair and averages 2 to 3 inches in length, with the longest collected being 15 inches. (Photos: Left, . Murphy; Right, Dr. H. Fahrenbach.)

439

*A frame from the Patterson/Gimlin 16mm film taken at Bluff Creek, California, in Octo-
ber 1967. The film started its "scientific journey" at the University of BC in that same
month. Remarkably, the images of the creature seen in the film are still the best ever
obtained. Unfortunately, neither the film nor other evidence has convinced the general
scientific community that intense research in warranted. Only a few intrepid scientists
have taken up the challenge. (Photo: R. Patterson, Public Domain.)*

A study performed by the author on the possible facial features of the creature seen in the Patterson/Gimlin film. I enlarged the head as seen in the first image and printed it on ordinary paper. I then used pastels to enhance the features. I closed the mouth to provide a more natural appearance. The images seen in the Patterson/Gimlin film are the best and most credible available to my knowledge for this kind of analysis. (Photo: C. Murphy.)

Robert Bateman's classic painting of a sasquatch. We see the creature as a "fleeting glimpse" in a misty rainforest which is very typical of sasquatch sightings in BC. Bateman's insights are highly significant in all of his as remarkable works, and I believe what we see here is likely very close to the actual appearance of the creature. (Artwork/Photo: R. Bateman.)

BC's First Nations people naturally have many legends. Top: Author at the Xa:Ytem Rock in Mission. The legend states: "When the Creator was walking this Earth putting things right, he met three Chiefs at this place. He gave them knowledge of the written language to share with the people. But when he came back, he found they had not done what he had instructed them to do, and so he threw them into a pile and changed them into that rock." (1991, Stolo Elder Bertha Peters' Story.) Many professionals put the sasquatch in the category of a legend. However, the sasquatch is seen by both First Nations and non-First Nations people in BC and elsewhere in North America.

Lower: Replicas of actual pit houses found near the above rock. The structures were used by aboriginal people in temperate zones (Northern Europe, Russia, and the Americas) for 2,000 years. The oldest remaining structure in North America is in Washington, USA, and is dated at over 6,500 years old. It is likely that stories of the sasquatch go back to the earliest human habitation of North America. (Photos: C. Murphy.)

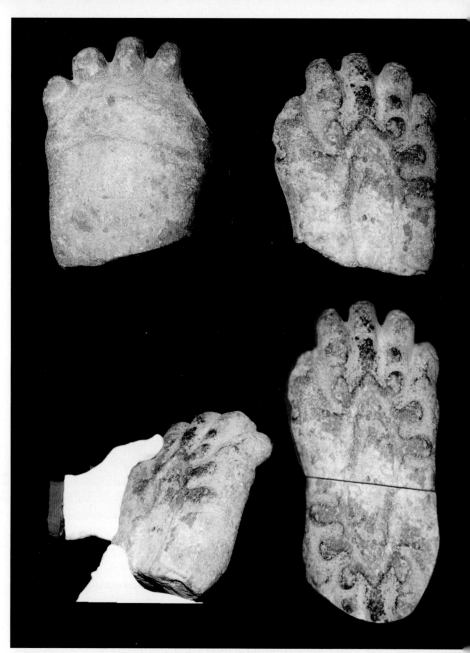

The "stone foot," shown from below and above in the first 2 images, and in the hands of a museum curator in the third image. I created the last image by .reversing the top half, less the toes, to reasonably complete the foot and thereby see how the artifact may have originally appeared. The foot is owned by the Vancouver Museum. (Photo: C. Murphy.)

The fearful D'sonoqua or "Wild Woman of the Woods," was both a good and bad omen in First Nations culture. From the above carving and art illustration, it is seen that D'sonoqua's body was used at major feasts as a receptacle for food.

The carving has a large bowl in the center that would have been used for meat or fish. It is seen that her knees are flat on top so they could be used for bowls of food.

The art illustration has bowls on her knees, also on her stomach, and there would be a bowl place where each of her breasts would have been. Also, her face lifted off and the hollow head below became another food receptacle.

In the background of the image, there are 2 heads facing each other on the ends of a little boat. The heads are also that of D'sonoqua, and the "boat" acted as a large food bowl.

Used in this way, D'sonoqua would probably have been regarded as a "giver." (Photos: C. Murphy.)

Kwakiutl First Nations people at Alert Bay (1926) with large carvings. The central carving is of D'sonoqua (also Dzunuk'wa) or "Wild Woman of the Woods." The information provided for the photograph states that the carvings are "feast dish covers." Another name for D'sonoqua is (in English) "Property Woman." She gives out her wealth and powers to a select few. Many tribes have an annual dance to celebrate or honor her. (Photo: Albert Paull, Vancouver Public Library, Image #1706.)

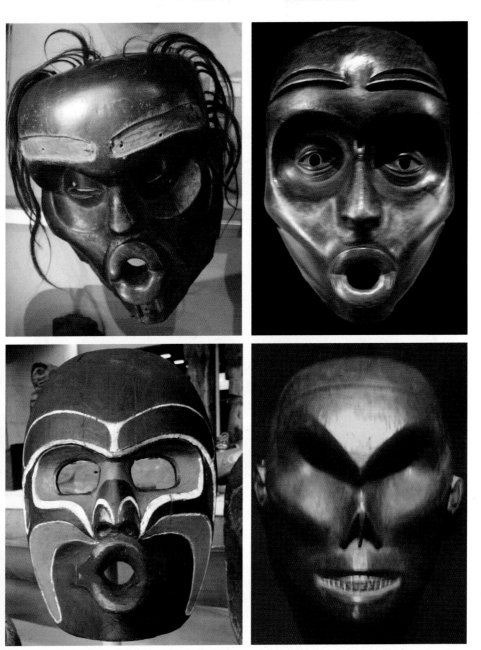

Four examples of sasquatch-related First Nations masks. The familiar "whistling lips" are seen in the first three, so they are depictions of D'snoqua (Wild Woman of the Woods). The last appears to show ape-like characteristics, so is considered sasquatch-related. The first three masks are in the Museum of Anthropology, University of BC, Vancouver. The last is in the Royal Museum, Victoria.

First Nations people call the creature represented by this mask a "Gagit." The mask was given to me on loan for my sasquatch exhibit by Robert Alley of Alaska. Alley has long been a sasquatch researcher and one day telephoned me and said he had 2 masks I could exhibit. One was definitely the "wild man of the woods," and the other this "Gagit" mask. Thomas Steenburg quickly pointed out to me that the mask was not sasquatch-related, and I initially agreed with him. However, Alley had said that another name (or perhaps English name?) for the Gagit was "land otter man." He pointed out that the spikes in the creature's lips are representations of sea urchin and fish dorsal spines, which the Gagit suffers when eating such food. As the sasquatch is noted for being a remarkable swimmer, I believe First Nations people would compare it to an otter. Furthermore, as the sasquatch is also a "land" creature, I can see it being called a "land otter man." (Photo: C. Murphy.)

(Above) Tony Healy from Australia with a display of sasquatch-related First Nations masks at the Museum of Anthropology, University of BC. Tony is one of the main "yowie" researchers in Australia. He visited with me in 2007. In 2009 Tony's friend and fellow researcher, Paul Cropper, visited me. From our combined research on the yowie and sasquatch, it appears that the creatures are so similar in nature as to be one and the same. Paul accompanied me to the Sasquatch Roundup in Yakima, Washington, and we later visited John Green at Harrison Hot Springs. Paul is seen in the adjacent photograph with John Green (left). (Photos: C. Murphy.)

449

The bases of 2 totem poles depicting D'sonoqua, or the "wild woman of the woods." The first (left) shows the creature with her offspring. This pole is on the grounds of the Royal Museum in Victoria. The second (right) appears to be a much older work. It is located inside at the Museum of Anthropology, University of BC, Vancouver. The repetition of the creature in First Nations art is quite remarkable. Equally remarkable is the fact that it is the only land creature in First Nations or native "lore" that has found its way into the world of non-natives. In other words, it is the only "mythical" land being that we (non-natives) claim to see in the flesh. (Photos: Left, J. Green; Right, C. Murphy.)

An elaborate sasquatch "transformer" mask. It is based on the Native belief that the sasquatch can "shape shift" or change its appearance at will (even change into other animals). (Photos: D. Hancock.)

451

Lynn Maranda of the Vancouver Museum holding the sasquatch mask created by Ambrose Point of the Chehalis Band. The size of the mask is clearly indicated. Point created it in the 1930s and it was donated to the Vancouver Museum later in that decade. It has been reasoned that Point based his work on an actual sighting. Kelsey Charlie, a First Nations sasquatch witness, agrees with this speculation. The mask facial features generally match those observed in actual sasquatch sightings, especially those seen in the Patterson/Gimlin film which was taken some 30 years after the mask was created. The fact that the mask was left unpainted also supports this contention. One might even muse that the large size of the mask is also directly related. In other words, Point created what could be called a "scale model." (Photo: C. Murphy.)

RobRoy Menzies captures the moment when Peter Nab fired a .22-caliber rifle at a sasquatch near Lumby in 1973 (see 1973/07/00 – Lumby. "Plinkers" Shoot at Sasquatch). Although it is hard to comprehend that even a small-caliber rifle would not inflict some damage to a sasquatch, the creature has been shot at with regular (high-caliber) hunting rifles with little or no effect. It has been reasoned that the muscle masses in its chest could stop bullets, and this appears to be the case. Nevertheless, it has allegedly seriously wounded by rifles, but in most cases manages to get away. Only in one case in BC did it appear to have been killed, and in this case another of its kind came to its aid and prevented its body from being retrieved (see 1905/YR/00 – Kitimat. Probable Sasquatch Killing). (Photo/artwork: RobRoy Menzies.)

453

Sasquatch sightings throughout North America often occur when the creature is either very near, or actually in, water. It has been observed swimming a considerable distance from shore, and also wading as if looking for something, as seen in this artwork by RobRoy Menzies. The fact that the creature has been sighted on small islands along the BC coast would indicate that swimming is a part of its normal daily existence.

Remarkably, according to some professionals, some sasquatch footprint casts might indicate that the creature has webbed feet. There is a highly controversial theory, proposed in 1942, that at one time there was an "aquatic ape" that measured in human evolution.

Whatever the case, we can reason that the size of a sasquatch would probably afford it protection from the cold water off BC's coast and in numerous rivers and lakes. For certain, sasquatch have been seen holding a fish, so we know they would have fishing skills.

454

William Munns is seen here with a model he created of a Gigantopithecus blacki, *a massive ape that became extinct about 300,000 years ago. Its existence is confirmed by teeth and jaw bones found in China and India. (Photo: W. Munns.)*

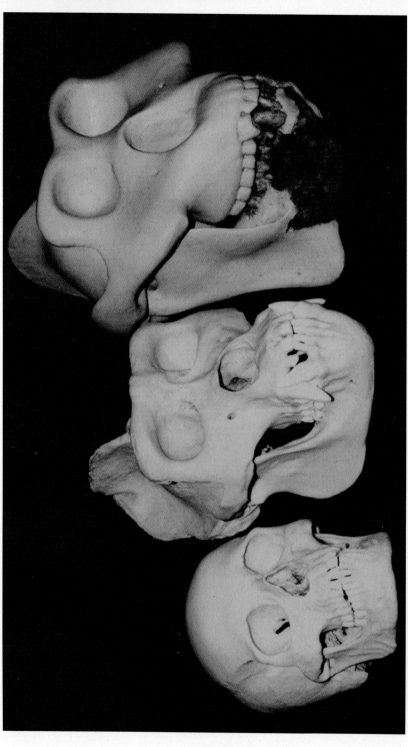

Left to right, model human, gorilla, and Gigantopithecus blacki skulls. The theory that an extinct ape, commonly called "Giganto," may have migrated from Asia to North America and there lived on to became the sasquatch was first offered by John Green. Dr. Grover Krantz constructed the Giganto skull based on a actual jawbone found in Asia and became the main propo-

Lynn Maranda (left) with Penny Birnam and the 4 sasquatch heads she created for my exhibit at the Vancouver Museum. The photograph was taken at Penny's studio. The heads have since traveled to an exhibit at a private museum in Seattle, Washington (Seattle Museum of Mysteries) and the Museum of Natural History in Pocatello, Idaho.

Penny gave me one of the heads (right) which I display at sasquatch-related functions. The head I have is around 18.5 inches high, so by my calculation it would represent that of about a 9-foot-tall sasquatch. This is based on my own findings that a sasquatch head is about one-sixth of its standing height (the average human head is one-eighth of standing height). (Photos: C. Murphy.)

457

Poster for the sasquatch exhibit at the Vancouver Museum.
(Photo: Author's collection.)

Author (left) with his son, Chris, daughter-in-law, Lisa, and some sort of wild man high in the mountains above Klein Lake, BC. The squinting carving overlooks a spectacular view of the region. There is no indication that I could see of who carved the figure, or when it was placed. The trail up to its presence is "vigorous," so I don't think very many people venture up that way.

Klein Lake is on the Sechelt Peninsula, which is the home of the Sechelt First Nations people. There have been a number of highly credible sasquatch sightings in this region. There is no road to the peninsula. The only way to get there is by ferry (about a 35-minute journey). Beyond Sechelt to the northwest, there are hundreds of miles of rugged mountains and dense rain forests. (Photo: T. Healy.)

Wanja Twan, René Dahinden's ex-wife, is seen here with the author at Wanja's little sasquatch museum in Hope (2003). The sasquatch wood carving is by Robert Forde, a local artist. The museum is now closed, but Wanja continues to support sasquatch research. (Photo: C. Murphy.)

The entrance to the village of Harrison Hot Springs with its "Sasquatch Country" sign. Many sasquatch conferences have been held at Harrison. Most (both here and elsewhere in BC) were organized by Stephen Harvey. All were well attended, with major researchers from both the US and Canada providing presentations on findings. Harrison continues to be the "mecca" of sasquatch research, even though some residents appear to wish otherwise. (Photos: T. Steenburg).

Stephen Harvey

The sasquatch/bigfoot entry in the 2006 Sand Sculpture Competition at Harrison Hot Springs. Sculptures were beyond amazing. The sand at Harrison was deemed to be the best sand in the Pacific Northwest for sand sculptures. John Green, a Lions Club member was responsible for bringing the event to Harrison and served as the first president of the Harrison Hot Springs Sand Sculpture Society. The annual event was recognized world wide and some record-breaking entries are in the Guinness Book of World Records. Thousands of people went to Harrison for the competitions, resulting in the largest crowds ever to visit the little village. The competition ran from 1986 to 2006. That Harrison was unable or unwilling to continue the event. was a great shame. (Photos: C. Murphy.)

A model of a giant (9 feet tall) human skeleton made by students at a university in Arizona, probably in the 1970s. It is made of iron and was used as a tourist attraction for many years at a motel in Castle Rock (near Quesnel). David Hancock purchased it in 2008 and had a student from Montreal, Ralph Goulet (left) and his friend Ben, drive to the motel and get it. Although the model is scaled to human proportions, one can get a bit of an appreciation as to what a 9-foot sasquatch skeleton would look like. (Photo: Hancock House.)

Sasquatch coin and presentation case. The tab showing a map of BC on the front pulls out. The case is 5 inches by 4 inches with the tab closed. The 25-cent coin is legal tender. It is much larger than a regular Canadian 25-cent piece. Indeed it is even larger than a Canadian 1-dollar or 2-dollar coin. (Photos: Royal Canadian Mint.)

Entries in Date Order

Note: Entries in **bold** print are either non-sighting/footprint incidents or they were excluded from the statistics previously provided.

000 – North America. Footprints In Stone (20)

00,000 BC – Asia. *Gigantopithecus blacki* (21)

D 500 or earlier – North America. Petroglyphs & Pictographs (22)

D 500 or earlier – Columbia River Valley, USA. Stone Heads (23)

D 500 or earlier – Lillooet. Stone Foot (23)

D 1700 and earlier – Pacific Coast. Totem Poles and Masks (24)

792/YR/00 – Nootka Sound (VI). First Nations People Fear Unusual Creatures (25)

811/01/07 – Jasper, Alberta. Unusual Footprints (27)

847/03/26 – Mt. St. Helens, Washington, USA. Race of a Different Species (28)

850/YR/00 – Nass River Area. The Monkey Mask (31)

864/YR/00 – Fraser River Canyon. The Caulfield Incident (33)

871/CI/00 – Harrison Mills (Chehalis Reservation). Abduction of Serephine Long (33)

875/12/00 – Stump Lake. Martin the Wild Man (35)

884/06/30 – Yale. Jacko the Ape-boy (36)

884/00/00 – Vancouver. Ape-boy Displayed? (40)

884/CI/00 – Spuzzum. Sasquatch Killed & Buried (41)

886/YR/00 – St. Alice Springs (now Harrison Hot Springs). Do Hot Springs Attract Sasquatch? (43)

895/YR/00 – Metchosin (VI). The Metchosin Monster—An Expensive Hoax (44)

898/CI/00 – Yale. A Strange Cave (45)

899/DE/00 – Vancouver. Loggers & Sasquatch (45)

900/CI/00 – Harrison Mills (Chehalis Reservation). First Report Known of Large Footprints in BC (47)

900/CI/00 – Lillooet. The Fossilized Man Hoax (48)

901/CI/00 – Campbell River (VI). The Monkey Man (49)

904/AU/00 – Port Alberni (VI). Some Sort of Wild Man—First Multiple Witness Sighting (50)

905/04/00 – Comox (VI). First Known Sasquatch Shooting (53)

905/06/18 – Little Qualicum (VI). First Urban Area Sighting (53)

905/07/29 – Cowichan (VI). Prospector Sees Wild Man (54)

905/10/00 – Comox (VI). Creatures Doing Sort of Sun Dance (54)

905/12/00 – Comox (VI). Carried a Lantern (56)

905/YR/00 – Nahanni Region (NWT). Skeletons Missing Their Heads (58)

905/YR/00 – Kitimat. Probable Sasquatch Killing (60)

906/08/05 – Alberni (VI). Wild Man No Figment of the Imagination (60)

906/09/00 – Vancouver. Wild Man—Beard Like Rip Van Winkle (61)

907/03/00 – Bishop Cove. Something Monkey-like (62)

909/05/00 – Harrison Mills (Chehalis Reservation). He Was In a Rage (63)

910/YR/00 – Harrison Mills (Chehalis Reservation). Same Type of Creature Seen (65)

911/YR/00 – Harrison Mills (Chehalis Reservation). Creatures Were a Man & a Woman (65)

1914/YR/00 – Hatzic. She Spoke the Douglas Tongue (66)
1915/YR/00 – Hope. Looked More Like a Human Being (68)
1919/YR/00 – Yale. Huge Nude Hairy Man (70)
1922/05/00 – Prince George. Primitive Cave or Tree-top Dweller (70)
1924/SU/00 – Toba Inlet. The Ostman Abduction (71)
1924/CI/00 – Bella Coola. Scientist Writes About the "Boqs" (75)
1926/CI/00 – Harrison Mills (Chehalis Reservation). The Word "Sasquatch" is Developed (76)
1927/09/00 – Agassiz. It Was Covered With Hair Like an Animal (76)
1928/YR/00 – Duncan (VI). Henry Napoleon Story (78)
1928/YR/00 – Lavington. An Odd Visitor (79)
1928/YR/00 – Bella Coola. Ape Seen & Shot (79)
1929/DE/00 – Windermere. Tall Human Skeletons (79)
1930/YR/00 – Swindle Island. Wading Sasquatch (80)
1934/03/23 – Harrison Mills (Chehalis Reservation). Rocks From Above (80)
1934/03/00 – Harrison Mills (Chehalis Reservation). Rock Bombardment (80)
1934/03/00 – Harrison Mills (Chehalis Reservation). Night Encounter (81)
1934/04/09 – Vancouver. First Official "Sasquatch Hunt" (81)
1934/05/00 – Harrison Mills. A Buzzing Sound (81)
1934/07/29 – Lincoln, Nebraska, USA. The Dickie Article (82)
1934/08/00 – Haney. "Human Face on a Fur-clad Body" (92)
1934/YR/00 – One Hundred Mile House. Quarter Mile of Tracks (94)
1935/10/00 – Harrison Hot Springs. Straddled a Log & Paddled with Hands & Feet (94)
1937/YR/00 – Lillooet. Eight-foot Skeleton (95)
1937/YR/00 – Osoyoos. Looked Just Like a Monkey (95)
1938/YR/00 – Harrison Mills (Chehalis Reservation). Bear & Sasquatch Fight (96)
1939/YR/00 – North Bend. Playful Sasquatch (96)
1939/DE/00 – Klemtu. Playing Around As They Walked (96)
1939/DE/00 – Harrison Mills (Chehalis Reservation). Life-like Mask (97)
1939/DE/00 – Alert Bay. D'sonoqua Seen (98)
1941/10/00 – Agassiz. Something Coming Out of the Woods (99)
1941/11/00 – Harrison Hot Springs. At Least 14 Feet Tall (100)
1943/CI/00 – Hope. Man Attacked by a Sasquatch (101)
1944/08/00 – Bella Coola. Bending Down In Water (101)
1945/04/00 – Swindle Island. Where the Apes Stay (103)
1945/CI/00 – Bella Coola. As Big As a Moose (103)
1945/CI/00 – Coombs (VI). Hair Streamed Out (103)
1945/CI/00 – Osoyoos. Faster Than a Man Could Run (103)
1947/YR/00 – Bella Coola. Not Bounding Like a Bear (104)
1947/YR/00 – Vancouver. Skin-clad Sasquatch? (104)
1948/YR/00 – Harrison Mills (Chehalis Reservation). Bicycle Chase (105)
1949/07/00 – Harrison Hot Springs. Horse Balked at Going Farther (105)
1949/10/00 – New Hazelton. Like a Bear (105)
1949/DE/00 – Agassiz. Huge Barefoot Tracks (106)

1949/CI/00 – Vanderhoof. Thought It Was a Gorilla (106)
1950/12/00 – Roderick Island. Tracks in Frost (107)
1950/CI/00 – Smithers. Trees Stripped (107)
1951/11/00 – McBride. Lean-to Slept In By Something (108)
1951/YR/00 – Swindle Island. Tracks Entered Lake Kitasu (108)
1952/YR/00 – Terrace. Watched the Man For a While (108)
1953/09/00 – Courtenay. "Thought It Was a Friend" (109)
1953/CI/00 – Esquimalt (VI). Massive Bone (109)
1954/YR/00 – Chilco Lake. Sasquatch Pass Named (109)
1955/10/00 – Tete Jaune Cache. Definitely Not a Bear (110)
1956/03/30 – Flood (or Floods). Sasquatch Couple (113)
1956/07/00 – Harrison Hot Springs. Sasquatch Hunt Planned (114)
1956/08/00 – Sechelt. Creature Standing In Doorway (115)
1956/08/00 – Sechelt. Rocks Thrown at US Visitor (115)
1956/CI/00 – Bella Coola. Eating Blueberries (115)
1957/03/00 – Harrison Hot Springs. Dahinden Speaks Out (115)
1957/04/10 – Harrison Hot Springs. A Search For the Sasquatch (116)
1957/04/21 – Victoria (VI). Provincial Archives Searched (117)
1957/04/24 – Harrison Hot Springs. Plans for Search Dampened (118)
1957/05/08 – Lillooet. Tired of Sasquatch Politics (119)
1957/08/25 – Vernon. Dahinden Returns After Month-long Search (120)
1957/YR/00 – East Sooke (VI). Large Bone With Foot (121)
1957/YR/00 – Agassiz. Green Takes Up the Search (121)
1957/CI/00 – East Sooke. (VI). Tore Trees Out By Roots (122)
1958/SU/00 – Harrison Hot Springs. A Hairy Monster (122)
1958/08/00 – New Hazelton. Ladies Report Sighting (124)
1958/08/00 – New Hazelton. Crossed the Road (125)
1958/10/29 – Yale. Residents Interviewed (125)
1958/12/08 – Bella Coola. Gaze Returned & Creature Runs Away (125)
1958/YR/00 – Vancouver. Oilman Needed (126)
1959/03/00 – Aristazabal Island. Strong Odor Sensed (127)
1959/08/00 – Kamloops. Queen Elizabeth II & Prince Philip Given Sasquatch Pres. (127)
1959/09/00 – Enderby. Nose Just a Flat Area With 2 Holes (128)
1959/AU/00 – Clinton. Massive Legs (129)
1959/DE/00 – Armstrong. Seven-foot Skeleton Uncovered (129)
1959/DE/00 – Rocky Mountains. Moved Like a Monkey (130)
1959/DE/00 – Hope. Sasquatch Caves (130)
1959/DE/00 – Mainland Bay. Sasquatch Scares Off Fisherman (131)
1959/DE/00 – Gilford Island. Digging for Clams (132)
1959/DE/00 – Peachland. Cook Sees Sasquatch (132)
1960/02/00 – Price Island. Size of a Small Man (132)
1960/02/00 – Roderick Island. Blood Found in Snow (132)
1960/02/00 – Langley. Four-and-a-half Feet Tall (132)

1960/08/00 – Nelson. A Great Beast (133)
1960/08/00 – Chilliwack. Stones Rain for About 2 Hours (134)
1961/04/19 – Harrison Hot Springs. Puzzling Print (136)
1961/09/00 – Cumberland. Seemed To Be Checking Where They Camped (136)
1961/10/00 – Nelson. Silhouetted In the Moonlight (136)
1961/10/00 – Swindle Island. Long String of Large Footprints (137)
1961/YR/00 – Moricetown. Prints Were Awe Inspiring (137)
1962/04/00 – Bella Coola. Four Sasquatch Seen Together (138)
1962/04/00 – Bella Coola. Mother & Child (138)
1962/06/08 – Kitimat. About 10 "Apes" (139)
1962/07/23 – Wells. Light Gray Animal (140)
1962/SU/00 – Aristazabal Island. Many Tracks Found (140)
11962/08/00 – Devastation Channel. Pace of 42 Inches (140)
1962/08/00 – Hixon. Hair-covered Man (141)
1962/08/00 – Quesnel. Small, Black Eyes (141)
1962/AU/00 – Swindle Island. Three Sets of Tracks (142)
1962/11/00 – Chilliwack. Eyes Glared (142)
1962/12/00 – Bella Coola. Four Sets of Footprints (144)
1962/CI/00 – Bella Coola. Million Dollar Reward (144)
1963/07/00 – Kemano. Titmus Sees Unusual Figures (145)
1963/SU/00 – Price Island. Sasquatch Roofed Nest (145)
1963/YR/00 – Bella Bella. Seen On the Shore (146)
1963/YR/00 – Minstrel Island. Monstrous Thing (146)
1964/WI/00 – Squamish. Very Large Tracks (146)
1964/04/00 – Turnour Island. House Moved (147)
1964/07/00 – Chilliwack. Four White Dots (147)
1964/CI/00 – Agassiz. Tracks With Short Toes (147)
1965/05/31 – Mission. Cows Stared At It (148)
1965/06/28 – Squamish. Something Dragged To an Ice Hole (148)
1965/07/00 – Butedale. Three Creatures Seen (151)
1965/07/00 – Hope. Lumbered Across Road (151)
1965/08/00 – Kitimat. Creature On the Shore (152)
1965/10/00 – Fort St. John. "Weetago" Seen (152)
1965/AU/00 – Harrison Mills. Wearing Fur Around Its Waist (152)
1965/11/00 – Bella Coola. White Streak Below Head (152)
1965/YR/00 – Agassiz. Sasquatch Incident Compilation (153)
1965/CI/00 – North Vancouver. White As a Sheet (153)
1965/CI/00 – Gold Bridge. Seen Through Rifle Scope (153)
1966/05/00 – Harrison Hot Springs. Sasquatch Not Welcome (154)
1966/05/00 – Spillimacheen. Strange Behavior (154)
1966/07/09 – Richmond. Fence Post Torn Out (155)
1966/07/14 – Richmond. Big Wooly Animal (155)
1966/07/21 & 22 – Richmond. Three Sightings in 24 Hours (156)

1966/07/00 – White Rock. It Fooled Around (157)

1966/08/00 – Penticton. Strange Shadow (158)

1966/CI/00 – Boston Bar. Beautiful Tawny Brown Coat (158)

1967/WI/00 – Bella Coola. Footprints & Yelling (158)

1967/02/00 – Hartley Bay. Screamed Like a Woman (159)

1967/02/00 – Quatna River. Ape Tracks (159)

1967/07/00 – Prince George. Legs Obscured By Long Hair (160)

1967/09/01 – Victoria (VI). Not a Subject for Mirth (160)

1967/09/19 – Victoria (VI). Government Gets Involved (162)

1967/09/19 – Agassiz. Getting Serious (163)

1967/09/28 – Victoria (VI). Keeping an Open Mind (163)

1967/09/00 – Comox (VI). Teenage Girls Encounter Sasquatch (163)

1967/09/30 – Victoria (VI). Government Impressed (164)

1967/10/23 – Vancouver. Screening Arranged for California Film (164)

1967/10/26 – Vancouver. The Webster Interview (165)

1967/10/26 – Vancouver. UBC Scientists First to See Film of Alleged Sasquatch (165)

1967/10/28 – Vancouver. The Killing Question (171)

1967/10/31 – Vancouver. A Crippled Giant (172)

1967/10/31 – Vancouver. Mining Promoter Considers Sasquatch Expedition (172)

1967/CI/00 – Swindle Island. Typical Ape Tracks (173)

1968/01/12 – Vancouver. Sasquatch Film Rights Bought by Canadian Researchers (173)

1968/WI/00 – Gosnell. Loggers Find Tracks (174)

1968/02/22 – Alert Bay. Ran As Fast As We Could (174)

1968/03/00 – Golden. Snow Survey – Tracks Found (175)

1968/05/17 – Sechelt. Ran Away Pretty Fast (175)

1968/06/01 – Mission. No Plan for a Hunt (177)

1968/06/25 – Vancouver. Hancock Weighs-in On Sasquatch Issue (177)

1968/07/21 – Lumby. Sasquatch Fever (179)

1968/08/00 – Stewart. Ran Up Steep Hill On "Hind Legs" (180)

1968/09/14 – Chetwynd. Game Guides Sighting (181)

1968/09/00 – Courtenay. Silent Watcher (181)

1968/11/16 – Lillooet. In & Out of the Bush (181)

1968/11/20 – Prince George. Absolutely Speechless (182)

1968/YR/00 – Kelowna. Walked Upright All the Time (182)

1968/YR/00 – Harrison Hot Springs. Sasquatch Provincial Park Named (183)

1969/01/10 – Hope. Turned Upper Body As Well (183)

1969/01/00 – Agassiz. First Authoritative Book On Sasquatch (184)

1969/02/00 – Alert Bay. Seen Looking at Them (185)

1969/02/00 – Butedale. Ran Off Screaming (185)

1969/03/00 – Powell River. Cabin Owner Shaken Up (186)

1969/04/12 – Vancouver. Ice Man's Canadian Tour Canceled (186)

1969/04/15 – Vancouver. Canadians "Cold Shoulder" Ice Man (187)

1969/06/01 – Merritt. Army Cadet Sighting (188)

1969/06/18 – Klemtu. Fifteen-inch Prints (190)

1969/06/00 – Squamish. Three-toed Tracks (190)

1969/07/29 – Agassiz. Green Convinced With US Sighting (191)

1969/09/20 – Rossland. Observed Movement for 200 Yards (191)

1969/10/25 – Bella Coola. Sitting by 2 Holes (192)

1969/11/11 – Harrison Hot Springs. Green Goes Forward With Book & Film (192)

1969/11/20 – Lytton. Wizened Old Man (193)

1969/12/23 – Harrison Hot Springs. Green Returns from Cross-Canada Trip (193)

1969/DE/00 – Turnour Island. Striking on House Posts (195)

1969/DE/00 – Whistler. Followed Her Husband (195)

1969/DE/00 – Harrison Mills (Chehalis Reservation). Unusual Prints in Snow (195)

1969/DE/00 – Invermere. White & Gray Creature (196)

1969/DE/00 – Nazko. Along the River (196)

1969/DE/00 – Remote Island. Heard Chest-beating (196)

1969/DE/00 – Nelson. Sasquatch Dumps Logs off Truck (197)

1969/CI/00 – Vancouver Island. Caver Sees Strange Creature (197)

1969/CI/00 – Gordon River (VI). Deer Heads Appeared Twisted Off (198)

1969/CI/00 – Keremeos. Green-blue Eyes Reflecting (198)

1970/01/07 – Squamish. It Was Carrying a Fish (198)

1970/02/00 – Klemtu. Parted the Salal With Its Hands (199)

1970/02/00 – Skidgate (QCI). "Small Beady Eyes" (200)

1970/03/00 – Juskatla. Lifted Up Its Arms (200)

1970/04/02 – Harrison Hot Springs. "Man or Gorilla"(200)

1970/04/23 – Klemtu. "Ape-like Face" (201)

1970/05/00 – Nazko. Watched It Go By (201)

1970/06/00 – Trail. Track in Dry Mud (202)

1970/08/05 – Harrison Hot Springs. Green & Dahinden At Odds (202)

1970/08/05 – Richmond. Dahinden Announces Pitt Lake Expedition (202)

1970/AU/00 – Clearwater. Tracks in Remote Area (203)

1970/11/00 – Princeton. Small Round Head (203)

1971/01/00 – Masset (QCI). Hairs More Than 7 Inches Long (204)

1971/WI/00 – Hope. Miles of Tracks (205)

1971/05/00 – Squamish. Tracks On a Sandbar (205)

1971/06/00 – Wells. Tracks Near Cave (205)

1971/07/00 – Lake Louise, Alberta. Shoulder-length Very Blond Hair (205)

1971/07/00 – Houston. Long White Hair (206)

1971/SU/00 – New Denver. Tracks Impressed 2 Inches Deep (206)

1971/09/00 – Princeton. Walked Slightly Stooped (207)

1971/AU/00 – Sechelt. Color of a Collie Dog (207)

1971/11/12 – Richmond. Dahinden Goes to Europe (207)

1972/01/15 – Yakima, Washington, USA. Roger Patterson Passes On (208)

1972/01/00 – Moscow, USSR. Dahinden Visits *Moscow News* (208)

1972/03/25 – Richmond. Dahinden Declares Trip to Europe Great Success (209)

972/07/00 – Hope. Came Out Of a Cave (210)
972/SU/00 – Houston. Something Seen & Filmed (210)
972/12/24 – Castlegar. 150 Yards of Tracks (210)
972/CI/00 – Stewart. Ten Feet Tall (211)
973/03/21 – Big Bay. Watched Taking Steps from Large Rock to Large Rock (211)
973/06/00 – Sechelt. Appeared to Have a Goatee (211)
973/07/10 – Ocean Falls. Rooting for Vegetation (212)
973/07/00 – Lumby. "Plinkers" Shoot at Sasquatch (213)
973/08/00 – Agassiz. I Saw a Sasquatch That Night (214)
973/11/26 – Vancouver. Stood About 10 Feet Tall (215)
973/YR/00 – Vancouver. Hunter with Dahinden Book Released (216)
974/01/00 – Terrace. Seen in Garbage (216)
974/01/00 – Jasper, Alberta. One Was a Step Ahead of the Other (217)
974/04/00 – Terrace. Eating Pine Needle Tips (217)
974/05/00 – Shawnigan Lake (VI). Five Young Girls See Sasquatch (217)
974/07/00 – Harrison Hot Springs. Sasquatch Seen at Youth Camp (218)
974/07/00 – Powell River. Pushing the Trees and Branches (220)
974/07/00 – Okanagan Lake. Standing On the Patio (220)
974/SU/00 – Hope. Inside the Town Limits (221)
974/08/25 – Chilliwack. Seen On a Rocky Cliff (221)
974/09/00 – Harrison Hot Springs. Went Into Shock (222)
974/09/00 – Cougar, Washington, USA. Canadians Meet With Morgan. (222)
974/10/00 – Mackenzie. Took a Shot Over Its Head (223)
974/10/00 – Princeton. King Kong or Something (224)
974/YR/00 – 100 Mile House. Crouched & Looked Back Down (225)
974/YR/00 – Castlegar. Not At All Bear-like (226)
974/YR/00 – Terrace. Odor Like a Camel (226)
974/CI/00 – Sicamous. Snow Angels (226)
975/01/00 – Kamloops. Three Teens See Sasquatch (227)
975/06/20 – Cherryville. Crossed 10 Feet in Front of Witness (228)
975/07/00 – Cranbrook. Strange Footprints (228)
975/SU/00 – Lake Cowichan (VI). Threw a Rock to Get Its Attention (228)
975/08/00 – Strathcona Provincial Park (VI). Black Human-like Form (229)
975/09/00 – Colwood (VI). Reflecting Yellow Eyes (230)
975/09/00 – Marble Canyon. Pronounced Long Neck (230)
975/09/00 – Kimberley. Did Not Make Any Noise (231)
975/10/00 – Princeton. Came Within 12 Feet (231)
975/10/17 – Agassiz. Ran On All Fours (232)
975/11/11 – Sooke (VI). Long Line of Tracks (232)
975/YR/00 – Natal. Seen Near Bridge (237)
975/CI/00 – Telegraph Cove (VI). Stones Scare Off Boys (237)
975/CI/00 – Kingcome Inlet. Unusual Nest (237)
976/01/22 – Victoria (VI). Shoulders 3 to 4 Feet Across (237)

1976/04/02 – Natal. Big Monkey (238)
1976/04/03 – Harmer Ridge. Webbed Feet (238)
1976/05/00 – Wycliffe. Taller Than the Horses (238)
1976/06/11 – Grand Forks. Kids See Sasquatch (239)
1976/06/00 – McBride. Over a Dozen Witnesses (239)
1976/07/17 – Terrace. Major Track Finding (239)
1976/07/00 – Bella Coola. Sasquatch Follows Witness (241)
1976/09/03 – Kimberley. Sasquatch Chased, Followed by Car (241)
1976/09/03 – Wycliffe. Polaroid Photo Taken (242)
1976/09/07 – Kimberley. It Simply Moved (242)
1976/09/09 – Kimberley. RCMP Reports 8 Sightings In a Week (243)
1976/09/29 – Deroche. Put Its Face In the Creek (243)
1976/11/29 – Bella Coola. Huge Man-type Tracks (243)
1976/11/00 – New Masset (QCI). Long Dark Hair (244)
1976/12/18 – New Masset (QCI). Same Individual? (244)
1976/12/21 – Bella Coola. Light-colored Chest (244)
1976/12/00 – Bella Coola. It Stinks (245)
1976/12/00 – Bella Coola. Beside Garbage Dump (245)
1976/YR/00 – Chilliwack. Smiling Sasquatch (245)
1976/YR/00 – Harrison Hot Springs. Sasquatch Carving History (245)
1976/CI/00 – Lillooet. Legs & Feet Seen In Trees (247)
1977/01/00 – West Kootenays. Splashing the Water (247)
1977/03/28 – Spuzzum. Witness White As a Ghost (247)
1977/05/15 – Mission. Misguided Misfits Pull Off Silly Hoax (249)
1977/05/19 – Victoria (VI). Sasquatch & Politics (250)
1977/06/17 – Vancouver. Titmus Says He Will Never Give Up (250)
1977/08/08 – Chetwynd. Four-toed Prints (251)
1977/08/12 – Victoria (VI). Government MLA Introduces Sasquatch Protection Bill (252)
1977/AU/00 – Prince George. Slightly Stooped Posture (253)
1977/AU/00 – Harrison Mills. Possible Sasquatch Corpse (253)
1977/YR/00 – Vancouver. UBC to Look at Sasquatch Evidence (255)
1977/CI/00 – Black Creek (VI). Sasquatch "Bower" (255)
1977/CI/00 – Hope. Like the Back of a Moose (255)
1978/02/16 – Berkley, California, USA. George Haas Passes On (256)
1978/05/05 – Vancouver. Sasquatch Poll (257)
1978/05/09 – Vancouver. "Anthropology of the Unknown" Conference (257)
1978/05/00 – Vancouver. Dr. Koffmann Reports on Russian Snowman (258)
1978/SU/00 – Quesnel. Face Like an Ape (260)
1978/08/00 – Rocky Mountains. Thump on Roof of Car (260)
1978/09/00 – Rock Bay (VI). Eye-watering Stench (261)
1978/12/00 – Kaslo. Large Footprints Very Convincing (261)
1978/YR/00 – Midway. Big Black Figure (261)
1979/04/28 – Little Fort. High-pitched Squeal (262)

1979/04/30 – Little Fort. Sasquatch Shot & Drops (262)

1979/05/20 – Haney. Tracks Old & New (265)

1979/06/18 – Salmon Arm. Standing In the Shadows (265)

1979/07/00 – Hope. Ten Feet Tall (265)

1979/07/00 – Port Alberni (VI). Eyes Gleaming Orange (266)

1979/YR/00 – Nelson. Whole Race of Them (266)

1979/YR/00 – Etolin Island, Alaska, USA. Fifteen Feet Away (266)

1979/DE/00 – Sooke (VI). Long Gray Hair On Its Chest (268)

1980/03/00 – Victoria (VI). Sasquatch Protection Put Aside (268)

1980/03/00 – Dallas, Texas, USA. Scientists a Virtual No-show (269)

1980/05/00 – Vavenby. Odd Figure at Water's Edge (269)

1980/07/21 – Cumberland (VI). Something Big Shaking the Alders (270)

1980/12/13 – Cumberland (VI). Like a Man In Torn Rain Gear (270)

1980/12/23 – Rossland. Poor Old Trapper (271)

1980/YR/00 – Harrison Mills. (Chehalis Reservation). Chehalis Band Adopts Sasquatch Logo (271)

1980/YR/00 – Vancouver. UBC Publishes Book on Humanoid Monsters (272)

1980/CI/00 – Kitimat. Appeared To Be Waving (275)

1981/06/14 – Vancouver. Dr. Halpin Provides Her Thoughts (276)

1981/11/00 – Squamish. Crouched By Water (277)

1981/YR/00 – Kaslo. Crossed Close to Car (277)

1982/01/30 – Saanichton (VI). Saw It & Ran (277)

1982/06/10 – Fernie. I Don't Think a Man Could Leap Like That (278)

1982/06/16 – Golden. Seemed To Be Looking At the Back Of the Train (279)

1982/10/00 – Vancouver. Scientific Proof Presented (280)

1983/07/00 – Hazelton. It Was Just Curious (282)

1983/SU/00 – Gibsons. Tall Figure & Roars (283)

1983/08/04 – Richmond. Dahinden Disputes Blue Mountain Footprints (284)

1983/08/00 – Nanaimo (VI). Up A Tree (285)

1983/AU/00 – Parksville (VI). Would Have Fired if Sure Not a Man (285)

1984/03/00 – Kelowna. Two Sets of Tracks (286)

1984/04/12 – Vancouver. Vietnam Vet Hunts Sasquatch (286)

1984/08/20 – Agassiz. Photos Snapped; Indistinct (286)

1985/07/30 – Burns Lake. Car Struck 2 or 3 Times by Sasquatch (287)

1985/07/00 – Revelstoke. Appeared To Look Both Ways (287)

1985/YR/00 – Queen Charlotte City (QCI). Deer Struck Down By Sasquatch (288)

1986/04/00 – Lone Butte. Trail Led To a Swampy Lake (288)

1986/SP/00 – Golden. A Gorilla, Dad! It was a Gorilla! (289)

1986/05/21 – Harrison Hot Springs. Hair Looked Stiff (289)

1986/06/15 – Gibsons. What Looked Like a Bed (290)

1986/07/00 – Coalmont. Appeared To Be Female (291)

1986/08/00 – Chilliwack. Sasquatch Snatches Fish Catch (293)

1986/10/00 – Bella Coola. Ripped the Screen Door Off (294)

1987/03/14 – Dawson Creek. Oil Rig Crew Sees Sasquatch (295)

1987/04/13 – Fort Nelson. It Was a "Nahganee" (297)

1987/08/00 – Golden. It Stayed On Two Legs the Whole Time (297)

1988/07/05 – Harrison Hot Springs. Showed Intelligence (298)

1988/08/24 – Wycliffe. Too Tall To Be a Man (299)

1988/09/08 – Woss (VI). Thought It Was His Son (299)

1988/10/00 – Strathcona Provincial Park (VI). Wildlife Biologist Finds Sasquatch Tracks (300)

1989/SP/00 – Harrison Hot Springs. Flexed Its Back (300)

1989/09/20 – Fort St. James. Strange Spur (301)

1989/11/00 – Bella Coola. Running Like a Human (302)

1989/YR/00 – Burnaby. BCSCC Formed (303)

1989/DE/00 – Sechelt. Students Attacked With Barrage of Rocks & Debris (303)

1990/10/00 – Ottawa, Ontario. Canadian "Sasquatch" Postage Stamp Issued (304)

1991/05/00 – Radium Hot Springs. Long Blond Hair (305)

1991/09/18 – Golden. Well Built (305)

1992/06/00 – Port Townsend, WA, USA. BC Could Be "Last Stand" For Sasquatch (306)

1992/SU/00 – Comox (VI). Night Sounds Frighten Campers (307)

1992/10/00 – Nelson. Half As High As the Trees (307)

1992/10/00 – Castlegar. Mystery For the Rest of My Life (307)

1992/AU/00 – Fort Nelson. In the Middle of Nowhere (307)

1992/YR/00 – Port McNeill (VI). All On 2 Legs (308)

1992/YR/00 – Pullman, Washington, USA. First Scientific Book On Sasquatch (309)

1992/CI/00 – Remote Island. Clam Diggers Attacked With Rocks & Driftwood (309)

1992/CI/00 – Port Alberni (VI). Camper Lifted (309)

1993/01/00 – Nakusp. Footprints In the Middle of Nowhere (310)

1993/03/30 – Harrison Hot Springs. Eating Devil's Club Roots (310)

1993/04/00 – Richmond. Murphy Teams Up With Dahinden (312)

1993/08/01 – Ainsworth. Cycling Encounter (313)

1993/08/00 – Fort Nelson. A Big Red Monkey (313)

1993/08/00 – Fort St. James. Somewhat Athletic Gait (313)

1993/09/00 – Cold Fish Lake. Flailing Its Arms (314)

1993/10/00 – Nakusp. Incident at Hot Springs (314)

1993/YR/00 – Pocatello, Idaho, USA. ISU Scientist Joins Search (315)

1994/05/00 – Remote mainland inlet. Loud Howling Concerns Workers (316)

1994/05/31 – Castlegar. On All Fours, But Odd (316)

1994/08/00 – Harrison Hot Springs. Footprint Mystery (316)

1994/09/11 – Vernon. Rifle Scope Crosshairs On Eyes (318)

1994/09/00 – Alert Bay. Makes You Gag... (318)

1994/AU/00 – Tofino (VI). Insistence of Tree Rapping (318)

1994/11/00 – Balfour. Four Paces to Cross Highway (319)

1994/YR/00 – Pullman, Washington, USA. Jacko Disappearance Theory (319)

1994/CI/00 – Gold River (VI). Straight Gray Hair (320)

1995/07/00 – Bella Coola. Talks With First Nations (321)

1995/08/03 – Harrison Mills (Chehalis Reservation). Footprints Found & Later a Sighting (321)

1995/08/27 – Kaslo. Eyes Were Horrifying (324)
1995/08/00 – Tofino. (VI). As If Walking On Air (325)
1995/AU/00 – Prince Rupert. Walked With Bent Knees (325)
1995/11/00 – Ainsworth. Crossed Road In 2 Paces (325)
1995/YR/00 – Harrison Hot Springs. Sasquatch Again On Welcome Sign (326)
1996/01/21 – McBride. Freezer Opened & Strange Footprints (327)
1996/SP/00 – Port Renfrew. (VI). Small Stones Thrown (327)
1996/YR/00 – Burnaby. Patty's Portrait (327)
1997/03/25 – Spuzzum. One Shape All the Way Down (328)
1997/05/00 – Chilliwack. Something On Tape (331)
1997/07/01 – Chilliwack. Bob Titmus Passes On (332)
1997/YR/00 – Vancouver. Dahinden & the Kokanee Commercials (336)
1998/06/00 – Hood River, Oregon, USA. NASI Report Released (337)
1998/09/00 – West Kootenays. Defies Conventional Description (338)
1998/YR/00 – Kemano. Standing On 2 Legs On Rocky Slope (338)
1998/YR/00 – Courtenay (VI). North America's Great Ape (339)
1999/01/00 – Hood River, Oregon, USA. NASI Folds (339)
1999/06/00 – Chilliwack. Hair Over Feet Like Bell-bottom Pants (339)
1999/09/14 – Hope. Like a Roll of Fur at Its Wrists (340)
1999/DE/00 – Vancouver. Woman Followed By Sasquatch (341)
1999/DE/00 – Sayward (VI). Dragging a Deer (341)
1999/DE/00 – Harrison Hot Springs. Ape-like Face In Window (341)
2000/WI/00 – Salt Spring Island. Bateman Paints a Sasquatch (342)
2000/SU/00 – Courtenay (VI). Wildlife Biologist Says It All (342)
2000/08/15 – Hope. Realized It Wasn't a Bear (356)
2000/09/00 – Harrison Hot Springs. Meandering Tracks (357)
2000/09/00 – Gifford Pinchot National Forest, Washington, USA. The Skookum Cast (359)
2001/03/30 – Radium Hot Springs. Running Like a Man (361)
2001/04/18 – Richmond. Farewell René—Fondly Remembered (361)
2001/05/00 – Hope. Near the Toll Gate (365)
2001/09/00 – Mission. Chattering Like a Bear (365)
2002/02/11 – Windermere. A "Bridge" Going Across the Foot (367)
2002/02/14 – Port Angeles, Wash., USA. Renowned Scientist Dr. Grover Krantz Passes On (368)
2002/07/00 – Grand Forks. We Know What We Saw (370)
2002/08/17 – Tete Jaune Cache. Thought It Was a Moose (371)
2002/08/00 – Port Alberni (VI). Thought It Was a Human With Down's Syndrome (371)
2002/09/00 – Mission. Steenburg Moves to BC (372)
2002/10/00 – Little Fort. It Bent Straight from the Hips (373)
2002/11/00. Tofino (VI). Real Big Orange Eyes (373)
2002/12/00 – Lillooet. Looked Like a Monkey Butt (374)
2002/YR/00 – Harrison Mills (Chehalis Reservation). You Will Never Catch a Sasquatch (374)
2003/03/00 – Burnaby. Film Site Model Provides Insights (375)
2003/06/00 – Duncan (VI). Prints in Strawberry Field (377)

2003/11/00 – Vancouver. **Sasquatch Heads Created (377)**

2003/12/18 – Vancouver. **First Nations Elders Provide Sasquatch Stories (379)**

2003/CI/00 – Bowen Island. Ladies Stared In Disbelief (382)

2004/05/10 – Merritt. Standing By a Pond (383)

2004/05/00 – Hope. Looked Like an Ape (384)

2004/06/17 – Vancouver. **Sasquatch Exhibit at Vancouver Museum (385)**

2004/07/22 – Canal Flats. Rocks Thrown (387)

2004/7/00 – Kincolith. As if Swimming, But Not Moving (388)

2004/08/15 – Sunshine Valley. Four or Five Steps to Cross Road (388)

2004/08/00 – Canal Flats. Walking Along a Ridge (389)

2004/11/30 – Tofino (VI). Stood Its Ground (389)

2005/02/13 – Squamish. Running In the Bushes (390)

2005/03/08 – Sechelt. Swinging Arms & Hunched Forward (391)

2005/06/10 – Hope. Glaring Eyes (392)

2005/07/28 – Victoria (VI). **Museum Anthropologist Don Abbot Passes On (393)**

2005/SU/00 – Cowichan (VI). Three Sightings (393)

2005/08/14 – Prince George. Primate-like Face (394)

2005/08/26 – Harrison Hot Springs. Tracks On Logging Road Appeared Very Fresh (394)

2005/YR/00 – Port Alberni (VI). Two See Sasquatch (397)

2006/06/20 – Mesachie Lake (VI). Gait Not the Same as a Human (398)

2006/07/16 – Pocatello, Idaho, USA. **Sasquatch Exhibit at Museum of Natural History (399)**

2006/10/29 – Dewdney. **Halloween Hoax (399)**

2007/01/00 – Harrison Hot Springs. Walking Along the Tree Line (400)

2007/02/13 – Harrison Mills (Chehalis Res.) Large Rock Lands Near Fishermen (400)

2007/03/30 – Ucluelet (VI). Laying On Log Over River (401)

2007/09/25 – Canal Flats. Odor Did Not Linger (401)

2007/YR/00 – Pocatello, Idaho, USA. **Footprints Recognized by Science (402)**

2008/02/00 – Vancouver. **"Quatchi" Made Olympics Mascot (404)**

2008/06/18 – Bridal Falls. Hands On Tree Stump (405)

2008/06/19 – Bridal Falls. Two Steps to Cross Road (406)

2008/06/19 – Chilliwack. Unusual Sounds & a Footprint (407)

2008/SU/00 – Enderby. Forest Service Worker Tells of Sasquatch Sightings (407)

2008/08/12 – Smithers. Leaning Against Tree (408)

2008/09/21 – Agassiz Something Glimpsed & Rock Thrown. (409)

2008/11/00 – Mission. Hair Like a Horse's Tail (411)

2009/04/04 – Langley. I Think Of It As a Gift (412)

2009/04/00 – Merritt. Possible Print—Cell Phone Photo (414)

2009/04/00 – Gibsons. Sasquatch Walking Insights (414)

2010/06/06 – Duncan (VI). An Extremely Loud Grunt (416)

2010/YR/00 – Port Mellon. Large Hand (417)

2011/02/08 – Hope. Something Walking In the Bush (418)

2011/04/09 – Harrison Hot Springs. **The Sasquatch Summit (420)**

2011/05/18 – Ottawa, Ontario. **Sasquatch Coin Issued (423)**

Entries in Location Order

100 Mile House:
>Crouched & Looked Back Down. 1974/YR/00 - (225)
>Quarter Mile of Tracks. 1934/YR/00 - (94)

Agassiz:
>First Authoritative Book on Sasquatch. 1969/01/00 - (184)
>Getting Serious. 1967/09/19 - (163)
>Green Convinced With US Sighting. 1969/07/29 - (191)
>Green Takes Up the Search. 1957/YR/00 - (121)
>Huge Barefoot Tracks. 1949/DE/00 - (106)
>I Saw a Sasquatch That Night. 1973/08/00 - (214)
>It Was Covered With Hair Like an Animal. 1927/09/00 - (76)
>Photos Snapped; Indistinct. 1984/08/20 - (286)
>Ran On All Fours. 1975/10/17 - (232)
>Sasquatch Incident Compilation. 1965/YR/00 - (153)
>Something Coming Out of the Woods. 1941/10/00 - (99)
>Something Glimpsed & Rock Thrown. 2008/09/21 - (409)
>Tracks With Short Toes. 1964/CI/00 - (147)

Ainsworth:
>Crossed Road In 2 Paces. 1995/11/00 - (325)
>Cycling Encounter. 1993/08/01 - (313)

Alberni (VI):
>Wild Man No Figment of the Imagination. 1906/08/05 - (60)

Alert Bay:
>D'sonoqua Seen. 1939/DE/00 - (98)
>"Makes you gag… 1994/09/00 - (318)
>Ran As Fast As We Could. 1968/02/22 - (174)
>Seen Looking at Them. 1969/02/00 - (185)

Aristazabal Island:
>Many Tracks Found. 1962/SU/00 - (140)
>Strong Odor Sensed. 1959/03/00 - (127)

Armstrong:
>Seven-foot Skeleton Uncovered. 1959/DE/00 - (129)

Asia:
>*Gigantopithecus blacki.* 300,000 BC - (21)

Balfour:
>Four Paces to Cross Highway. 1994/11/00 - (319)

Bella Bella:
>Seen On Shore. 1963/YR/00 - (146)

Bella Coola:

Ape Seen & Shot. 1928/YR/00 - (79)

As Big As a Moose. 1945/CI/00 - (103)

Bending Down in Water. 1944/08/00 - (101)

Beside Garbage Dump. 1976/12/00 - (245)

Eating Blueberries. 1956/CI/00 - (115)

Footprints & Yelling. 1967/WI/00 - (158)

Four Sasquatch Seen Together. 1962/04/00 - (138)

Four Sets of Footprints. 1962/12/00 - (144)

Gaze Returned & Creature Runs Away. 1958/12/08 - (125)

Huge Man-type Tracks. 1976/11/29 - (243)

It Stinks. 1976/12/00 - (245)

Light-colored Chest. 1976/12/21 - (244)

Million Dollar Reward. 1962/CI/00 - (144)

Mother & Child. 1962/04/00 - (138)

Not Bounding Like a Bear. 1947/YR/00 - (104)

Ripped the Screen Door Off. 1986/10/00 - (294)

Running Like a Human. 1989/11/00 - (302)

Sasquatch Follows Witness. 1976/07/00 - (241)

Scientist Writes About the "Boqs." 1924/CI/00 - (75)

Sitting by 2 Holes. 1969/10/25 - (192)

Talks With First Nations. 1995/07/00 - (321)

White Streak Below Head. 1965/11/00 - (152)

Berkley, California, USA:

George Haas Passes On. 1978/02/16 - (256)

Big Bay:

Watched Taking Steps from Large Rock to Large Rock. 1973/03/21 - (211)

Bishop Cove:

Something Monkey-like. 1907/03/00 - (62)

Black Creek (VI):

Sasquatch "Bower." 1977/CI/00 - (255)

Boston Bar:

Beautiful Tawny Brown Coat. 1966/CI/00 - (158)

Bowen Island:

Ladies Stared In Disbelief. 2003/CI/00 - (382)

Bridal Falls:

Hands On Tree Stump. 2008/06/18 - (405)

Two Steps to Cross Road. 2008/06/19 - (406)

Burnaby:

BCSCC Formed. 1989/YR/00 - (303)

Film Site Model Provides Insights. 2003/03/00 - (375)

Patty's Portrait. 1996/YR/00 - (327)

Burns Lake:

Car Struck 2 or 3 Times by Sasquatch. 1985/07/30 - (287)

Butedale:

Ran Off Screaming. 1969/02/00 - (185)

Three Creatures Seen. 1965/07/00 - (151)

Campbell River (VI):

The Monkey Man. 1901/CI/00 - (49)

Canal Flats:

Odor Did Not Linger. 2007/09/25 - (401)

Rocks Thrown. 2004/07/22 - (387)

Walking Along a Ridge. 2004/08/00 - (389)

Castlegar:

Mystery For the Rest of My Life. 1992/10/00 - (307)

Not At All Bear-like. 1974/YR/00 - (226)

On All Fours, But Odd. 1994/05/31 - (316)

150 Yards of Tracks. 1972/12/24 - (210)

Cherryville:

Crossed 10 Feet in Front of Witness. 1975/06/20 - (228)

Chetwynd:

Four-toed Prints. 1977/08/08 - (251)

Game Guides Sighting. 1968/09/14 - (181)

Chilco Lake:

Sasquatch Pass Named. 1954/YR/00 - (109)

Chilliwack:

Bob Titmus Passes On. 1997/07/01 - (332)

Eyes Glared. 1962/11/00 - (142)

Four White Dots. 1964/07/00 - (147)

Hair Over Feet Like Bell-bottom Pants. 1999/06/00 - (339)

Sasquatch Snatches Fish Catch. 1986/08/00 - (293)

Seen On a Rocky Cliff. 1974/08/25 - (221)

Smiling Sasquatch. 1976/YR/00 - (245)

Something On Tape. 1997/05/00 - (331)

Stones Rain For About 2 Hours. 1960/08/00 - (134)

Unusual Sounds & a Footprint. 2008/06/19 - (407)

Clearwater:

Tracks in Remote Area. 1970/AU/00 - (203)

Clinton:

Massive Legs. 1959/AU/00 - (129)

Coalmont:

Appeared To Be Female. 1986/07/00 - (291)

Cold Fish Lake:

Flailing Its Arms. 1993/09/00 - (314)

Columbia River Valley, USA:
>   Stone Heads. AD 500 or earlier - (23)

Colwood (VI):
>   Reflecting Yellow Eyes. 1975/09/00 - (230)

Comox (VI):
>   Carried a Lantern. 1905/12/00 - (56)
>   Creatures Doing Sort of Sun Dance. 1905/10/00 - (54)
>   First Known Sasquatch Shooting. 1905/04/00 - (53)
>   Night Sounds Frighten Campers. 1992/SU/00 - (307)
>   Teenage Girls Encounter Sasquatch. 1967/09/00 - (163)

Coombs (VI):
>   Hair Streamed Out. 1945/CI/00 - (103)

Cougar, Washington, USA:
>   Canadians Meet With Morgan. 1974/09/00 - (222)

Courtenay (VI):
>   Silent Watcher. 1968/09/00 - (181)
>   Thought it was a Friend. 1953/09/00 - (109)
>   North America's Great Ape. 1998/YR/00 – (339)
>   Wildlife Biologist Says It All. 2000/SU/00 - (342)

Cowichan (VI):
>   Prospector Sees Wild Man. 1905/07/29 - (54)
>   Strange Footprints. 1975/07/00 - (228)
>   Threw Rock to Get Its Attention. 1975/SU/00 - (228)
>   Three Sightings. 2005/SU/00 - (393)

Cumberland (VI):
>   Like a Man In Torn Rain Gear. 1980/12/13 - (270)
>   Something Big Shaking the Alders. 1980/07/21 - (270)
>   Seemed To Be Checking Where They Camped. 1961/09/00 - (136)

Dallas, Texas, USA:
>   Scientists a Virtual No-show. 1980/03/00 - (269)

Dawson Creek:
>   Oil Rig Crew Sees Sasquatch. 1987/03/14 - (295)

Deroche:
>   Put Its Face In the Creek. 1976/09/29 - (243)

Devastation Channel:
>   Pace of 42 Inches. 1962/08/00 - (140)

Dewdney:
>   Halloween Hoax. 2006/10/29 - (399)

Duncan (VI):
>   An Extremely Loud Grunt. 2010/06/06 - (416)
>   Henry Napoleon Story. 1928/YR/00 - (78)
>   Prints in Strawberry Field. 2003/06/00 - (377)

480

East Sooke (VI):

Large Bone With Foot. 1957/YR/00 - (121)

Tore Trees Out By Roots. 1957/CI/00 - (122)

Enderby:

Forest Service Worker Tells of Sasquatch Sightings. 2008/SU/00 - (407)

Nose Just a Flat Area With 2 Holes. 1959/09/00 - (128)

Esquimalt (VI):

Massive Bone. 1953/CI/00 - (109)

Etolin Island, Alaska, USA:

Fifteen Feet Away. 1979/YR/00 - (266)

Fernie:

I Don't Think a Man Could Leap Like That. 1982/06/10 - (278)

Flood (or Floods):

Sasquatch Couple. 1956/03/30 - (113)

Fort Nelson:

A Big Red Monkey. 1993/08/00 - (313)

In the Middle of Nowhere. 1992/AU/00 - (307)

It was a "Nahganee." 1987/04/13 - (297)

Fort St. James:

Somewhat Athletic Gait. 1993/08/00 - (313)

Strange Spur. 1989/09/20 - (301)

Fort St. John:

"Weetago" Seen. 1965/10/00 - (152)

Fraser River Canyon:

The Caulfield Incident. 1864/YR/00 - (33)

Gibsons:

Sasquatch Walking Insights. 2009/04/00 - (414)

Tall Figure & Roars. 1983/SU/00 - (283)

What Looked Like a Bed. 1986/06/15 - (290)

Gifford Pinchot National Forest, Washington, USA

The Skookum Cast. 2000/09/00 - (359)

Gilford Island:

Digging for Clams. 1959/DE/00 - (132)

Gold Bridge:

Seen Through Rifle Scope. 1965/CI/00 - (153)

Gold River (VI):

Straight Gray Hair. 1994/CI/00 - (320)

Golden:

A Gorilla, Dad! It was a Gorilla! 1986/SP/00 - (289)

It Stayed On 2 Legs the Whole Time. 1987/08/00 - (297)

Seemed to be Looking at the Back of the Train. 1982/06/16 - (279)

Snow Survey—Tracks Found. 1968/03/00 - (175)

Well Built. 1991/09/18 - (305)

Gordon River (VI):
> Deer Heads Appeared Twisted Off. 1969/CI/00 -(198)

Gosnell:
> Loggers Find Tracks. 1968/WI/00 - (174)

Grand Forks:
> Kids See Sasquatch. 1976/06/11 - (239)
> We Know What We Saw. 2002/07/00 - (370)

Haney:
> Human Face On a Fur-clad Body. 1934/08/00 - (92)
> Tracks Old & New. 1979/05/20 - (265)

Harmer Ridge:
> Webbed Feet. 1976/04/03 - (238)

Harrison Hot Springs:
> A Hairy Monster. 1958/SU/00 - (122)
> A Search For the Sasquatch. 1957/04/10 - (116)
> Ape-like Face in the Window. 1999/DE/00 - (341)
> At Least 14 Feet Tall. 1941/11/00 - (100)
> Dahinden Speaks Out. 1957/03/00 - (115)
> Eating Devil's Club Roots. 1993/03/30 - (310)
> Flexed Its Back. 1989/SP/00 - (300)
> Footprint Mystery. 1994/08/00 - (316)
> Green & Dahinden at Odds. 1970/08/05 - (202)
> Green Goes Forward With Book & Film. 1969/11/11 - (192)
> Green Returns from Cross-Canada Trip. 1969/12/23 - (193)
> Hair Looked Stiff. 1986/05/21 - (289)
> Horse Balked at Going Farther. 1949/07/00 - (105)
> Man or Gorilla. 1970/04/02 - (200)
> Meandering Tracks. 2000/09/00 - (357)
> Plans for Search Dampened. 1957/04/24 - (118)
> Puzzling Print. 1961/04/19 - (136)
> Sasquatch Again on Welcome Sign. 1995/YR/00 - (326)
> Sasquatch Carving History. 1976/YR/00 - (245)
> Sasquatch Hunt Planned. 1956/07/00 - (114)
> Sasquatch Not Welcome. 1966/05/00 - (154)
> Sasquatch Provincial Park Named. 1968/YR/00 - (183)
> Sasquatch Seen at Youth Camp. 1974/07/00 - (218)
> Showed Intelligence. 1988/07/05 - (298)
> Straddled a Log & Paddled With Hands & Feet. 1935/10/00 - (94)
> The Sasquatch Summit. 2011/04/09 - (420)
> Tracks On Logging Road Appeared Very Fresh. 2005/08/26 - (394)
> Walking Along the Tree Line. 2007/01/00 - (400)
> Went into Shock. 1974/09/00 - (222)

Harrison Mills:

A Buzzing Sound. 1934/05/00 - (81)
Abduction of Serephine Long. 1871/CI/00 - (33)
Bear & Sasquatch Fight. 1938/YR/00 - (98)
Bicycle Chase. 1948/YR/00 - (105)
Chehalis Band Adopts Sasquatch Logo. 1980/YR/00 - (271)
Creatures Were a Man & a Woman. 1911/YR/00 - (65)
First Report Known of Large Footprints in BC. 1900/CI/00 - (47)
Footprints Found & Later a Sighting. 1995/08/03 - (321)
He was in a Rage. 1909/05/00 - (63)
Large Rock Lands Near Fishermen. 2007/02/13 - (400)
Life-like Mask. 1939/DE/00 - (97)
Night Encounter. 1934/03/00 - (81)
Possible Sasquatch Corpse. 1977/AU/00 - (253)
Rock Bombardment. 1934/03/00 - (80)
Rocks From Above. 1934/03/23 - (80
Same Type of Creature Seen. 1910/YR/00 - (65)
The Word "Sasquatch" is Developed. 1926/CI/00 - (76)
Unusual Prints in Snow. 1969/DE/00 - (195)
Wearing Fur Around Its Waist. 1965/AU/00 - (152)
You Will Never Catch a Sasquatch. 2002/YR/00 - (374)

Hartley Bay:

Screamed Like a Woman. 1967/02/00 - (159)

Hatzic:

She Spoke the Douglas Tongue. 1914/YR/00 - (66)

Hazelton:

It Was Just Curious. 1983/07/00 - (282)

Hixon:

Hair-covered Man. 1962/08/00 - (141)

Hood River, Oregon, USA:

NASI Report Released. 1998/06/00 - (337)
NASI Folds. 1999/01/00 - (339)

Hope:

Came Out Of a Cave. 1972/07/00 - (210)
Glaring Eyes. 2005/06/10 - (392)
Inside the Town Limits. 1974/SU/00 - (221)
Like a Roll of Fur at Its Wrists. 1999/09/14 - (340)
Like the Back of a Moose. 1977/CI/00 - (255)
Looked Like an Ape. 2004/05/00 - (384)
Looked More Like a Human Being. 1915/YR/00 - (68)
Lumbered Across the Road. 1965/07/00 - (151)
Man Attacked by a Sasquatch. 1943/CI/00 - (101)

Miles of Tracks. 1971/WI/00 - (205)

Near the Toll Gate. 2001/05/00 - (365)

Realized It Wasn't a Bear - 2000/08/15 - (356)

Sasquatch Caves. 1959/DE/00 - (130)

Something Walking In the Bush. 2011/02/08 - (418)

Ten Feet Tall. 1979/07/00 - (211)

Turned Upper Body As Well. 1969/01/10 - (183)

Houston:

Long White Hair. 1971/07/00 - (206)

Something Seen & Filmed. 1972/SU/00 - (210)

Invermere

White & Gray Creature. 1969/DE/00 - (196)

Jasper, Alberta:

One Was a Step Ahead of the Other. 1974/01/00 - (217)

Unusual Footprints. 1811/01/07 - (27)

Juskatla:

Lifted Up Its Arms. 1970/03/00 - (200)

Kamloops:

Queen Elizabeth II & Prince Philip Given Sasquatch Presentation. 1959/08/00 - (127)

Three Teens See Sasquatch. 1975/01/00 - (227)

Kaslo:

Crossed Close to Car. 1981/YR/00 - (277)

Eyes Were Horrifying. 1995/08/27 - (324)

Large Footprints Very Convincing. 1978/12/00 - (261)

Kelowna:

Two Sets of Tracks. 1984/03/00 - (286)

Walked Upright All the Time. 1968/YR/00 - (182)

Kemano:

Standing on 2 Legs On Rocky Slope. 1998/YR/00 - (338)

Titmus Sees Unusual Figures. 1963/07/00 - (145)

Keremeos:

Green-blue Eyes Reflecting. 1969/CI/00 - (198)

Kimberley:

Did Not Make Any Noise. 1975/09/00 - (231)

It Simply Moved. 1976/09/07 - (242)

RCMP Reports 8 Sightings In a Week. 1976/09/09 - (243)

Sasquatch Chased; Followed by Car. 1976/09/03 - (241)

Kincolith:

As if Swimming, But Not Moving. 2004/7/00 - (388)

Kingcome Inlet:

Unusual Nest. 1975/CI/00 - (237)

Kitimat:

    About 10 "Apes." 1962/06/08 - (139)

    Appeared To Be Waving. 1980/CI/00 - (275)

    Creature On the Shore. 1965/08/00 - (152)

    Probable Sasquatch Killing. 1905/YR/00 - (60)

Klemtu:

    Ape-like Face. 1970/04/23 - (201)

    Fifteen-inch Prints. 1969/06/18 - (190)

    Parted the Salal with Its Hands. 1970/02/00 - (199)

    Playing Around as They Walked. 1939/DE/00 - (96)

Lake Louise, Alberta:

    Shoulder Length Very Blond Hair. 1971/07/00 - (205)

Langley:

    Four-and-a-half Feet Tall. 1960/02/00 - (132)

    I Think Of It As a Gift. 2009/04/04 - (412)

Lavington:

    An Odd Visitor. 1928/YR/00 - (79)

Lillooet:

    Eight-foot Skeleton. 1937/YR/00 - (95)

    In & Out the Bush. 1968/11/16 - (181)

    Legs & Feet Seen in Trees. 1976/CI/00 - (247)

    Looked Like a Monkey Butt. 2002/12/00 - (374)

    Stone Foot. AD 500 or earlier - (23)

    The Fossilized Man Hoax. 1900/CI/00 - (48)

    Tired of Sasquatch Politics. 1957/05/08 - (119)

Lincoln, Nebraska, USA:

    The Dickie Article. 1934/07/29 - (82)

Little Fort:

    High Pitched Squeal. 1979/04/28 - (262)

    It Bent Straight From the Hips. 2002/10/00 - (373)

    Sasquatch Shot & Drops. 1979/04/30 - (262)

Little Qualicum (VI):

    First Urban Area Sighting. 1905/06/18 - (53)

Lone Butte:

    Trail Led to a Swampy Lake. 1986/04/00 - (288)

Lumby:

    "Plinkers" Shoot at Sasquatch. 1973/07/00 - (213)

    Sasquatch Fever. 1968/07/21 - (179)

Lytton:

    Wizened Old Man. 1969/11/20 - (193)

Mackenzie:

    Took a Shot Over Its Head. 1974/10/00 - (223)

Mainland Bay:
      Sasquatch Scares Off Fisherman. 1959/DE/00 - (131)
Marble Canyon:
      Pronounced Long Neck. 1975/09/00 - (230)
Masset (QCI):
      Hairs More Than 7 Inches Long. 1971/01/00 - (204)

McBride:
      Freezer Opened & Strange Footprints. 1996/01/21 - (327)
      Lean-to Slept In By Something. 1951/11/00 - (108)
      Over a Dozen Witnesses. 1976/06/00 - (239)
Merritt:
      Army Cadet Sighting. 1969/06/01 - (188)
      Possible Print—Cell Phone Photo. 2009/04/00 - (414)
      Standing by a Pond. 2004/05/10 - (383)
Mesachie Lake (VI):
      Gait Not the Same As a Human. 2006/06/20 - (398)
Metchosin (VI):
      The Metchosin Monster—An Expensive Hoax. 1895/YR/00 - (44)
Midway:
      Big Black Figure. 1978/YR/00 - (245)
Minstrel Island:
      Monstrous Thing. 1963/YR/00 - (146)
Mission:
      Chattering Like a Bear. 2001/09/00 - (365)
      Cows Stared At It. 1965/05/31 - (148)
      Hair Like a Horse's Tail. 2008/11/00 - (411)
      Misguided Misfits Pull Off Silly Hoax. 1977/05/15 - (249)
      No Plan for a Hunt. 1968/06/01 - (177)
      Steenburg Moves to BC. 2002/09/00 - (372)
Moricetown:
      Prints Were Awe Inspiring. 1961/YR/00 - (137)
Moscow, USSR:
      Dahinden Visits Moscow News. 1972/01/00 - (208)
Mt. St. Helens, Washington, USA:
      Race of a Different Species. 1847/03/26 - (28)
Nahanni Region (NWT):
      Skeletons Missing Their Heads. 1905/YR/00 - (58)
Nakusp:
      Footprints In the Middle of Nowhere. 1993/01/00 - (310)
      Incident at Hot Springs. 1993/10/00 - (314)
Nanaimo (VI):
      Up A Tree. 1983/08/00 - (285)

Nass River Area:
    The Monkey Mask. 1850/YR/00 - (31)
Natal:
    Big Monkey. 1976/04/02 - (238)
    Seen Near Bridge. 1975/YR/00 - (237)
Nazko:
    Along the River. 1969/DE/00 - (196)
    Watched It Go By. 1970/05/00 - (201)
Nelson:
    A Great Beast. 1960/08/00 - (133)
    Half As High As the Trees. 1992/10/00 - (307)
    Sasquatch Dumps Logs Off Truck. 1969/DE/00 - (197)
    Silhouetted In the Moonlight. 1961/10/00 - (136)
    Whole Race of Them. 1979/YR/00 - (266)
New Denver:
    Tracks Impressed 2 Inches Deep. 1971/SU/00 - (206)
New Hazelton:
    Crossed the Road. 1958/08/00 - (125)
    Ladies Report Sighting. 1958/08/00 - (124)
    Like a Bear. 1949/10/00 - (105)
New Masset (QCI):
    Long Dark Hair. 1976/11/00 - (244)
    Same Individual? 1976/12/18 - (244)
Nootka Sound (VI):
    First Nations People Fear Unusual Creatures. 1792/YR/00 - (25)
North America:
    Footprints In Stone. 0000 - (20)
    Petroglyphs & Pictographs. AD 500 or earlier - (22)
North Bend:
    Playful Sasquatch. 1939/YR/00 - (96)
North Vancouver:
    White As a Sheet. 1965/CI/00 - (153)
Ocean Falls:
    Rooting for Vegetation. 1973/07/10 - (212)
Okanagan Lake:
    Standing On the Patio. 1974/07/00 - (220)
Osoyos:
    Faster Than a Man Could Run. 1945/CI/00 - (103)
    Looked Just Like a Monkey. 1937/YR/00 - (95)
Ottawa, Ontario:
    Canadian "Sasquatch" Postage Stamp Issued. 1990/10/00 - (304)
    Canadian "Sasquatch" Coin Issued. 2011/05/18 - (423)

Pacific Coast:
>   Totem Poles and Masks. 1700 AD and earlier - (24)

Parksville (VI):
>   Would Have Fired if Sure Not a Man. 1983/AU/00 - (285)

Peachland:
>   Cook Sees Sasquatch. 1959/DE/00 - (132)

Penticton:
>   Strange Shadow. 1966/08/00 - (158)

Pocatello, Idaho, USA:
>   Footprints Recognized by Science. 2007/YR/00 - (402)
>   ISU Scientist Joins the Search. 1993/YR/00 - (315)
>   Sasquatch Exhibit at Museum of Natural History. 2006/07/16 - (399)

Port Alberni (VI):
>   Camper Lifted. 1992/CI/00.- (309)
>   Eyes Gleaming Orange. 1979/07/00 - (266)
>   Some Sort of Wild Man—First Multiple Witness Sighting. 1904/AU/00 - (50)
>   Thought It Was a Human With Down's Syndrome. 2002/08/00 - (371)
>   Two See Sasquatch. 2005/YR/00 - (397)

Port Angeles, Washington, USA:
>   Renowned Scientist Dr. Grover Krantz Passes On. 2002/02/14 - (368)

Port McNeill (VI):
>   All On Two Legs. 1992/YR/00 - (308

Port Mellon:
>   Large Hand. 2010/YR/00 - (417)

Port Renfrew (VI):
>   Small Stones Thrown. 1996/SP/00 - (327)

Port Townsend, Washington, USA:
>   BC Could Be "Last Stand" For Sasquatch. 1992/06/00 - (306)

Powell River:
>   Cabin Owner Shaken Up. 1969/03/00 - (186)
>   Pushing the Trees & Branches. 1974/07/00 - (220)

Price Island:
>   Sasquatch Roofed Nest. 1963/SU/00 - (145)
>   Size Of a Small Man. 1960/02/00 - (132)

Prince George:
>   Absolutely Speechless. 1968/11/20 - (182)
>   Legs Obscured By Long Hair. 1967/07/00 - (160)
>   Primate-like Face. 2005/08/14 - (394)
>   Primitive Cave or Tree-top Dweller. 1922/05/00 - (70)
>   Slightly Stooped Posture. 1977/AU/00 - (253)

Prince Rupert:
>   Walked With Bent Knees. 1995/AU/00 - (325)

Princeton:

    Came Within 12 Feet. 1975/10/00 - (231)

    King Kong or Something. 1974/10/00 - (224)

    Small Round Head. 1970/11/00 - (203)

    Walked Slightly Stooped 1971/09/00 - (207)

Pullman, Washington, USA:

    First Scientific Book On Sasquatch. 1992/YR/00 - (309)

    Jacko Disappearance Theory. 1994/YR/00 - (319)

Quatna River:

    Ape Tracks. 1967/02/00 - (159)

Queen Charlotte City (QCI)

    Deer Struck Down By Sasquatch. 1985/YR/00 - (288)

Quesnel:

    Face Like an Ape. 1978/SU/00 - (260)

    Small, Black Eyes. 1962/08/00 - (141)

Radium Hot Springs:

    Long Blond Hair. 1991/05/00 - (305

    Running Like a Man. 2001/03/30 - (361)

Remote Island

    Clam Diggers Attacked With Rocks & Driftwood. 1992/CI/00 - (309)

    Heard Chest-beating. 1969/DE/00 - (196)

Remote mainland inlet:

    Loud Howling Concerns Workers. 1994/05/00 - (316)

Revelstoke:

    Appeared to Look Both Ways. 1985/07/00 - (287)

Richmond:

    Big Wooly Animal. 1966/07/14 - (155)

    Dahinden Announces Pitt Lake Expedition. 1970/08/05 - (202)

    Dahinden Declares Trip to Europe Great Success. 1972/03/25 - (209)

    Dahinden Disputes Blue Mountain Footprints. 1983/08/04 - (284)

    Dahinden Goes to Europe. 1971/11/12 - (207)

    Farewell René—Fondly Remembered. 2001/04/18 - (361)

    Fence Post Torn Out. 1966/07/09 - (155)

    Murphy Teams Up With Dahinden. 1993/04//00 - (312)

    Three Sightings in 24 Hours. 1966/07/21 & 22 - (156)

Rock Bay (VI):

    Eye-Watering Stench. 1978/09/00 - (261)

Rocky Mountains:

    Moved Like a Monkey. 1959/DE/00 - (130)

    Thump on Roof of Car. 1978/08/00 - (260)

Roderick Island:

    Blood Found in Snow. 1960/02/00 - (132)

    Tracks in Frost. 1950/12/00 - (107)

Rossland:
>Observed Movement for 200 Yards. 1969/09/20 - (191)
>Poor Old Trapper. 1980/12/23 - (271)

Saanichton (VI):
>Saw It & Ran. 1982/01/30 - (277)

Salmon Arm:
>Standing In the Shadows. 1979/06/18 - (265)

Salt Spring Island:
>Bateman Paints a Sasquatch. 2000/WI/00 - (342)

Sayward (VI):
>Dragging a Deer. 1999/DE/00 - (341)

Sechelt:
>Appeared to Have a Goatee. 1973/06/00 - (211)
>Color of a Collie Dog. 1971/AU/00 - (207)
>Creature Standing In Doorway. 1956/08/00 - (115)
>Ran Away Pretty Fast. 1968/05/17 - (175)
>Rocks Thrown at US Visitor. 1956/08/00 - (115)
>Students Attacked With Barrage of Rocks & Debris. 1989/DE/00 - (303)
>Swinging Arms & Hunched Forward. 2005/03/08 - (391)

Shawingan Lake (VI):
>Five Young Girls See Sasquatch. 1974/05/00 - (217)

Sicamous:
>Snow Angels. 1974/CI/00 - (226)

Skidgate (QCI):
>Small Beady Eyes. 1970/02/00 - (200)

Smithers:
>Leaning Against a Tree. 2008/08/12 - (408)
>Trees Stripped. 1950/CI/00 - (107)

Sooke (VI):
>Long Gray Hairs On Its Chest. 1979/DE/00 - (268)
>Long Line of Tracks. 1975/11/11 - (232)

Spillimacheen:
>Strange Behavior. 1966/05/00 - (154)

Spuzzum:
>One Shape All the Way Down. 1997/03/25 - (328)
>Sasquatch Killed & Buried. 1884/CI/00 - (41)
>Witness White as a Ghost. 1977/03/28 - (247)

Squamish:
>Crouched By the Water. 1981/11/00 - (277)
>It Was Carrying a Fish. 1970/01/07 - (198)
>Running In the Bushes. 2005/02/13 - (390)
>Something Dragged To an Ice Hole. 1965/06/28 - (148)

Three-toed Tracks. 1969/06/00 - (190)

Tracks on a Sandbar. 1971/05/00 - (205)

Very Large Tracks. 1964/WI/00 - (146)

St. Alice Springs (now Harrison Hot Springs):

Do Hot Springs Attract Sasquatch? 1886/YR/00 - (43)

Stewart:

Ran Up Steep Hill On "Hind Legs." 1968/08/00 - (180)

Ten Feet Tall. 1972/CI/00 - (211)

Strathcona Provincial Park (VI):

Black Human-like Form. 1975/08/00 - (229)

Wildlife Biologist Finds Sasquatch Tracks. 1988/10/00 - (300)

Stump Lake:

Martin the Wild Man. 1875/12/00 - (35)

Sunshine Valley:

Four or Five Steps to Cross Road. 2004/08/15 - (388)

Swindle Island:

Long String of Large Footprints. 1961/10/00 - (137)

Three Sets of Tracks. 1962/AU/00 - (142)

Tracks Entered Lake Kitasu. 1951/YR/00 - (108)

Typical Ape Tracks. 1967/CI/00 - (173)

Wading Sasquatch. 1930/YR/00 - (80)

Where the Apes Stay. 1945/04/00 - (103)

Telegraph Cove (VI):

Stones Scare Off Boys. 1975/CI/00 - (237)

Terrace

Eating Pine Needle Tips. 1974/04/00 - (217)

Major Track Finding. 1976/07/17 - (239)

Odor Like a Camel. 1974/YR/00 - (226)

Seen in Garbage. 1974/01/00 - (216)

Watched the Man For a While. 1952/YR/00 - (108)

Tete Jaune Cache:

Definitely Not a Bear. 1955/10/00 - (110)

Thought It Was a Moose at First. 2002/08/17 - (371)

Toba Inlet:

The Ostman Abduction. 1924/SU/00 - (71)

Tofino (VI):

As If Walking On Air. 1995/08/00 - (325)

Insistence of Tree Rapping. 1994/AU/00 - (318)

Real Big Orange Eyes - (373)

Stood Its Ground. 2004/11/30 - (389)

Trail:

Track in Dry Mud. 1970/06/00 - (202)

Turnour Island:
House Moved. 1964/04/00 - (147)
Striking on House Posts. 1969/DE/00 - (195)
Ucluelet (VI):
Laying On Log Over River. 2007/03/30 - (401)
Vancouver Island (No specific location):
Caver Sees Strange Creature. 1969/CI/00 - (197)
Vancouver:
A Crippled Giant. 1967/10/31 - (172)
"Anthropology of the Unknown" Conference. 1978/05/09 - (257)
Ape-boy Displayed? 1884/00/00 - (40)
Canadians "Cold Shoulder" Ice Man. 1969/04/15 - (187)
Dahinden & the Kokanee Commercials. 1997/YR/00 - (336)
Dr. Halpin Provides Her Thoughts. 1981/06/14 - (276)
Dr. Koffmann Reports on Russian Snowman. 1978/05/00 - (243)
First Nations Elders Provide Sasquatch Stories. 2003/12/18 - (379)
First Official Sasquatch Hunt. 1934/04/09 - (81)
Hancock Weighs-in On Sasquatch Issue. 1968/06/25 - (177)
Hunter with Dahinden Book Released. 1973/YR/00 - (216)
Ice Man's Canadian Tour Canceled. 1969/04/12 - (186)
Loggers & Sasquatch. 1899/DE/00 - (45)
Mining Promoter Considers Sasquatch Expedition. 1967/10/31 - (172)
Oilman Needed. 1958/YR/00 - (126)
"Quatchi" Made Olympics Mascot. 2008/02/00 - (404)
Sasquatch Exhibit at Vancouver Museum. 2004/06/17 - (385)
Sasquatch Film Rights Bought by Canadian Researchers. 1968/01/12 - (173)
Sasquatch Heads Created. 2003/11/00 - (377)
Sasquatch Poll. 1978/05/05 - (257)
Scientific Proof Presented. 1982/10/00 - (280)
Screening Arranged for California Film. 1967/10/23 - (164)
Skin-clad Sasquatch? 1947/YR/00 - (104)
Stood About 10 Feet Tall. 1973/11/26 - (215)
The Killing Question. 1967/10/28 - (171)
The Webster Interview. 1967/10/26 - (165)
Titmus Says He Will Never Give Up. 1977/06/17 - (250)
UBC Publishes Book on Humanoid Monsters. 1980/YR/00 - (272)
UBC Scientists First to See Film of Alleged Sasquatch. 1967/10/26 - (165)
UBC to Look at Sasquatch Evidence. 1977/YR/00 - (255)
Vietnam Vet Hunts Sasquatch. 1984/04/12 - (286)
Wild Man—Beard Like Rip Van Winkle. 1906/09/00 - (61)
Woman Followed by Sasquatch. 1999/DE/00 - (341)
Vanderhoof:
Thought It Was a Gorilla. 1949/CI/00 - (106)

Vavenby:

Odd Figure at Water's Edge. 1980/05/00 - (269)

Vernon:

Dahinden Returns After Month-long Search. 1957/08/25 - (120)

Rifle Scope Crosshairs on Eyes. 1994/09/11 - (318)

Victoria (VI):

Government Gets Involved. 1967/09/19 - (162)

Government Impressed. 1967/09/30 - (164)

Government MLA Introduces Sasquatch Protection Bill. 1977/08/12 - (252)

Keeping an Open Mind. 1967/09/28 - (163)

Museum Anthropologist Don Abbot Passes On. 2005/07/28 - (393)

Not a Subject for Mirth. 1967/09/01 - (160)

Provincial Archives Searched. 1957/04/21 - (117)

Sasquatch & Politics. 1977/05/19 - (250)

Sasquatch Protection Put Aside. 1980/03/00 - (253)

Shoulders 3 to 4 Feet Across. 1976/01/22 - (237)

Wells:

Light Gray Animal. 1962/07/23 - (140)

Tracks Near Cave. 1971/06/00 - (205)

West Kootenays:

Defies Conventional Description. 1998/09/00 - (338)

Splashing the Water. 1977/01/00 - (247)

Whistler:

Followed Her Husband. 1969/DE/00 - (195)

White Rock:

It Fooled Around. 1966/07/00 - (157)

Windermere:

A "Bridge" Going Across the Foot. 2002/02/11 - (367)

Tall Human Skeletons. 1929/DE/00 - (79)

Woss (VI):

Thought It Was His Son. 1988/09/08 - (299)

Wycliffe:

Polaroid Photo Taken. 1976/09/03 - (242)

Taller Than the Horses. 1976/05/00 - (238)

Too Tall To Be a Man. 1988/08/24 - (299)

Yakima, Washington, USA:

Roger Patterson Passes On. 1972/01/15 - (208)

Yale:

A Strange Cave. 1898/CI/00 - (45)

Huge Nude Hairy Man. 1919/YR/00 - (70)

Jacko the Ape-boy. 1884/06/30 - (36)

Residents Interviewed. 1958/10/29 - (125)

Entries in Title Order

Bending Down in the Water.1944/08/00 - Bella Coola. (101)

Beside Garbage Dump. 1976/12/00 - Bella Coola. (245)

Bicycle Chase. 1948/YR/00 - Harrison Mills (Chehalis Reservation). (105)

Big Black Figure. 1978/YR/00 - Midway. (261)

Big Monkey. 1976/04/02 - Natal. (238)

Big Wooly Animal. 1966/07/14 - Richmond. (155)

Black Human-like Form. 1975/08/00 - Strathcona Provincial Park (VI). (229)

Blood Found in Snow. 1960/02/00 - Roderick Island. (132)

Bob Titmus Passes On. 1997/07/01 - Chilliwack. (332)

Cabin Owner Shaken Up. 1969/03/00 - Powell River. (186)

Came Out Of a Cave. 1972/07/00 - Hope. (210)

Came Within 12 Feet. 1975/10/00 - Princeton. (231)

Camper Lifted. 1992/CI/00 - Port Alberni (VI). (309)

Canadian "Sasquatch" Coin Issued. 2011/05/18 - Ottawa, Ontario. (423)

Canadian "Sasquatch" Postage Stamp Issued. 1990/10/00 - Ottawa, Ontario. (304)

Canadians "Cold Shoulder" Ice Man. 1969/04/15 - Vancouver. (187)

Canadians Meet with Morgan. 1974/09/00 - Cougar, Washington, USA. (222)

Car Struck 2 or 3 Times by Sasquatch. 1985/07/30 - Burns Lake. (287)

Carried a Lantern. 1905/12/00 - Comox (VI). (56)

Caver Sees Strange Creature. 1969/CI/00 - Vancouver Island. (197)

Chattering Like a Bear. 2001/09/00 - Mission. (365)

Chehalis Band Adopts Sasquatch Logo. 1980/YR/00 - Harrison Mills. (Chehalis Reservation). (271)

Clam Diggers Attacked With Rocks & Driftwood. 1992/CI/00 - Remote Island. (309)

Color of a Collie Dog. 1971/AU/00 - Sechelt. (207)

Cook Sees Sasquatch. 1959/DE/00 - Peachland. (132)

Cows Stared at It. 1965/05/31 - Mission. (148)

Creature On the Shore. 1965/08/00 - Kitimat. (152)

Creature Standing In Doorway. 1956/08/00 - Sechelt. (115)

Creatures Doing Sort Of Sun Dance. 1905/10/00 - Comox (VI). (54)

Creatures Were a Man & a Woman. 1911/YR/00 - Harrison Mills (Chehalis Reservation). (65)

Crossed Close to Car. 1981/YR/00 - Kaslo. (277)

Crossed Road In 2 Paces. 1995/11/00 - Ainsworth (325)

Crossed the Road. 1958/08/00 - New Hazelton. (125)

Crossed 10 Feet in Front of Witness. 1975/06/20 - Cherryville. (228)

Crouched & Looked Back Down. 1974/YR/00 - 100 Mile House. (225)

Crouched By the Water. 1981/11/00 - Squamish. (277)

Cycling Encounter. 1993/08/01 - Ainsworth. (313)

D'sonoqua Seen.1939/DE/00 - Alert Bay. (98)

Dahinden & the Kokanee Commercials. 1997/YR/00 - Vancouver. (336)

Dahinden Announces Pitt Lake Expedition. 1970/08/05 - Richmond. (202)

Dahinden Declares Trip to Europe was Great Success. 1972/03/25 - Richmond. (209)

Dahinden Disputes Blue Mountain Footprints. 1983/08/04 - Richmond. (284)

Footprint Mystery. 1994/08/00 - Harrison Hot Springs. (316)
Footprints & Yelling. 1967/WI/00 - Bella Coola. (158)
Footprints Found & Later a Sighting. 1995/08/03 - Harrison Mills (Chehalis Reservation). (321)
Footprints In Stone. 0000 - North America. (20)
Footprints In the Middle of Nowhere. 1993/01/00 - Nakusp. (310)
Footprints Recognized by Science. 2007/YR/00 - Pocatello, Idaho, USA. (402)
Forest Service Worker Tells of Sasquatch Sightings. 2008/SU/00 - Enderby. (407)
Four-and-a-half Feet Tall. 1960/02/00 - Langley. (132)
Four or Five Steps to Cross Road. 2004/08/15 - Sunshine Valley. (388)
Four Paces to Cross Highway. 1994/11/00 - Balfour. (319)
Four Sasquatch Seen Together. 1962/04/00 - Bella Coola. (138)
Four Sets of Footprints. 1962/12/00 - Bella Coola. (144)
Four White Dots. 1964/07/00 - Chilliwack. (147)
Four-toed Prints. 1977/08/08 - Chetwynd. (251)
Freezer Opened & Strange Footprints. 1996/01/21 - McBride. (327)
Gait Not the Same as a Human. 2006/06/20 - Mesachie Lake (VI). (398)
Game Guides Sighting. 1968/09/14 - Chetwynd. (181)
Gaze Returned & Creature Runs Away 1958/12/08 - Bella Coola. (125)
George Haas Passes. 1978/02/16 - Berkley, California, USA. (256)
Getting Serious. 1967/09/19 - Agassiz. (163)
Gigantopithecus blacki. 300,000 BC - Asia. (21)
Glaring Eyes. 2005/06/10 - Hope. (392)
Government Gets Involved. 1967/09/19 - Victoria (VI). (162)
Government Impressed. 1967/09/30 - Victoria (VI). (164)
Government MLA Introduces Sasquatch Protection Bill. 1977/08/12 - Victoria (VI). (252)
Green & Dahinden at Odds. 1970/08/05 - Harrison Hot Springs. (202)
Green Convinced With US Sighting. 1969/07/29 - Agassiz. (191)
Green Goes Forward With Book & Film. 1969/11/11 - Harrison Hot Springs. (192)
Green Returns from Cross-Canada Trip. 1969/12/23 - Harrison Hot Springs. (193)
Green Takes Up the Search. 1957/YR/00 - Agassiz. (121)
Green-blue Eyes Reflecting. 1969/CI/00 - Keremeos. (198)
Hair Covered Man. 1962/08/00 - Hixon. (141)
Hair Like a Horse's Tail. 2008/11/00 - Mission. (411)
Hair Looked Stiff. 1986/05/21 - Harrison Hot Springs. (289)
Hair Over Feet Like Bell-Bottom Pants. 1999/06/00 - Chilliwack. (339)
Hair Streamed Out. 1945/CI/00 - Coombs (VI). (103)
Hairs More Than 7 Inches Long. 1971/01/00 - Masset (QCI). (204)
Half as High as the Trees. 1992/10/00 - Nelson. (307)
Halloween Hoax. 2006/10/29 - Dewdney. (399)
Hancock Weighs-in on Sasquatch Issue. 1968/06/25 - Vancouver. (177)
Hands on a Tree Stump. 2008/06/18 - Bridal Falls. (405)
He Was In a Rage. 1909/05/00 - Harrison Mills (Chehalis Reservation). (63)

Heard Chest-beating. 1969/DE/00 - Remote Island. (196)

Henry Napoleon Story. 1928/YR/00 - Duncan (VI). (78)

High Pitched Squeal. 1979/04/28 - Little Fort. (262)

Horse Balked at Going Farther. 1949/07/00 - Harrison Hot Springs. (105)

House Moved. 1964/04/00 - Turnour Island. (147)

Huge Barefoot Tracks. 1949/DE/00 - Agassiz. (106)

Huge Man-type Tracks. 1976/11/29 - Bella Coola. (243)

Huge Nude Hairy Man. 1919/YR/00 - Yale. (70)

Human Face on a Fur-clad Body. 1934/08/00 - Haney. (92)

Hunter with Dahinden Book Released. 1973/YR/00 - Vancouver. (216)

I Don't Think a Man Could Leap Like That. 1982/06/10 - Fernie. (278)

I Saw a Sasquatch That Night. 1973/08/00 - Agassiz. (214)

I Think of It As a Gift. 2009/04/04 - Langley. (412)

Ice Man's Canadian Tour Canceled. 1969/04/12 - Vancouver. (186)

In & Out the Bush. 1968/11/16 - Lillooet. (181)

In the Middle of Nowhere. 1992/AU/00 - Fort Nelson. (307)

Incident at Hot Springs. 1993/10/00 - Nakusp. (314)

Inside the Town Limits. 1974/SU/00 - Hope. (221)

Insistence of Tree Rapping. 1994/AU/00 - Tofino (VI). (318)

ISU Scientist Joins the Search. 1993/YR/00 - Pocatello, Idaho, USA. (315)

It Bent Straight from the Hips. 2002/10/00 - Little Fort. (373)

It Fooled Around. 1966/07/00 - White Rock. (157)

It Simply Moved. 1976/09/07 - Kimberley. (242)

It Stayed On 2 Legs the Whole Time. 1987/08/00 - Golden. (297)

It Stinks. 1976/12/00 - Bella Coola. (245)

It Was a "Nahganee." 1987/04/13 - Fort Nelson. (297)

It Was Carrying a Fish. 1970/01/07 - Squamish. (198)

It Was Covered With Hair Like an Animal. 1927/09/00 - Agassiz. (76)

It Was Just Curious. 1983/07/00 - Hazelton. (282)

Jacko Disappearance Theory. 1994/YR/00 - Pullman, Washington, USA. (159)

Jacko the Ape-boy. 1884/06/30 - Yale. (36)

Keeping an Open Mind. 1967/09/28 - Victoria (VI). (163)

Kids See Sasquatch. 1976/06/11 - Grand Forks. (239)

King Kong or Something. 1974/10/00 - Princeton. (224)

Ladies Report Sighting. 1958/08/00 - New Hazelton. (124)

Ladies Stared In Disbelief. 2003/CI/00 - Bowen Island. (382)

Large Bone with Foot. 1957/YR/00 - East Sooke (VI). (121)

Large Footprints Very Convincing. 1978/12/00 - Kaslo. (261)

Large Hand. 2010/YR/00 - Port Mellon. (417)

Large Rock Lands Near Fishermen. 2007/02/13 - Harrison Mills (Chehalis Res.). (400)

Laying On Log Over River. 2007/03/30 - Ucluelet (VI). (401)

Leaning Against a Tree. 2008/08/12 - Smithers. (408)

Lean-to Slept In By Something. 1951/11/00 - McBride. (108)

Legs & Feet Seen In Trees. 1976/CI/00 - Lillooet. (247)

Legs Obscured by Long Hair. 1967/07/00 - Prince George. (160)

Life-like Mask. 1939/DE/00 - Harrison Mills (Chehalis Reservation). (97)

Lifted Up Its Arms. 1970/03/00 - Juskatla. (200)

Light Colored Chest. 1976/12/21 - Bella Coola. (244)

Light Gray Animal. 1962/07/23 - Wells. (140)

Like a Bear. 1949/10/00 - New Hazelton. (105)

Like a Man In Torn Rain Gear. 1980/12/13 - Cumberland (VI). (270)

Like a Roll of Fur at Its Wrists. 1999/09/14 - Hope. (340)

Like the Back of a Moose. 1977/CI/00 - Hope. (255)

Loggers & Sasquatch. 1899/DE/00 - Vancouver. (45)

Loggers Find Tracks. 1968/WI/00 - Gosnell. (174)

Long Blond Hair. 1991/05/00 - Radium Hot Springs. (305)

Long Dark Hair. 1976/11/00 - New Masset (QCI). (244)

Long Gray Hairs On Its Chest. 1979/DE/00 - Sooke (VI). (268)

Long Line of Tracks. 1975/11/11 - Sooke (VI). (232)

Long String of Large Footprints. 1961/10/00 - Swindle Island. (137)

Long White Hair. 1971/07/00 - Houston. (206)

Looked Just Like a Monkey. 1937/YR/00 - Osoyoos. (95)

Looked Like a Monkey Butt. 2002/12/00 - Lillooet. (374)

Looked Like an Ape. 2004/05/00 - Hope. (384)

Looked More Like a Human Being. 1915/YR/00 - Hope. (68)

Loud Howling Concerns Workers. 1994/05/00 - Remote mainland inlet. (316)

Lumbered Across the Road. 1965/07/00 - Hope. (151)

Major Track Finding. 1976/07/17 - Terrace. (239)

Makes You Gag... 1994/09/00 - Alert Bay. (318)

Man Attacked By a Sasquatch. 1943/CI/00 - Hope. (101)

Man or Gorilla. 1970/04/02 - Harrison Hot Springs. (200)

Many Tracks Found. 1962/SU/00 - Aristazabal Island. (140)

Martin the Wild Man. 1875/12/00 - Stump Lake. (35)

Massive Bone. 1953/CI/00 - Esquimalt, (VI). (109)

Massive Legs. 1959/AU/00 - Clinton. (129)

Meandering Tracks. 2000/09/00 - Harrison Hot Springs. (357)

Miles of Tracks. 1971/WI/00 - Hope. (205)

Million Dollar Reward. 1962/CI/00 - Bella Coola. (144)

Mining Promoter Considers Sasquatch Expedition. 1967/10/31 - Vancouver. (172)

Misguided Misfits Pull Off Silly Hoax. 1977/05/15 - Mission. (249)

Monstrous Thing. 1963/YR/00 - Minstrel Island. (146)

Mother & Child. 1962/04/00 - Bella Coola. (138)

Moved Like a Monkey. 1959/DE/00 - Rocky Mountains. (130)

Murphy Teams Up with Dahinden. 1993/04//00 - Richmond. (312)

Museum Anthropologist Don Abbott Passes On. 2005/07/28 - Victoria (VI). (393)

Mystery For the Rest of My Life. 1992/10/00 - Castlegar. (307)

NASI Folds. 1999/01/00 - Hood River, Oregon, USA. (339)

NASI Report Released. 1998/06/00 - Hood River, Oregon, USA. (337)

Near the Toll Gate. 2001/05/00 - Hope. (365)

Night Encounter. 1934/03/00 - Harrison Mills (Chehalis Reservation). (81)

Night Sounds Frighten Campers. 1992/SU/00 - Comox (VI). (307)

No Plan for a Hunt. 1968/06/01 - Mission. (177)

North America's Great Ape. 1998/YR/00 - Courtenay (VI). (339)

Nose Just a Flat Area With 2 Holes. 1959/09/00 - Enderby. (128)

Not a Subject for Mirth. 1967/09/01 - Victoria (VI). (160)

Not At All Bear-like. 1974/YR/00 - Castlegar. (226)

Not Bounding Like a Bear. 1947/YR/00 - Bella Coola. (104)

Observed Movement for 200 Yards. 1969/09/20 - Rossland. (191)

Odd Figure at Water's Edge. 1980/05/00 - Vavenby. (269)

Odor Did Not Linger. 2007/09/25 - Canal Flats. (401)

Odor Like a Camel. 1974/YR/00 - Terrace. (226)

Oil Rig Crew Sees Sasquatch. 1987/03/14 - Dawson Creek. (295)

Oilman Needed. 1958/YR/00 - Vancouver. (126)

On All Fours, But Odd. 1994/05/31 - Castlegar. (316)

One Shape All the Way Down. 1997/03/25 - Spuzzum. (328)

One Was a Step Ahead of the Other. 1974/01/00 - Jasper, Alberta. (217)

Over a Dozen Witnesses. 1976/06/00 - McBride. (239)

Pace of 42 Inches. 1962/08/00 - Devastation Channel. (140)

Parted the Salal With Its Hands. 1970/02/00 - Klemtu. (199)

Patty's Portrait. 1996/YR/00 - Burnaby. (327)

Petroglyphs & Pictographs. 500 AD or earlier - North America (22)

Photos Snapped; Indistinct. 1984/08/20 - Agassiz. (286)

Plans for Search Dampened. 1957/04/24 - Harrison Hot Springs. (118)

Playful Sasquatch. 1939/YR/00 - North Bend. (96)

Playing Around as They Walked. 1939/DE/00 - Klemtu. (96)

"Plinkers" Shoot at Sasquatch. 1973/07/00 - Lumby. (213)

Polaroid Photo Taken. 1976/09/03 - Wycliffe. (242)

Poor Old Trapper. 1980/12/23 - Rossland. (271)

Possible Print—Cell Phone Photo. 2009/04/00 - Merritt. (414)

Possible Sasquatch Corpse. 1977/AU/00 - Harrison Mills. (253)

Primate-like Face. 2005/08/14 - Prince George. (394)

Primitive Cave or Tree-top Dweller. 1922/05/00 - Prince George. (70)

Prints in Strawberry Field. 2003/06/00 - Duncan (VI). (377)

Prints Were Awe Inspiring. 1961/YR/00 - Moricetown. (137)

Probable Sasquatch Killing. 1905/YR/00 - Kitimat. (60)

Pronounced Long Neck. 1975/09/00 - Marble Canyon. (230)

Prospector Sees Wild Man. 1905/07/29 - Cowichan (VI). (54)
Provincial Archives Searched. 1957/04/21 - Victoria. (117)
Pushing the Trees & Branches. 1974/07/00 - Powell River. (220)
Put Its Face In the Creek. 1976/09/29 - Deroche. (243)
Puzzling Print. 1961/04/19 - Harrison Hot Springs. (136)
Quarter Mile of Tracks. 1934/YR/00 - One Hundred Mile House. (404)
"Quatchi" Made Olympics Mascot. 2008/02/00 - Vancouver. (373)
Queen Elizabeth II & Prince Philip Given Sasquatch Presentation. 1959/08/00 - Kamloops. (127)
Race of a Different Species. 1847/03/26 - Mt. St. Helens, Washington, USA. (28)
Ran As Fast As We Could. 1968/02/22 - Alert Bay. (174)
Ran Away Pretty Fast. 1968/05/17 - Sechelt. (175)
Ran Off Screaming. 1969/02/00 - Butedale. (185)
Ran On All Fours. 1975/10/17 - Agassiz. (232)
Ran Up Steep Hill On "Hind Legs." 1968/08/00 - Stewart. (180)
RCMP Reports 8 Sightings In a Week. 1976/09/09 - Kimberley. (243)
Real Big Orange Eyes. 2002/11/00. Tofino (VI). (373)
Realized It Wasn't a Bear. 2000/08/15 - Hope. (356)
Reflecting Yellow Eyes. 1975/09/00 - Colwood (VI). (230)
Renowned Scientist Dr. Grover Krantz Passes On. 2002/02/14 - Port Angeles, Washington, USA. (368)
Residents Interviewed. 1958/10/29 - Yale. (125)
Rifle Scope Crosshairs on Eyes. 1994/09/11 - Vernon. (318)
Ripped Off the Screen Door. 1986/10/00 - Bella Coola. (294)
Rock Bombardment. 1934/03/00 - Harrison Mills (Chehalis Reservation). (80)
Rocks From Above. 1934/03/23 - Harrison Mills (Chehalis Reservation). (80)
Rocks Thrown at US Visitor. 1956/08/00 - Sechelt. (115)
Rocks Thrown. 2004/07/22 - Canal Flats. (387)
Roger Patterson Passes On. 1972/01/15 - Yakima, Washington, USA. (208)
Rooting for Vegetation. 1973/07/10 - Ocean Falls. (212)
Running Like a Man. 2001/03/30 - Radium Hot Springs. (390)
Running In the Bushes. 2005/02/13 - Squamish. (302)
Running Like a Human. 1989/11/00 - Bella Coola. (361)
Same Type of Creature Seen. 1910/YR/00 - Harrison Mills (Chehalis Reservation). (244)
Same Individual? 1976/12/18 - New Masset (QCI). (65)
Sasquatch "Bower." 1977/CI/00 - Black Creek (VI). (255)
Sasquatch Again on "Welcome Sign. 1995/YR/00 - Harrison Hot Springs. (250)
Sasquatch & Politics. 1977/05/19 - Victoria (VI). (326)
Sasquatch Carving History. 1976/YR/00 - Harrison Hot Springs. (245)
Sasquatch Caves. 1959/DE/00 - Hope. (130)
Sasquatch Chased; Followed by Car. 1976/09/03 - Kimberley. (241)
Sasquatch Couple. 1956/03/30 - Flood (or Floods). (113)
Sasquatch Dumps Logs Off Truck. 1969/DE/00 - Nelson. (197)
Sasquatch Exhibit at Museum of Natural History. 2006/07/16 - Pocatello, Idaho, USA. (399)

Sasquatch Exhibit at Vancouver Museum. 2004/06/17 - Vancouver. (385)

Sasquatch Fever. 1968/07/21 - Lumby. (179)

Sasquatch Film Rights Bought by Canadian Researchers. 1968/01/12 - Vancouver. (173)

Sasquatch Follows Witness. 1976/07/00 - Bella Coola. (241)

Sasquatch Heads Created. 2003/11/00 - Vancouver. (377)

Sasquatch Hunt Planned. 1956/07/00 - Harrison Hot Springs. (114)

Sasquatch Incident Compilation. 1965/YR/00 - Agassiz. (153)

Sasquatch Killed & Buried. 1884/CI/00 - Spuzzum. (41)

Sasquatch Not Welcome. 1966/05/00 - Harrison Hot Springs. (154)

Sasquatch Pass Named. 1954/YR/00 - Chilco Lake. (109)

Sasquatch Poll. 1978/05/05 - Vancouver. (257)

Sasquatch Protection Put Aside. 1980/03/00 - Victoria (VI). (268)

Sasquatch Provincial Park Named. 1968/YR/00 - Harrison Hot Springs. (183)

Sasquatch Roofed Nest. 1963/SU/00 - Price Island. (145)

Sasquatch Scares Off Fisherman. 1959/DE/00 - Mainland Bay. (131)

Sasquatch Seen at Youth Camp. 1974/07/00 - Harrison Hot Springs. (218)

Sasquatch Shot & Drops. 1979/04/30 - Little Fort. (262)

Sasquatch Snatches Fish Catch. 1986/08/00 - Chilliwack. (293)

Sasquatch Walking Insights. 2009/04/00 - Gibsons. (414)

Saw It & Ran. 1982/01/30 - Saanichton (VI). (277)

Scientific Proof Presented. 1982/10/00 - Vancouver. (280)

Scientist Writes About the "Boqs." 1924/CI/00 - Bella Coola. (75

Scientists a Virtual No-show. 1980/03/00 - Dallas, Texas, USA. (269)

Screamed Like a Woman. 1967/02/00 - Hartley Bay. (159)

Screening Arranged for California Film. 1967/10/23 - Vancouver. (164)

Seemed To Be Checking Where They Camped. 1961/09/00 - Cumberland. (136)

Seemed To Be Looking At the Back Of the Train. 1982/06/16 - Golden. (279)

Seen in Garbage. 1974/01/00 - Terrace. (216)

Seen Looking at Them. 1969/02/00 - Alert Bay. (185)

Seen Near Bridge. 1975/YR/00 - Natal. (237)

Seen On a Rocky Cliff. 1974/08/25 - Chilliwack. (221)

Seen On Shore. 1963/YR/00 - Bella Bella. (146)

Seen Through Rifle Scope. 1965/CI/00 - Gold Bridge. (153)

Seven Foot Skeleton Uncovered. 1959/DE/00 - Armstrong. (129)

She Spoke the Douglas Tongue. 1914/YR/00 - Hatzic. (66)

Shoulder Length Very Blond Hair. 1971/07/00 - Lake Louise, Alberta. (205)

Shoulders 3 to 4 Feet Across. 1976/01/22 - Victoria (VI). (237)

Showed Intelligence. 1988/07/05 - Harrison Hot Springs. (298)

Silent Watcher. 1968/09/00 - Courtenay. (181)

Silhouetted in the Moonlight. 1961/10/00 - Nelson. (136)

Sitting by 2 Holes. 1969/10/25 - Bella Coola. (192)

Size of a Small Man. 1960/02/00 - Price Island. (132)

Skeletons Missing Their Heads. 1905/YR/00 - Nahanni Region (NWT). (58)

Skin-clad Sasquatch? 1947/YR/00 - Vancouver. (104)

Slightly Stooped Posture. 1977/AU/00 - Prince George. (253)

Small Beady Eyes. 1970/02/00 - Skidgate (QCI). (200)

Small Round Head. 1970/11/00 - Princeton. (203)

Small Stones Thrown. 1996/SP/00 - Port Renfrew. (VI). (327)

Small, Black Eyes. 1962/08/00 - Quesnel. (141)

Smiling Sasquatch. 1976/YR/00 - Chilliwack. (245)

Snow Angels. 1974/CI/00 - Sicamous. (226)

Snow Survey - Tracks Found. 1968/03/00 - Golden. (175)

Some Sort of Wild Man—First Multiple Witness Sighting. 1904/AU/00 - Port Alberni (VI). (50)

Something Big Shaking the Alders. 1980/07/21 - Cumberland (VI). (270)

Something Coming Out of the Woods. 1941/10/00 - Agassiz. (99)

Something Dragged to an Ice Hole. 1965/06/28 - Squamish. (148)

Something Glimpsed & Rock Thrown. 2008/09/21 - Agassiz. (409)

Something Monkey-like. 1907/03/00 - Bishop Cove. (62)

Something on Tape. 1997/05/00 - Chilliwack. (331)

Something Seen & Filmed. 1972/SU/00 - Houston. (210)

Something Walking in the Bush. 2011/02/08 - Hope. (418)

Somewhat Athletic Gait. 1993/08/00 - Fort St. James. (313)

Splashing the Water. 1977/01/00 - West Kootenays. (247)

Standing By a Pond. 2004/05/10 - Merritt. (383)

Standing In the Shadows. 1979/06/18 - Salmon Arm. (265)

Standing On the Patio. 1974/07/00 - Okanagan Lake. (220)

Standing on 2 Legs On a Rocky Slope. 1998/YR/00 - Kemano. (338)

Steenburg Moves to BC. 2002/09/00 - Mission. (372)

Stone Foot. AD 500 or earlier - Lillooet. (23)

Stone Heads. AD 500 or earlier - Columbia River Valley, USA. (23)

Stones Rain For About 2 Hours. 1960/08/00 - Chilliwack. (134)

Stones Scare Off Boys. 1975/CI/00 - Telegraph Cove (VI). (237)

Stood About 10 Feet Tall. 1973/11/26 - Vancouver. (215)

Stood Its Ground. 2004/11/30 - Tofino (VI). (389)

Straddled a Log & Paddled With Hands & Feet. 1935/10/00 - Harrison Hot Springs. (94)

Straight Gray Hair. 1994/CI/00 - Gold River (VI). (320)

Strange Behavior. 1966/05/00 - Spillimacheen. (154)

Strange Footprints. 1975/07/00 - Cranbrook. (228)

Strange Shadow. 1966/08/00 - Penticton. (158)

Strange Spur. 1989/09/20 - Fort St. James. (301)

Striking on House Posts. 1969/DE/00 - Turnour Island. (195)

Strong Odor Sensed. 1959/03/00 - Aristazabal Island. (127)

Students Attacked With Barrage of Rocks & Debris. 1989/DE/00 - Sechelt. (303)

Swinging Arms & Hunched Forward. 2005/03/08 - Sechelt. (391)

Talks with First Nations. 1995/07/00 - Bella Coola. (321)

Tall Figure & Roars. 1983/SU/00 - Gibsons. (266)

Tall Human Skeletons. 1929/DE/00 - Windermere. (79)

Taller Than the Horses. 1976/05/00 - Wycliffe. (238)

Teenage Girls Encounter Sasquatch. 1967/09/00 - Comox (VI). (163)

Ten Feet Tall. 1972/CI/00 - Stewart. (211)

Ten Feet Tall. 1979/07/00 - Hope. (265)

The Caulfield Incident. 1864/YR/00 - Fraser River Canyon. (33)

The Dickie Article. 1934/07/29 - Lincoln, Nebraska, USA. (82)

The Fossilized Man Hoax. 1900/CI/00 - Lillooet. (48)

The Killing Question. 1967/10/28 - Vancouver. (171)

The Metchosin Monster—An Expensive Hoax. 1895/YR/00 - Metchosin (VI). (44)

The Monkey Man. 1901/CI/00 - Campbell River (VI). (49)

The Monkey Mask. 1850/YR/00 - Nass River Area. (31)

The Ostman Abduction. 1924/SU/00 - Toba Inlet. (71)

The Sasquatch Summit. 2011/04/09 - Harrison Hot Springs. (420)

The Skookum Cast. 2000/09/00 – Gifford Pinchot National Forest, Washington, USA. (359)

The Webster Interview. 1967/10/26 - Vancouver. (165)

The Word "Sasquatch" is Developed. 1926/CI/00 - Harrison Mills (Chehalis Reservation). (76)

Thought It Was a Friend. 1953/09/00 - Courtenay. (109)

Thought It Was a Gorilla. 1949/CI/00.- Vanderhoof. (106)

Thought It Was a Human with Down's Syndrome. 2002/08/00 - Port Alberni (VI). (371)

Thought It Was a Moose at First. 2002/08/17 - Tete Jaune Cache. (371)

Thought It Was His Son. 1988/09/08 - Woss (VI). (299)

Three Creatures Seen. 1965/07/00 - Butedale. (151)

Three Sets of Tracks.1962/AU/00 - Swindle Island. (142)

Three Sightings in 24 Hours. 1966/07/21 & 22 - Richmond. (156)

Three Sightings. 2005/SU/00 - Cowichan (VI). (393)

Three Teens See Sasquatch. 1975/01/00 - Kamloops. (227)

Three-toed Tracks. 1969/06/00 - Squamish. (190)

Threw a Rock to Get Its Attention. 1975/SU/00 - Lake Cowichan (VI). (228)

Thump On Roof of Car. 1978/08/00 - Rocky Mountains. (260)

Tired of Sasquatch Politics. 1957/05/08 - Lillooet. (119)

Titmus Says He Will Never Give Up. 1977/06/17 - Vancouver. (250)

Titmus Sees Unusual Figures. 1963/07/00 - Kemano. (145)

Too Tall To Be a Man, 1988/08/24 - Wycliffe. (299)

Took a Shot Over Its Head. 1974/10/00 - Mackenzie. (223)

Tore Trees Out By their Roots. 1957/CI/00 - East Sooke (VI). (122)

Totem Poles & Masks. 1700 AD and earlier - Pacific Coast. (24)

Track in Dry Mud. 1970/06/00 - Trail. (202)

Tracks Entered Lake Kitasu. 1951/YR/00 - Swindle Island. (108)

Tracks Impressed 2 Inches Deep. 1971/SU/00 - New Denver. (206)

Tracks in Frost. 1950/12/00 - Roderick Island. (107)

Tracks in Remote Area. 1970/AU/00 - Clearwater. (203)

Tracks Near Cave. 1971/06/00 - Wells. (205)

Tracks Old & New. 1979/05/20 - Haney. (265)

Tracks On a Sandbar. 1971/05/00 - Squamish. (205)

Tracks On Logging Road; Appeared Very Fresh. 2005/08/26 - Harrison Hot Springs. (394)

Tracks With Short Toes. 1964/CI/00 - Agassiz. (147)

Trail Led To a Swampy Lake. 1986/04/00 - Lone Butte. (288)

Trees Stripped. 1950/CI/00 - Smithers. (107)

Turned Upper Body As Well. 1969/01/10 - Hope. (183)

Two See Sasquatch. 2005/YR/00 - Port Alberni (VI). (397)

Two Sets of Tracks. 1984/03/00 - Kelowna. (286)

Two Steps to Cross the Road. 2008/06/19 - Bridal Falls. (406)

Typical Ape Tracks. 1967/CI/00 - Swindle Island. (173)

UBC Publishes Book on Humanoid Monsters. 1980/YR/00 - Vancouver. (272)

UBC Scientists First to See Film of Alleged Sasquatch. 1967/10/26 - Vancouver. (165)

UBC to Look at Sasquatch Evidence. 1977/YR/00 - Vancouver. (255)

Unusual Footprints. 1811/01/07 - Jasper, Alberta. (27)

Unusual Nest. 1975/CI/00 - Kingcome Inlet. (237)

Unusual Prints in Snow. 1969/DE/00 - Harrison Mills (Chehalis Reservation). (195)

Unusual Sounds & a Footprint. 2008/06/19 - Chilliwack. (407)

Up A Tree. 1983/08/00 - Nanaimo (VI). (285)

Very Large Tracks. 1964/WI/00 - Squamish. (146)

Vietnam Vet Hunts Sasquatch. 1984/04/12 - Vancouver. (286)

Wading Sasquatch. 1930/YR/00 - Swindle Island. (80)

Walked Slightly Stooped. 1971/09/00 - Princeton. (207)

Walked Upright All the Time. 1968/YR/00 - Kelowna. (182)

Walked With Bent Knees. 1995/AU/00 - Prince Rupert. (325)

Walking Along a Ridge. 2004/08/00 - Canal Flats. (389)

Walking Along the Tree Line. 2007/01/00 - Harrison Hot Springs. (400)

Watched It Go By. 1970/05/00 - Nazko. (201)

Watched Taking Steps from Large Rock to Large Rock. 1973/03/21 - Big Bay. (211)

Watched the Man For a While. 1952/YR/00 - Terrace. (108)

We Know What We Saw. 2002/07/00 - Grand Forks. (370)

Wearing Fur Around Its Waist. 1965/AU/00 - Harrison Mills. (152)

Webbed Feet. 1976/04/03 - Harmer Ridge. (238)

Weetago Seen. 1965/10/00 - Fort St. John. (152)

Well Built. 1991/09/18 - Golden. (305)

Went Into Shock. 1974/09/00 - Harrison Hot Springs. (222)

What Looked Like a Bed. 1986/06/15 - Gibsons. (290)

Where the Apes Stay. 1945/04/00 - Swindle Island. (103)

White & Gray Creature. 1969/DE/00 - Invermere. (196)

White as a Sheet. 1965/CI/00 - North Vancouver. (153)

White Streak Below Its Head. 1965/11/00 - Bella Coola. (152)

Whole Race of Them. 1979/YR/00 - Nelson. (266)

Wild Man—Beard Like Rip Van Winkle.1906/09/00 - Vancouver. (61)

Wild Man No Figment of the Imagination. 1906/08/05 - Alberni (VI). (60)

Wildlife Biologist Finds Sasquatch Tracks. 1988/10/00 - Strathcona Provincial Park (VI). (300)

Wildlife Biologist Says it All. 2000/SU/00 - Courtenay (VI). (342)

Witness White as a Ghost. 1977/03/28 - Spuzzum. (247)

Wizened Old Man. 1969/11/20 - Lytton. (193)

Woman Followed by Sasquatch. 1999/DE/00 - Vancouver. (341)

Would Have Fired if Sure Not a Man. 1983/AU/00 - Parksville (VI). (285)

Yale. Residents Interviewed. 1958/10/29 - (125)

You Will Never Catch a Sasquatch. 2002/YR/00 - Harrison Mills (Chehalis Reservation). (374)

BIBLIOGRAPHY

Books

Bayanov, Dmitri. 1996. *In the Footsteps of the Russian Snowman.* Moscow, Russia: Crypto Logos Publishers. (Available from Hancock House Publishers, Surrey, BC)

———. 1997. *America's Bigfoot: Fact Not Fiction.* Moscow, Russia: Crypto Logos Publishers. (Available from Hancock House Publishers, Surrey, BC)

———. 2001. *Bigfoot: To Kill or to Film, The Problem of Proof.* Burnaby, BC: Pyramid Publications.

Bindernagel, John A. 1998. *North America's Great Ape: The Sasquatch.* Courtenay, BC: Beachcomber Books.

———. 2010. *The Discovery of the Sasquatch.* Courtenay, BC: Beachcomber Books.

Bord, Janet and Colin. 1982. *The Bigfoot Casebook.* London, UK: Granada Publishing Ltd.

Cherrington, John A. 1992. *The Fraser Valley: A History.* Madeira Park, BC: Harbour Publishing.

Coil Suchy, Linda. 2009. *Who's Watching You? An Exploration of the Bigfoot Phenomenon in the Pacific Northwest.* Surrey, BC: Hancock House Publishers.

Colombo, John Robert. 1989. *Mysterious Canada.* Toronto, ON: Doubleday Canada Ltd.

———. 2004. *The Monster Book of Canadian Monsters.* Shelburne, ON: The Battered Silicon Dispatch Box.

Cusick, David. 1828. *Sketches of Ancient History of the Six Nations*, Tuscarora Village, Lewiston, Niagara County, New York, USA.

Glickman, J. 1998. *Toward a Resolution of the Bigfoot Phenomenon.* Hood River, OR: North American Science Institute.

Green, John. 1973. *The Sasquatch File.* Agassiz, BC: Cheam Publishing.

———. 1981. *Sasquatch, the Apes Among Us.* Surrey, BC: Hancock House Publishers.

Halpin, Marjorie and Michael M. Ames. 1980. *Manlike Monsters on Trial, Early Records and Modern Evidence.* Vancouver, BC: University of British Columbia Press.

Hunter, Don with René Dahinden. 1993. *Sasquatch/Bigfoot, The Search for North America's Incredible Creature.* Toronto, ON: McClelland and Stewart.

Kane, Paul. 1925. *Wanderings of an Artist Among the Indians of North America.* Toronto, ON: The Radisson Society of Canada Ltd.

Krantz, Grover S. 1992. *Big Footprints: A Scientific Inquiry into the Reality of Sasquatch.* Boulder, CO: Johnson Printing Co.

———. 1999. *Bigfoot/Sasquatch Evidence.* Surrey, BC: Hancock House Publishers.

Laforet, Andrea and Annie York. 1998. *Spuzzum: Fraser Canyon Histories, 1808–1939.* Vancouver, BC: University of British Columbia Press.

Markotic, Vladimir and Grover Krantz (editors). 1984. *The Sasquatch and other Unknown Hominoids.* , Calgary, AB: Western Publishers.

McClean, Scott. 2005. *Big News Prints.* Self-published.

Mocine, Jose Mariano. 1792. *Noticias de Nutka: An Account of Nootka Sound in 1792.* Toronto/Montreal: McClelland and Steward Ltd.

Murphy, Christopher L., with John Green and Thomas Steenburg. 2004. *Meet the Sasquatch.* Surrey, BC: Hancock House Publishers.

———. 2008. *Bigfoot Film Journal.* Surrey, BC: Hancock House Publishers.

———. 2010. *Know the Sasquatch: Sequel and Update to Meet the Sasquatch.* Surrey, BC: Hancock House Publishers.

Napier, John. 1972. *Bigfoot: Startling Evidence of Another Form of Life on Earth Now.* New York, NY: Berkley Publishing.

NASI. 1997. Bigfoot Phenomenon Anecdotal Reports "A" Classification, Volume 1. North American Science Institute publication.

Sleigh, Daphne. 1990. *The People of the Harrison.* Abbotsford, BC: (self published).

Patterson, R.M. 1954. *The Dangerous River.* London, UK: George Allen & Unwin Ltd.

Steenburg, Thomas. 1990. *Sasquatch/Bigfoot: The Continuing Mystery.* Surrey, BC: Hancock House Publishers.

———. 2000. *In Search of Giants: Bigfoot Sasquatch Encounters.* Surrey, BC: Hancock House Publishers.

Thompson, David. 1916. *David Thompson: Narrative of his Explorations in Western America, 1784–1812.* Westport, CT: Greenwood Press.

Magazine Articles

Bindernagel, John. 2000. "Sasquatches in Our Woods." *Beautiful BC* magazine. Summer.

Burns, John W. 1929. "Introducing B.C.'s Hairy Giants." *MacLean's* magazine. April 1.

———. 1954. "My Search for B.C.'s Giant Indians." *Liberty* magazine. December.

Hancock, David, 1968. "The Sasquatch Returns." *Weekend* magazine. The *Sun.* No. 2, June 25.

Murphy, Christopher L. 2003. "Bigfoot Film Site Insights." *Fate* magazine. March.

Paterson, T.W. 1967. "Wanted, Dead or Alive Sasquatch." *Real West* magazine. July.

———. 1973. "Not All Quiet on the Western Front," published in *Pursuit* magazine. Volume 7, No. 4.

Other References:

All other references, together with the above, are provided in the "Source" section that follows each book entry.

General Index

Note on Inclusions: This index covers all applicable references in the main text of the book. To facilitate research, it also includes extensive coverage of pertinent information is the "Source" section that follows each book entry.

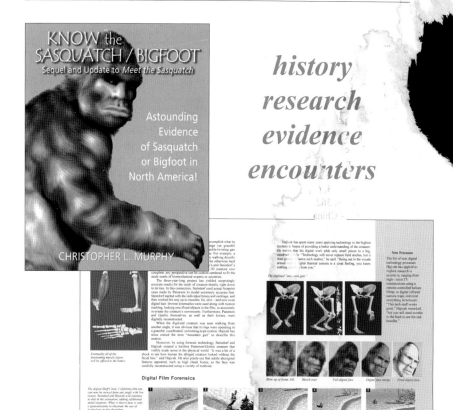

Other **HANCOCK HOUSE** *cryptozoology titles*

Best of Sasquatch Bigfoot
John Green
0-88839-546-9
8½ x 11, sc, 144 pages

Bigfoot Discovery Coloring Book
Michael Rugg
0-88839-592-2
8½ x 11, sc, 24pages

Bigfoot Encounters in Ohio
C. Murphy, J. Cook, G. Clappison
0-88839-607-4
5½ x 8½, sc, 152 pages

Bigfoot Encounters in New York & New England
Robert Bartholomew
Paul Bartholomew
978-0-88839-652-5
5½ x 8½, sc, 176 pages

Bigfoot Film Controversy
Roger Patterson, Christopher Murphy
0-88839-581-7
5½ x 8½, sc, 240 pages

Bigfoot Film Journal
Christopher Murphy
0-88839-658-7
5½ x 8½, sc, 106 pages

Bigfoot Research: The Russian Vision
Dmitri Bayanov
978-0-88839-706-5
5½ x 8½, sc, 422 pages

Bigfoot Sasquatch Evidence
Dr. Grover S. Krantz
0-88839-447-0
5½ x 8½, sc, 348 pages

Giants, Cannibals & Monsters
Kathy Moskowitz Strain
0-88839-650-3
8½ x 11, sc, 288 pages

Hoopa Project
David Paulides
0-88839-653-2
5½ x 8½, sc, 336 pages

In Search of Giants
Thomas Steenburg
0-88839-446-2
5½ x 8½, sc, 256 pages

In Search of Ogopogo
Arlene Gaal
0-88839-482-9
5½ x 8½, sc, 208 pages

The Locals
Thom Powell
0-88839-552-3
5½ x 8½, sc, 272 pages

Meet the Sasquatch
Christopher Murphy, John Green, Thomas Steenburg
0-88839-574-4
8½ x 11, hc, 240 pages

Raincoast Sasquatch
J. Robert Alley
978-0-88839-508-5
5½ x 8½, sc, 360 pages

Rumours of Existence
Matthew A. Bille
0-88839-335-0
5½ x 8½, sc, 192 pages

Sasquatch: The Apes Among Us
John Green
0-88839-123-4
5½ x 8½, sc, 492 pages

Sasquatch Bigfoot
Thomas Steenburg
0-88839-312-1
5½ x 8½, sc, 128 pages

Sasquatch/Bigfoot and the Mystery of the Wild Man
Jean-Paul Debenat
978-0-88839-685-3
5½ x 8½, sc, 428 pages

Shadows of Existence
Matthew A. Bille
0-88839-612-0
5½ x 8½, sc, 320 pages

Strange Northwest
Chris Bader
0-88839-359-8
5½ x 8½, sc, 144 pages

Tribal Bigfoot
David Paulides
978-0-88839-687-7
5½ x 8½, sc, 336 pages

UFO Defense Tactics
A.K. Johnstone
0-88839-501-9
5½ x 8½, sc, 152 pages

Who's Watching You?
Linda Coil Suchy
0-88839-664-8
5½ x 8½, sc, 408 pages

Yale & the Strange Story of Jacko the Ape-boy
Christopher Murphy with Barry Blount
0-88839-712-6
5½ x 8½, sc, 48 pages

View all **HANCOCK HOUSE** *titles at* **hancockhouse.com**